Electrical Systems Design

Electrical Systems Design

Third Edition

M.K. Giridharan
Former Dean (Academics) & Professor
Department of Electrical & Electronics Engineering
Mar Baselios College of Engineering & Technology
Thiruvananthapuram, Kerala, India

TECHSAR™
NEW DELHI

Published by

Techsar Pvt. Ltd.
4435-36/7, Ansari Road, Daryaganj
New Delhi-110 002 (India)
E-mail: tech_info@techsarworld.com
Website: www.techsarworld.com

ISBN 978-93-90620-13-5

Reprint 2025

Printed by Shree Maitrey Printech Pvt. Ltd. Noida

Dedicated to my beloved wife

Latha

the Electrical Engineers in my family

Snita, Prejith, Nishi and Rajesh

and my grandchildren

Neha, Pratik and Shyamak

Preface to the Third Edition

This third edition of the well accepted book continues in its tradition of giving a comprehensive treatment of Electrical Systems Design. It covers on one hand the guiding principles of electrical system design, lighting design, designing for household electricity installation, industrial electrical system installation, exterior lighting, cable sizing, earthing and so on, on the other hand, it also discusses the challenges that an electrical system designer faces to improve power quality, power factor, renewable sources and energy efficiency. Keeping in view the current advances in renewable sources of energy and environmental consciousness, the book also covers solar PV installation and emergency/standby generation installation. The book also carries an introductory chapter on the relevant laws and compliance codes that an electrical system designer should be aware of, namely, Electricity Act of 2003 and National Electrical Code (NEC) 2011.

Only a basic knowledge of electrical engineering is required to understand the concepts of systems design. Even though the current practice is to use software tools for every design process, this book provides the background information to help the users to understand how to use electricity efficiently, safely and economically. Although safety is the primary objective of a good electrical system design, the information given in this book is not intended to be a substitute for the national or manufacturer's safety guidelines.

With its coverage and approach, the book will be very useful to the engineering degree students as well as practising engineers.

M.K. Giridharan

PREFACE TO THE FIRST EDITION

Electrical industry worldwide is undergoing rapid and widespread changes. These changes have been largely driven by market oriented restructuring policies of the respective countries and increasing global environmental concerns. In India also these issues have been foreseen and the result is the new **"Energy Conservation Act, 2001"** and **"The Indian Electricity Act, 2003"**, enacted by the Indian Parliament.

The progressive features like de-licensing of power generation, open access in power transmission, decentralized management of distribution, etc., envisaged in the new Acts have necessitated the electrical system designers to design systems which are more energy efficient, environment-friendly , reliable and with reduced greenhouse gas emissions.

A very well planned and carefully installed electrical system can be a pleasure to operate. They reward us with many years of safe, efficient and reliable service. On the other hand a poorly designed, badly executed electrical system can be dangerous to human lives and property, unreliable and a never ending source of head ache and extra expenses.

The subject of electrical system design is only marginally treated in the undergraduate programmes of many universities. Being a field oriented topic, many of the teachers are also not aware of the new trends in electrical systems design. Electrical systems design can never be learnt overnight. It can be mastered only by continuous practice and updating to the current level of technology. It also requires a detailed knowledge of the relevant statutes. The proposed book addresses these problems in the light of the Electricity Act 2003 and National Electrical Code-2011 and will try to provide the readers a better insight into the subject and thereby motivate the engineers to be more socially committed and environmentally concerned citizens.

This text is primarily intended to provide design assistance to electrical systems commonly found in residential, recreational and industrial premises. It emphasizes the design of electrical systems in accordance with the requirements of the National Electrical code (NEC-2011). Statutory requirements as contained in the NEC are explained to ensure that the students appreciate the relevance of the NEC as well as the systems are designed to ensure public safety and well-being.

Chapter 1 provides an overview of the statutes that are applicable to electrical system design. It gives a brief description of the Indian Electricity Act 2003, Indian Electricity Rules 1956, and National Electrical Code, etc. It also highlights the standard voltages, its tolerances and various electrical safety aspects to be included in the system design. A brief list of graphic symbols used in electrical systems design is also included in this chapter

Chapter 2 introduces the basic principles of design of electrical installations. The classification of the building services, design aspects of various building services are highlighted in this chapter.

Chapter 3 discusses the lighting design aspects of buildings and covers the various light sources with its characteristics. Three types of lighting calculations are discussed in detail with examples. The computation of coefficient of utilization and light loss factors are also discussed in detail. Exact methods of lighting design based on IES standards are also discussed with examples. The quality aspects of lighting also are dealt with in this chapter.

Chapter 4 highlights the design aspects of electrical installations for domestic purposes. Estimation of the load requirements, selection of types of wiring and wiring accessories are also covered in this chapter. Requirements of standby generation and earthing are discussed with specific reference to the NEC. Special features of high rise apartment buildings are also discussed and typical schematic wiring diagrams of a high rise apartment building is discussed.

Chapter 5 focusses on the electrical installations of industrial buildings. The classifications and general characteristics are discussed. Selection of supply voltage, design of transformer substations and selection of drives are detailed with examples. The chapter also covers the short circuit calculations and the design of earthing scheme for transformer substations. Different types of switchgear elements are also explained with its application to electrical systems.

Chapter 6 details the different types of cables and the calculation of voltage drops in the cables. The derating of cables for different environmental conditions and different types of laying are discussed. Simplified approach of selection of cables is also discussed.

Chapter 7 provides an introduction to exterior lighting. Definition of terms, classification of luminaries, etc., are discussed in this chapter. The classification of roads as per NEC is detailed .The design principle for road lighting and area lighting are discussed with examples.

Chapter 8 discusses the application of power factor correction capacitors to power distribution systems. The location and the optimal compensation methodologies are discussed. Power factor correction for induction motors are also discussed in this chapter. The chapter concludes with the discussion of control and protective measures and the impact of voltage transients due to power factor correction capacitors.

Chapter 9 addresses the issues relating to power quality. The impact of harmonics and calculation of harmonic distortion are discussed in detail with examples. The problem of resonance with power factor improving capacitors is also discussed along with its remedy, i.e., D-tuned capacitors. IEEE standard 519-1992 is also discussed in this chapter.

Chapter 10 discusses the energy economic aspects like time value of money in different models like single payment compound model, uniform series compound model, uniform series present worth. The impact of depreciation and different forms of calculating depreciation are explained with examples. Tax considerations are also discussed in this chapter. The impact of energy efficient retrofits is also discussed with examples.

Chapter 11 explains the different forms of power system earthing and its special features. Selection criteria for power system earthing are explained. The importance of earth leakage protection is also highlighted in this chapter.

Chapter 12 deals with the basics of solar PV systems and their application to domestic buildings. Brief descriptions of different types of solar panels are provided with typical design problem for a domestic building. Details of associated components like solar inverter, battery charge controller, protective systems, etc., are also explained.

Chapter 13 gives a brief description of standby generation systems in electrical systems. Basic guidelines for the selection of the standby generating systems are explained with statutory requirements.

This book will be useful as a textbook at undergraduate level in engineering colleges offering degree program in Electrical and Electronics Engineering and also to practising engineers.

M.K. Giridharan

Acknowledgements

Many persons have helped me in the preparation of this book. I would like to put on record my sincere thanks to the following for their support and encouragement. His Beatitude Moran Mor Baselios Cleemis Catholicos, Major Archbishop of Trivandrum and the Patron of the College; Rev. Father John Thundiyath, the Bursar and Prof. K.M. Raju, Former Principal of Mar Baselios College of Engineering & Technology, Thiruvananthapuram. I am also thankful to Dr. T.M. George, the Principal, MBCET for his encouragement in this project.

I also wish to express my gratitude to Prof. N.K. Radhakrishnan Nair, Former Professor of Electrical Engineering, College of Engineering, Thiruvananthapuram, for initiating me into the field of consultancy services. My sincere thanks to all my erstwhile teachers and colleagues who have made possible a highly satisfying and rewarding teaching and consultancy career for me.

My special thanks are due to Ms. Elizabeth Varghese, Ms. Jomole Joseph P., Ms. Beena Paul and all other colleagues of the Electrical & Electronics Engineering Department of Mar Baselios College of Engineering & Technology. Ms. Prabha and Mr. Arun, Technical support staff and Ms. Linumole and Ms. Sindhu of the Central Computing Facility also deserve special mention for their assistance in typing and drafting the manuscript.

My wife and children also have provided untiring support and encouragement during the preparation of this book. I wish to thank them for their patience and constant inspiration.

M.K. Giridharan

CONTENTS

Preface to the Third Edition *vii*
Preface to the First Edition *ix*
Acknowledgements *xiii*

1. Role of Statutes in Electrical Systems Design **1**
1.1 The Electricity Act, 2003 [3] 1
1.2 Central Electricity Authority (CEA) [4] 3
1.3 National Electrical Code (NEC-2011) [1] 6
1.4 International Electrotechnical Commission (IEC) 8
1.5 Standard Values of Voltages 10
1.6 Voltage Limits for Alternating Current Systems 10
1.7 Safety Aspects of Electrical System Design 11
1.8 Graphic Symbols for Diagrams (NEC-2011) 12
1.9 Some Important BIS Specifications 16
1.10 Questions 16

2. Design Aspects of Electrical Systems **17**
2.1 Classification of Building Services 17
2.2 Design Aspects of Lighting Services 18
2.3 Design Aspects of Ventilation 19
2.4 Design Aspects of Climate Control 22
2.5 Design Aspects of Vertical Transportation 23
2.6 Design Aspects of Minor Building Services 23
2.7 Questions 24

3. Principles of Lighting Design **25**
3.1 Light Sources 25
3.2 Lighting Calculations 44
3.3 Point by Point Method 45
3.4 Average Lumen Method 46
3.5 Lumen 47
3.6 Light Loss Factor (LLF) 48
3.7 Quality of Lighting 50
3.8 Design Procedures 52
3.9 Arrangement of Fixtures 55
3.10 Questions 65

4. Electrical Installations in Domestic Buildings **67**
4.1 Classification 67
4.2 Estimation of Load Requirements 75
4.3 Selection of Type of Wiring 88
4.4 Special Features Applicable for High-Rise Apartment Buildings 93
4.5 Precommissioning Tests 100
4.6 Questions 102

5. Industrial Installations **107**
5.1 Introduction 108
5.2 Classification of Industrial Buildings 108
5.3 General Characteristics 109
5.4 Selection of Distribution Architecture 113
5.5 Selection of Transformer Substations 126
5.6 Selection of Drives 138
5.7 Short Circuit Studies 167
5.8 Selection of Switchgears 180
5.9 Switchboards in MV Installations 216
5.10 Questions 222

6. Cable Sizing and Voltage Drop Calculations **225**
6.1 Introduction 225
6.2 Voltage Designation 226
6.3 Choice of Materials used in Cables 226
6.4 Current Carrying Capacity 229
6.5 Short-Circuit Ratings 232
6.6 Voltage Drop Calculation 232
6.7 General Methods of Cable Laying 251
6.8 Simplified Approach for Cable Selection 256
6.9 Questions 259

7. Exterior Lighting **261**
7.1 Introduction 261
7.2 General Considerations 261
7.3 Road Lighting 266
7.4 Area Lighting 272
7.5 High Mast Lighting 274
7.6 Questions 274

8. Power Factor Improvement **277**
8.1 Nature of Reactive Energy 277
8.2 Power Factor 278
8.3 How to Improve the Power Factor 280
8.4 Location of Capacitors 282
8.5 Optimal Compensation 283
8.6 Power Factor Correction of Induction Motors 287
8.7 Protection and Control 288
8.8 Voltage Transients 290
8.9 Switching Considerations 290
8.10 Questions 294

9. Power Quality Issues and Resonance Problems in System Design **295**
9.1 Introduction 295
9.2 Overview of Power Quality Issues 297
9.3 Harmonics 297
9.4 Sources of Harmonics 302
9.5 Disturbances Caused by Harmonics 305
9.6 Methods to Reduce the Impact of Harmonics 322
9.7 Design of Detuned Capacitor Bank (Harmonic Filter) 325
9.8 IEEE Standard 519-1992 334
9.9 IEEE 519-1992 Voltage Distortion Limits 339
9.10 Summary of Detuned Harmonic Filter Design 339
9.11 Questions 340

10. Energy Economics in System Design **343**
10.1 Introduction 343
10.2 Time Value of Money 343
10.3 Single Payment Compound Amount Model (SPCA) 344
10.4 Uniform Series Compound Amount Model (USCA) 344
10.5 Uniform Series Present Worth Model (USPW) 345
10.6 Depreciation 349
10.7 Tax Considerations 353
10.8 After Tax Analysis 353
10.9 Conclusion 357
10.10 Questions 358

11. Power System Earthing **359**
11.1 Introduction 359
11.2 Types of System Earthing 360
11.3 TN System 361

11.4 TT System 364
11.5 IT System 365
11.6 Main Characteristics of the Earthing Systems 367
11.7 Selection Criteria for Earthing 368
11.8 Earth Leakage Protection 369
11.9 Questions 375

12. Solar PV Systems for Household Applications 377

12.1 Introduction 377
12.2 Types of PV Electrical Systems 378
12.3 System Components 380
12.4 Typical Design for Home Applications 386
12.5 Estimating the Wh/Day that can be used for a Particular Size of Panel 396
12.6 Questions 400

13. Emergency and Standby Generation 401

13.1 Introduction 401
13.2 Ratings of the Standby Generating Sets 402
13.3 Starting and Stopping of Generator Sets 404
13.4 Stand-alone Operation 405
13.5 Parallel Operation with Utility Supply 405
13.6 General Requirements of Installation of MV Standby Generators [2] 408
13.7 Guidelines for Generators Running in Parallel 409
13.8 Guidelines for Generators Installed Aboveground Level 410
13.9 Guidelines for Portable Generators 410
13.10 Questions 411

Appendix 413

Bibliography 423

Index 425

CHAPTER 1

ROLE OF STATUTES IN ELECTRICAL SYSTEMS DESIGN

OBJECTIVES

At the end of this chapter you will

- Understand the various statutes that are governing the Electricity Industry in our country.
- Have an idea about the significance of Central Electricity Authority.
- Appreciate the relevance of National Electrical Code.
- Know about the International Electrotechnical Commission.
- Have a knowledge of the common system voltages used in Electrical Industry.
- Understand the use of standard symbols used in schematic representations.
- Understand the safety aspects of system design.

1.1 THE ELECTRICITY ACT, 2003 [3]

The Electricity Supply Industry in our country was governed by three major enactments viz.

1. The Indian Electricity Act, 1910
2. The Electricity (Supply) Act, 1948 and
3. The Electricity Regulatory Commission Act, 1998.

The Indian Electricity Act, 1910 created the basic framework for the development of electric supply industry in India. This Act envisaged growth of the Electricity Industry which was in its infancy, through Private Licensees. It also created the legal framework for laying down of conductors and other work relating to the supply of Electricity.

The Electricity Supply Act, 1948 mandated the creation of State Electricity Boards with the responsibility of making available the electricity in the respective States. Soon after the Independence, the need for providing Electricity for the overall development was felt and this responsibility was vested with the State Electricity Boards. Accordingly, the State Electricity Boards, through the successive five-year plans, undertook rapid expansion programmes by utilizing plan funds. However, there was a gradual deterioration in the performance of the State Boards and several State Electricity Boards were in the red due to factors most of which are the ills that were the factors contributing to the poor performance of public sector undertakings. Even though the power to fix the tariff was vested with the State Electricity Boards, they were unable to take decision on tariff in a professional and independent manner because of the ruling Government's intervention on political mileage. This led to a grim situation where cross subsidies reached unsustainable levels.

To address these issues and to provide for distancing of the Government from determination of tariffs, The Electricity Regulatory Commission Act was enacted by the Parliament in 1998 with enabling provision through which the State Governments can create State Electricity Regulatory Commissions. Accordingly many States have created State Electricity Regulatory Commissions.

Way back in 1995, starting with the State of Orissa, several State Governments have realised the need for restructuring the functioning of State Electricity Boards for improving their performance. They have enacted reforms which involved unbundling of the State Electricity Board into separate Generation, Transmission and Distribution Companies through transfer of assets and staff. Orissa (1995), Haryana (1997), Andhra Pradhesh (1998), Uttar Pradhesh (1999), Karnataka (1999), Rajasthan (1999), Delhi (2000) and Madhya Pradhesh (2000) are some of the forerunners in this reform process.

The economic liberalisation policies of 1990's have opened up a plethora of opportunities for massive industrialisation and this has resulted in an ever increasing gap between the supply and demand for electric power. This being a major threat to industrial development, and with the policy adopted for private participation in generation, transmission and distribution and with the objective of distancing the regulatory responsibilities from the Government, the need for a more harmonious and rational law was inevitable. Therefore it became necessary to enact a new legislation for regulating the Electricity Supply industry in the country. The main objectives were,

- **To replace the existing laws while preserving the core features other than those relating to the mandatory existence of State Electricity Boards and the responsibilities of the State Government with respect to regulating licensees.**
- **To provide for introducing new concepts like power trading, open access, etc.**
- **To obviate the requirement of each State Government to pass its own Reform Acts**
- **To include progressive features and endeavours to strike a right balance, given the current realities of the Power Sector of India and**
- **To give every State Government enough flexibility to develop their Power Sector in the manner they consider appropriate.**

Extensive consultations were carried out with the State Governments, other Stakeholders and Experts. The result was "The Electricity Act, 2003" which seeks to replace Indian Electricity Act 1910, The Electricity (Supply) Act 1948 and The Electricity Regulatory Commission Act, 1998. The Electricity Act, 2003 is an Act promulgated by the Parliament to consolidate the laws relating to Generation, Transmission, Distribution, Trading and use of Electricity and aims at,

- Promoting measures conducive to the development of Electricity Industry
- Promoting competition
- Protecting the interest of consumers

- Providing electric supply to all areas
- Providing transparent policies on subsidies
- Promoting efficient and environment-friendly policies
- Constitution of Central Electricity Authority and Regulatory Commission
- Establishment of Appellate Tribunals and
- For all matters connected therewith and incidental thereto.

This Act has progressive features and endeavours to strike the right balance given to the current realities of the power sector in India. It gives the States enough flexibility to develop their own sector in the manner they consider appropriate.

1.2 CENTRAL ELECTRICITY AUTHORITY (CEA) [4]

The Central Electricity Authority (CEA) is a statutory organization constituted under section 3 of the repealed Electricity (Supply) Act, 1948. It was established as a part-time body in the year 1951 and was subsequently made a full-time body in the year 1975.

The Electricity Act, 2003 (No. 36 of 2003) has been enacted by the Indian Parliament with the objective of reforming the Power Sector in India and the provisions of this Act have been brought into force with effect from 10th June, 2003. With the coming into force of The Electricity Act, 2003; Indian Electricity Act, 1910, Electricity (Supply) Act, 1948 and Electricity Regulatory Commissions Act, 1998 stand repealed. As per section 73 of The Electricity Act, 2003, the Central Electricity Authority shall perform such functions and duties as the Central Government may prescribe or direct, and in particular to:

(a) Advise the Central Government on the matters relating to the National Electricity Policy, formulate short-term and perspective plans for the development of the electricity system and coordinate the activities of the planning agencies for the optimal utilization of resources to subserve the interests of the national economy and to provide reliable and affordable electricity to all consumers;
(b) Specify the technical standards for the construction of electrical plants, electric lines and connectivity to the grid;
(c) Specify the safety requirements for construction, operation and maintenance of electrical plants and electric lines;
(d) Specify the Grid Standards for operation and maintenance of transmission lines;
(e) Specify the conditions for installation of meters for transmission and supply of electricity;
(f) Promote and assist in the timely completion of schemes and projects for improving and augmenting the electricity system;
(g) Promote measures for advancing the skills of persons engaged in electricity industry;
(h) Advise Central Government on any matter on which its advice is sought or make recommendation to that Government on any matter if, in the opinion of the Authority, the recommendation would help in improving the generation, transmission, trading, distribution and utilization of electricity;
(i) Collect and record the data concerning the generation, transmission, trading, distribution and utilization of electricity and carry out studies relating to cost, efficiency, competitiveness and such like matters;

(j) Make public, from time to time, the information secured under this Act, and provide for the publication of reports and investigations;
(k) Promote research in matters affecting the generation, transmission, distribution and trading of electricity;
(l) Carry out, or cause to be carried out, any investigation for the purpose of generating or transmitting or distributing electricity;
(m) Advise any State Government, licensees or the generating companies on such matters which shall enable them to operate and maintain the electricity system under their ownership or control in an improved manner and where necessary, in coordination with any other Government, licensee or the generating company owning or having the control of another electricity system;
(n) Advise the Appropriate Government and the Appropriate Commission on all technical matters relating to generation, transmission and distribution of electricity; and
(o) Discharge such other functions as may be provided under this Act.

In addition to the above functions and duties, CEA has to perform the following functions also in terms of the undermentioned sections of the Electricity Act, 2003 [3]:

Section 3–National Electricity Policy and Plan

(1) The Central Government shall, from time to time, prepare the National Electricity Policy and tariff policy, in consultation with the State Governments and the Authority for development of the power system based on optimal utilization of resources, such as coal, natural gas, nuclear substances or materials, hydro and renewable sources of energy.
(2) The Central Government shall publish the National Electricity Policy and tariff policy from time to time.
(3) The Central Government may, from time to time, in consultation with the State Governments and the Authority, review or revise the National Electricity Policy referred to in sub-section (1).
(4) The Authority shall prepare a National Electricity Plan in accordance with the National Electricity Policy and notify such plan once in five years.

PROVIDED that the Authority while preparing the National Electricity Plan shall publish the draft National Electricity Plan and invite suggestions and objections thereon from licensees, generating companies and the public within such time as may be prescribed;

PROVIDED FURTHER that the Authority shall:

(a) Notify the plan after obtaining the approval of the Central Government;
(b) Revise the plan incorporating therein directions, if any, given by the Central Government while granting approval under clause (a).

(5) The Authority may review or revise the National Electricity Plan in accordance with the National Electricity Policy.

Section 8–Hydroelectric Generation

(1) Any generating company intending to set up a hydro-generating station shall prepare and submit to the Authority for its concurrence, a scheme estimated to involve a capital expendi-

ture exceeding such sum, as may be fixed by the Central Government, from time to time, by notification.

Section 53–Provision Relating to Safety and Electricity Supply

The Authority may in consultation with the State Governments, specify suitable measures for-

a) Protecting the public (including the person engaged in the generation, transmission or distribution or trading) from dangers arising from the generation, transmission or distribution or trading of electricity, or use of electricity supplied or installation, maintenance or use of any electric line of electrical plant.
b) Eliminating or reducing the risks of personal injury to any person, or damage to property of any person or interference with use of such property.
c) Prohibiting the supply or transmission of electricity except by means of a system which conforms to the specification as may be specified.
d) Giving a notice in the specified form to the Appropriate Commission and the Electrical Inspector, of accidents and failures of supplies or transmission of electricity.
e) Keeping by a generating company or licensee, the maps, plant and sections relating to supply or transmission of electricity.
f) Inspection of maps, plans and sections by any person authorized by it or by Electrical Inspector or by any person on payment of specified fee.
g) Specifying action to be taken in relation to any electric line or electrical plant, or any electrical appliance under the control of a consumer for the purpose of eliminating or reducing the risk of personal injury or damage to property or interference with its use.

Section 55–Use, Etc. of Meters

(2) For proper accounting and audit in the generation, transmission and distribution or trading of electricity, the Authority may direct the installation of meters, by a generating company or licensee at such stages of generation, transmission or distribution or trading of electricity and at such locations of generation, transmission or distribution or trading, as it may deem necessary.

Section 177–Powers of Authority to Make Regulations

1) The Authority may, by notification, make regulations consistent with this Act and the rules generally to carry out the provisions of this Act.
2) In particular and without prejudice to the generality of the power conferred in sub-section (1), such regulations may provide for all or any of the following matters, namely:

   a) The Grid Standards under section 34;
   b) Suitable measures relating to safety and electricity supply under section 53;
   c) The installation and operation of meters under section 55;
   d) The rules of procedure for transaction of business under sub-section (9) of section 70;
   e) The technical standards for construction of electrical plants and electric lines and connectivity to the grid under clause (b) of section 73;

f) The form and manner in which and the time at which the State Government and licensees shall furnish statistics, returns or other information under section 74;

g) Any other matter which is to be, or may be, specified.

(3) All regulations made by the Authority under this Act shall subject to the conditions of previous publication.

The CEA plays a lead role in promoting the integrated operations of the regional power grids and the evolution of a state-of-the-art national grid. The eastern, western and north-eastern grids have been integrated and are operating in a synchronous mode. The eastern grid is connected to the northern as well as southern grid through HVDC back to back links. The western grid is also connected to the northern and southern grid through similar arrangements. The CEA facilitates exchange of power within the country from surplus to deficit regions and with neighbouring countries for mutual benefits.

Chief Electrical Inspectorate Division (CEI) is a part of the Power System Wing of Central Electricity Authority. CEI Division has been given the responsibility to frame regulation and measures relating to Safety and Electric Supply under section 53 and 177 of Electricity Act, 2003. In line with that, CEA Regulation, on Measures relating to Safety and Electric Supply, was notified on 24th September, 2010. As a consequence, **IE Rules, 1956 got repealed** with the notification of these Regulations under section 185 (2) (c) of EA, 2003 **(CEA Regulation on Measures relating to Safety and Electric Supply 2010)**. This Division is located in NRPC Building, Katwaria Sarai, New Delhi. This Division has five Regional Inspectorial Organisations (RIOs) across India. RIO (North) is located in NRPC Building, Katwaria Sarai, New Delhi, RIO (West) in Mumbai, RIO (South) in Chennai, RIO (East) in Kolkata and RIO (North-East) in Shillong.

1.3 NATIONAL ELECTRICAL CODE (NEC-2011) [1]

The economic development of a country largely depends on the Electric power generation capabilities of the country. Hence a greater emphasis is to be made on the implementation of programmes for Power generation, Transmission and Distribution. The designs and practices that were followed by various agencies involved in the Electric Power Industry were very irrational. As technology advanced with new products and processes, a more rational and unified code became a necessity. The National Electric Code was first introduced in 1985 and it unifies the practices, procedures and safety requirements to be complied within the Design, Execution, Inspection and Maintenance of Electrical Installations in the Country. The IE Rules, 1956 contains specific regulations to be adhered to, in the supply and use of Electricity in the interest of Safety. The NEC is formulated primarily to elaborate the IE Rules which are statutory in nature and serve as a compilation document on Electrical practices in our country.

The objective of NEC is to complement IE Rules by way of elaborating as well as recommending practices to comply with their requirements. NEC provides information in a consolidated form to the Electrical Engineers and Contractors who are concerned with the design and operation of Electrical Installation. The information in NEC is presented in seven parts. They are:

1. General and common aspects
2. Electrical installations in standby generating stations and substations
3. Electrical installations in non-industrial buildings
4. Electrical installations in industrial buildings

5. Outdoor installations
6. Electrical installations in agricultural premises
7. Electrical installations in hazardous areas.

The information relating to each type of installation identified above is further classified and presented in the following manner:

1. Assessment of the general characteristics of occupancy from the point of view of electrical installation.
2. Heavy current installation in the occupancy.
    - Power supply and distribution system.
    - Cables and accessories for power distribution.
    - Protective equipment.
    - Metering of energy consumption.
    - Emergency supply requirements.
    - Reactive power compensation.
    - Guidance on electrical aspects of building services like lighting, air conditioning, lifts, etc.
3. Light current installations in the occupancy.
    - Electric bells and clock system.
    - Electrical audio system.
    - Fire protection signalling.
    - Cable TV.
    - Networking, etc.
4. Specific requirements for protection and safety.

The approach of NEC is primarily to cater to System Engineers. This code excludes the requirements under the purview of Utilities. Product details are also excluded. This code also does not cover the requirements of underground mines.

There is no one-to-one correspondence between the building classification of National Building Code. NBC classification of buildings is mainly based on fire safety considerations. NEC takes a modified approach and takes into account the following:

1. Classification should take into account the skill and capabilities of the occupants utilizing the installation.
2. Classification also should be based on the degree of sophistication of the electrical installation of the building, rather than the type of building.

NEC is basically a compilation document taking assistance from detailed IS specifications already available. The code is intended to be advisory in nature. Its provisions will not be mandatory, but will serve as a model for adoption in the interest of safety and economy and with the intent to keep our electrical installation practices at par with the best practices in the world.

During the formulation of the National Electrical Code in 1985, it was realized that the referred codes, for example, those on wiring practice, earthing, lightning protection, etc. need to be revised in line with the practice and technology available at that time. It had also been planned that after the relevant codes are revised, the National Electrical Code would also need to be revised. After the publication of NEC 1985, the referred Codes were revised. However, the task of revision of NEC could not be taken up in earnest immediately after the revision of various codes of practice. Over the years, there have been yet more changes in the technology; new practices have evolved and got modified. There have been tremendous socio-economic changes, and corresponding change in the pattern of the usage of electricity. Electricity Act 2003 has been notified and power sector reforms have been firmly established. During the Ninth Plan, it was realized that it is necessary to have an Energy Conservation Act. Accordingly, the Government has enacted the Energy Conservation Act, 2001 to meet the legal requirement needed to enforce energy efficiency and conservation measures. Due to all such changes, the present scenario is at great variance with that of 1985, when the Code was first formulated. Therefore, an urgent need was felt to revise the NEC at the earliest to maintain its relevance in the present context.

Also, there have been tremendous socio-economic changes in the country after the economic liberalization of 1990's and corresponding changes in the usage of electricity. Moreover, over the years, there have been changes in the technology, new products and practices have been evolved and got modified. In order to keep pace with the technological advancements and to incorporate the additional knowledge gained through the implementation of the Code, an urgent need was felt to revise the NEC to maintain its relevance in the present context. The revised version follows the earlier structure of NEC, with modifications and additions being incorporated in tune with IEC 60364 series on electrical installations as well as the changes and developments that took place since the publication in 1985. The revised **NEC 2011** has eight parts which are further divided into sections. The eighth part is **SOLAR PHOTOVOLTAIC (PV) POWER SUPPLY SYSTEMS**.

The Code excludes guidance on tariff. Product details are also excluded from the Code, as separate product standards are available for these. When these standards are revised subsequent to the revision of NEC, there could be instances where the latest Codes differ from the revised NEC. It is, therefore, recommended to follow the provisions of the latest standards/codes of practice. In order to avoid instances where the Indian Standards and provisions of this Code differ, attention has been drawn to the relevant standard. However, for certain provisions in this Code, the relevant requirements from corresponding Indian Standards have been extracted and reproduced. In all cases, for detailed guidance, reference should be made to the individual standard and should any contradiction be observed between the provisions in individual standard and those reproduced herein, the provisions of the former shall be considered accurate. As a general rule, technological innovations, such as better materials or new and better method also proved as 'good practice', would first be introduced in the individual standard as appropriate as in the National Electrical Code. In order to keep pace with such changes and to incorporate the additional knowledge that will be gained through the implementation of the Code, a continuous review is envisaged. Thus, the users of this Code are encouraged to bring to the notice of Bureau of Indian Standards, need of modifications that may be required in the light of changes in technology or other factors, as this is a continuous process.

1.4 INTERNATIONAL ELECTROTECHNICAL COMMISSION (IEC)

The International Electrotechnical Commission (IEC) is a non-profit making, non-governmental international standards organization that prepares and publishes International Standards for all electrical, electronic and related technologies - collectively known as "electrotechnology". IEC standards cover a vast range of technologies from power generation, transmission and distribution to home appliances and office equipment, semiconductors, fibre optics, batteries, solar energy, nanotechnology and marine energy as well as many others. The IEC also manages three global conformity assessment systems that certify whether equipment, system or components conform to its International Standards.

The IEC charter embraces all electrotechnologies including energy production and distribution, electronics, magnetics and electromagnetics, electro acoustics, multimedia and telecommunication, as well as associated general disciplines such as terminology and symbols, electromagnetic compatibility (by its Advisory Committee on Electromagnetic Compatibility, ACEC), measurement and performance, dependability, design and development, safety and the environment.

The International Electro technical Commission (IEC) was officially founded in June 1906 in London, based on a resolution of the delegates of The International Electrical Congress held in St. Louis, USA on 15th September 1904. By 1914, IEC had formed four technical committees to deal with nomenclature, symbols, ratings of electrical machinery and prime movers. The Commission also issued a first list of terms and definitions covering electrical machinery and apparatus, a list of international letter symbols for quantities and signs for names of units. The first world war interrupted the work of IEC and was resumed only in 1919. By 1923, the number of technical committees increased to 10. In 1930, IEC established the units for the following electrical quantities.

Hertz	frequency
Oersted	magnetic field strength
Gauss	magnetic flux density
Maxwell	magnetic flux
Gilbert	magneto motive force
Var	reactive power
Weber	magnetic flux

It was also decided to extend the existing series of practical units into a comprehensive system of physical units, which became "Georgi System" in honour of the Italian scientist Giovanni Georgi (1871–1950). This system has been elaborated and is now commonly known as "System International" or SI units.

Between the First and Second world wars, a number of International organisations came into being and the IEC recognised the need for cooperation between them to avoid overlapping of efforts. In 1938, IEC produced the first edition of International Electrotechnical Vocabulary (IEV), now known as Electropedia and freely available to common man. The activities of the IEC came to a standstill from 1939 due to the Second World War and did not resume for the next six years. In 1948, the central office of IEC was shifted to Geneva, Switzerland.

From 1948 to 1980 the number of technical committees grew from 34 to 80 and began to include new technologies in semiconductor devices, electrical equipments in medical practice, maritime navigation and radio communications. In 1974, IEC created Technical Committee 76, to address the standards relating to lasers. The last two decades of the century saw the IEC continue to address new technologies as they emerged, creating new technical committees to prepare standards like lightning protection, fibre optics, ultrasonic, wind turbines, etc.

In 1995, IEC created the Lord Kelvin Award. A maximum of three recipients are chosen each year to pay tribute to their outstanding contributions to global electro technical standardisation over a number of years.

In keeping pace with the new technological developments of the 21st century, IEC has most recently created new committees for fuel cell technology, avionics etc. In 2005 IEC published the most recent edition of IEC multilingual dictionary which now contain 19,400 electrotechnical definitions in French, English and equivalent terms in 13 other languages. In 2006, IEC celebrated its 100th anniversary and is set forth on its journey of service to the mankind.

Today the IEC carries out its work using the most contemporary tools and the most up-to-date ideas for serving its market. Participation has always been a key aspect of the IEC. Whoever wants to be involved is given the opportunity to do so. In 2001, the IEC took this idea to a new level by creating the concept of Affiliated Country Programme, which helps industrialising countries to get involved in IEC work at no cost. Many countries including India have joined this program and are getting benefited from this truly international organization.

1.5 STANDARD VALUES OF VOLTAGES

The standard values of voltages to be used in the electrical systems as per the NEC are as detailed in Table 1.1.

The standard distribution voltage for direct current system shall be 220/440 V.

Table 1.1 Standard voltages for three phase systems

Standard voltages
240 V (voltage to neutral)
415 V (Line to line voltage)
3.3 kV
6.6 kV
11 kV
22 kV
33 kV
66 kV
110 kV
132 kV
220 kV
400 kV

1.6 VOLTAGE LIMITS FOR ALTERNATING CURRENT SYSTEMS

As per statutory requirements the supply authorities are required to maintain the voltages on the system under normal condition within the tolerances specified below:

1. 6% in the case of low and medium voltage installations
2. 6% on the higher side or 9% on the lower side in the case of high voltage systems
3. 12.5% in the case of extra high voltage systems.

1.7 SAFETY ASPECTS OF ELECTRICAL SYSTEM DESIGN

Two important aspects of electrical system design are:

- To assure proper functioning of the installation for the use intended by the designer
- To provide safety to persons, livestock and property against dangers and damages that may arise in the use of electrical installations.

Every electrical installation exposes the end users to two major types of risks. One is the risk of an electric shock current and the other is the risk of very high temperature due to sparking and likely to cause burns, fires and other injurious effects. When there is a current flow through the human body, the electric current itself is the source of danger which has a negative effect when it passes through the body (ventricular fibrillation, muscle paralysis), or changes the body substance (coagulation, burn).

Therefore the system designers shall incorporate adequate protective measures in the system to provide protection against the following:

1. Protection against direct contact
2. Protection against indirect contact
3. Protection against thermal effects
4. Protection against over current
5. Protection against fault current and
6. Protection against over voltages.

1.7.1 Protection against direct contact

The primary aim of this type of protection is to prevent persons and live stock coming in direct contact with live parts of the installation. This can be achieved by preventing a current from passing through the body of any person or live stock or by limiting the magnitude of the current which passes through the body to a value lower than the perceptible shock current.

1.7.2 Protection against indirect contact

In this case the persons and live stock are protected against the possible dangers from a contact with the exposed conductive parts of the electrical installation. Many electrical gadgets are manufactured with external bodies that are conductive in nature. For example, the electric motors have cast iron stator body which under normal conditions is at zero potential. In the event of an insulation failure, the cast iron body becomes partially conductive and will be a threat to the users. The protection against indirect contact can be achieved by

- Preventing the fault current from passing through the body of any person or livestock.
- Limiting the magnitude of the fault current which can pass through the body to a value lower than the perceptible shock current.
- By automatic disconnection of the supply on the occurrence of a fault likely to cause a current to flow through a body in contact with exposed conductive parts, where the value of the current is equal or greater than the perceptible shock current.

1.7.3 Protection against thermal effects

The electrical installations shall be so designed that there is no risk of ignition of flammable materials due to the high temperature of the conductive parts or due to an electric arc. Also, the design shall guarantee that during normal operation, there shall be no risk of persons suffering from burns due to thermal effects.

1.7.4 Protection against over current

Overloading and thereby an over current cannot be eliminated from an electrical system. However, persons and live stock shall be protected against injury and property shall be protected against damage due to excessive temperature or electromechanical stress caused by over current. This can be achieved by one of the following methods:

- Automatic disconnection of the supply in the event of an over current before it reaches a dangerous value taking into account its duration.
- By limiting the maximum value of the over current to a safe value and duration.

1.7.5 Protection against fault currents

Faults are likely events on an electrical system and therefore adequate measures shall be taken to see that all conductors, other than live conductors, and any other parts intended to carry a fault current shall be capable of carrying the fault current without reaching excessive temperature levels.

1.7.6 Protection against over/under voltage

There are several instances when the system experiences an over voltage. Lightning, switching or faults between live parts etc. can cause an over voltage on the system. There shall be measures incorporated in the design to protect persons and livestocks against the harmful effects of over voltages.

The present day scenario is that the public distribution lines are so overloaded or poorly designed that the customers most often experience the problem of under voltages. The under voltage may not be much of a problem in lighting applications. But when it comes to industrial use a situation of under voltage can lead to overloading of the motors or even unwanted tripping of motor starters, causing production loss. Hence most of the modern starting devices are equipped with under voltage protective elements also.

1.8 GRAPHIC SYMBOLS FOR DIAGRAMS (NEC-2011)

Table 1.2 lists some of the graphic symbols to be used in preparing the schematic wiring diagrams in electrical systems design.

Table 1.2 Graphic symbols

No.	Name	Symbol
1.8.1	Direct Current	———

Continued. . .

No.	Name	Symbol
1.8.2	Alternating Current, General Symbol	
	a) Alternating Current, Single-Phase 50 Hz	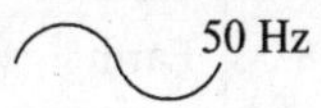
	b) Alternating Current, 3-Phase 415 V 50 Hz	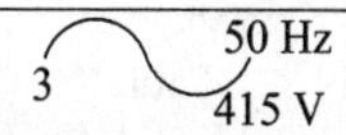
	c) Alternating Current, 3-Phase with Neutral	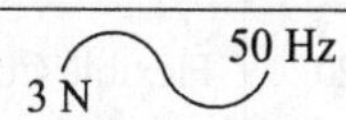
1.8.3	Neutral	N
1.8.4	Positive Polarity	+
1.8.5	Negative Polarity	–
1.8.6	Direct Current, 2 Conductors 110 V	2——110 V
1.8.7	Direct Current, 3 Cond: including Neutral	2 N——220 V
1.8.8	Underground Cable	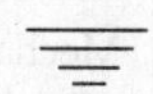
1.8.9	Overhead Line	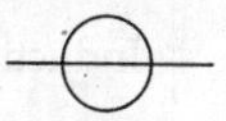
1.8.10	Winding, Delta	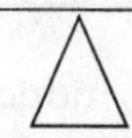
1.8.11	Winding, Star	
1.8.12	Terminals	
1.8.13	Resistance, Resistor, Variable Resistor	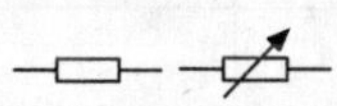
1.8.14	Impedance	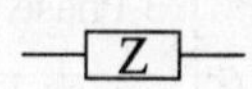
1.8.15	Inductance, Inductor	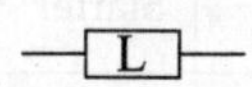
1.8.16	Winding	

Continued. . .

No.	Name	Symbol
1.8.17	Capacitance, Capacitor	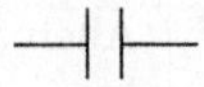
1.8.18	Earth	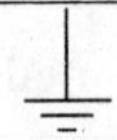
1.8.19	Fault	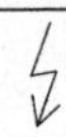
1.8.20	Flexible Conductor	
1.8.21	Generator	
	a) AC Generator	
	b) DC Generator	
1.8.22	Motor	
1.8.23	Synchronous Motor	
1.8.24	Mechanically Coupled Machines	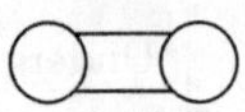
1.8.25	Induction Motor 3-Phase, Squirrel Cage	
	Induction Motor with Wound Rotor	
1.8.26	Transformers with Two Separate Windings	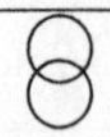
1.8.27	Auto-Transformer	
1.8.28	3-Phase T_f with 3 Separate Windings-Star-Star-Delta	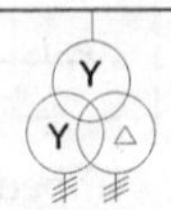
1.8.29	Starter	
1.8.30	Direct-on-Line Starter for Reversing Motor	

Continued...

No.	Name	Symbol
1.8.31	Star-Delta Starter	
1.8.32	Auto-Transformer Starter	
1.8.33	Rheostatic Starter	
1.8.34	Switch	
1.8.35	Contactor	
1.8.36	Circuit-Breaker	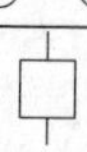
1.8.37	Isolator	
1.8.38	Fuse	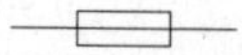
1.8.39	Signal Lamp	
1.8.40	Link	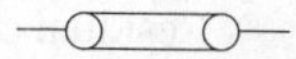
1.8.41	Distribution Board, Cubicle Box, Main Fuse Board with Switches	
1.8.42	Socket Outlet, 5A	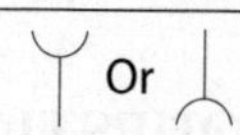
1.8.43	Socket Outlet, 15A	
1.8.44	Plug	

1.9 SOME IMPORTANT BIS SPECIFICATIONS

S. No.	Title	ISS. No.
1	Code of practice for electrical wiring installations	732:89
2	Code of practice for Earthing	3043:87
3	Code of Practice for Industrial lighting	6665:72
4	Code of practice for Hospital lighting	4347:67
5	Guide for electrical layout in residential buildings	4648:68
6	Guide for short circuit calculation of 3 phase AC systems	5728:70
7	Calculation of the effects of short circuit currents	13235:91
8	Code of practice for selection, installation and maintenance of transformers-3 parts	10028:85
9	Guide for testing of three phase induction motors	4029:67
10	Guide for testing synchronous machines	7132:73
11	Guide for testing insulation resistance of rotating machines	7816:75
12	Degree of protections of enclosures	13234:92
13	Designation of methods of cooling of electrical machines	6262:71
14	LV switch gear and control gear-4 parts	13947:93
15	Dry type transformers	11171:85
16	Guide for safety procedures and practices in electrical work	5216:82
17	Miniature circuit breaker boards for voltages up to and including 1000 V.AC	13032:91
18	PVC insulated (heavy duty) electric cables-2 parts	1554:88
19	Recommended current rating of cables-5 parts	3961:67
20	Terminal marking and direction of rotation of electrical machines	4728:75
21	Type of duty and class of rating of rotating machines	12824:89
22	XLPE insulated PVC sheathed cables-2 parts	7098:88
23	Residual current operated circuit-breakers	12640:88
24	Current transformers-3 parts	2705:92

1.10 QUESTIONS

1. What is the relevance of The Electricity Act 2003?
2. Explain briefly the significance of The Central Electricity Authority.
3. What are the standard voltages used in Electrical Systems in India?
4. Explain why voltage tolerances are permitted in the supply systems.
5. What are the important design aspects in Electrical systems?
6. Explain the different types of safety aspects to be incorporated in system design.
7. How is NEC different from CEA regulations 2010?
8. What are the roles of NEC and NBC (National Building Code) in building design?
9. What is the significance of standard symbols in electrical system design?
10. What is the role of International Electrotechnical Commission in the standardisation of Electro technology?

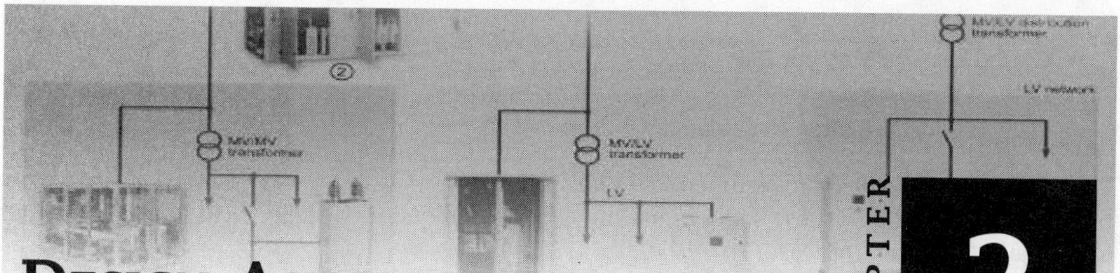

CHAPTER 2

DESIGN ASPECTS OF ELECTRICAL SYSTEMS

OBJECTIVES

At the end of this chapter you will,

- Understand the basis on which building services are classified.
- Have an idea about the design aspects of major building services like lighting, ventilation, climate control and vertical transportation.
- Know the design requirements of minor building services.
- Know the importance of building services in system design.
- Understand the necessity for coordination among the various building service providers.

2.1 CLASSIFICATION OF BUILDING SERVICES

Building services are those aspects of building design that makes the building worthy of its purpose for which they are designed. These services include:

- Lighting and ventilation
- Air-conditioning
- Lifts and escalators.

These three services are called the major building services, because the amount of electricity required for these services is considerably high. There are several other building services which consume lesser amount of power and these services are referred to as minor building services. The types and number of minor building services will depend on the type and occupancy of the building. Minor building services include:

- Electric audio systems
- Electric call bell systems
- Electric clock systems

- Fire alarm system
- Closed circuit TV system
- Cable TV network
- Data networking intercom etc.,

These services are referred to as functional/safety services and are very important for proper functioning of the building. These services use only very small quantity of electricity when compared to the major building services.

Since all building services mentioned above require electricity for their normal functioning, the electrical system designer shall have a clear understanding of the requirements of these services and shall be capable of including these requirements in the electrical system. Most of the building services are designed and executed by specialists in those areas and proper co-ordination is essential between the electrical engineer and the other agencies to arrive at a perfect solution. Therefore, it is the responsibility of the electrical engineer who designs the electrical system, to assess the requirements of the electrical energy required by the other building services and to incorporate these requirements in the electrical distribution network. Then only the other building services will get uninterrupted power for their normal functioning.

Lighting and ventilation services are absolute requirements of any type of buildings. It is essential that the buildings are designed in such a way that, during day time sufficient quantity of natural light and ventilation is made available inside the buildings. The National Building Code gives extensive guidelines in the orientation of the building so as to provide physically and psychologically comfortable living inside the buildings by creating conditions that will ward off the undesirable effects of severe weather conditions. The Architects and the Civil engineers have a great role in selecting the orientation of the building so as to guarantee that the building requires a minimum amount of artificial lighting and ventilation during day time.

2.2 DESIGN ASPECTS OF LIGHTING SERVICES

Good lighting is an absolute necessity for all kinds of building and has three primary aims:

- To promote the work and other activities carried out within the building.
- To promote the safety of the people using the building.
- To create in conjunction with the structure and decoration, a pleasing environment.

In order to realise these aims, the lighting designer will have to be involved in the following aspects of lighting:

- Careful planning of the brightness and colour pattern within the work space and the surroundings so that attention is drawn naturally to the important areas, details are seen quickly and accurately, and the room is free from monotony.
- Using directional lighting, wherever appropriate, to assist the perception of task detail and to give good modelling.
- Controlling the direct and indirect glare from the light sources to eliminate the visual discomfort.
- Minimising the flicker from discharge lamps and paying attention to the colour rendering properties of the light sources.

- Correlating the lighting throughout the building to prevent excessive differences between adjacent areas so as to reduce the risk of accidents.
- Installing emergency lighting systems where necessary.

Good lighting is also the process of providing the right quantity of light at the right quality. The quantity of lighting or the level of illumination for a particular application depends on the following factors:

- Adequacy for preventing both strain in seeing and liability to accidents due to poor visibility
- Adequacy for realising visual comfort
- Adequacy for performance of the task at high efficiency
- Adequacy for pleasantness.

The recommended levels of illumination are specified in IS 3646 (Part-2). Therefore we can see that a good lighting design will involve:

1. Planning the brightness pattern from the point of view of visual performance, safety and amenity.
2. Creating form and texture in the task area and the surroundings.
3. Controlling glare, flicker and stroboscopic effects.
4. Selecting the right kind of light sources to provide correct colour rendering.
5. Maintaining the right level differences in lighting in the adjoining areas to prevent accidents.
6. Providing emergency lighting services.
7. Including effective maintenance and easy installation features.
8. Assuring energy effectiveness.

The complete guidelines on good lighting are given in IS 3646 (Part-1).

2.3 DESIGN ASPECTS OF VENTILATION

Ventilation systems are means of providing a building with the required quantity of fresh air in order to improve the general environment of the building. Proper ventilation means the creation of an environment that stimulates the people to higher efficiency levels.

The important considerations to be looked into while providing the ventilation for a building are:

- To provide fresh air for respiration of the occupants.
- To remove the products of combustion or other contaminants like carbon dioxide, air borne chemicals, pollutants etc., from the inside air.
- To dilute the inside air to prevent vitiation by body odour.
- To maintain the heat balance of the human body.

If these factors are considered, it can guarantee a healthy and comfortable living environment within the building. Ventilation can be arranged in any one of the following methods:

1. Natural supply and natural exhaust of air
2. Natural supply and mechanical exhaust of air
3. Mechanical supply and natural exhaust of air
4. Mechanical supply and mechanical exhaust of air.

2.3.1 Natural supply and natural exhaust

This form of ventilation is perhaps the simplest form of providing ventilation. In this case the role of electrical engineer is practically nil. The Architect/Civil Engineer who designs the building shall have a thorough knowledge about the climatic conditions in order to take full advantage of this type of ventilation. The building orientation is fixed in such a way that natural winds enter the building and sweeps through the building providing the required level of ventilation. The orientation of the building and the proper positioning of the doors, windows and ventilators will ensure a satisfactory level of ventilation. The effects of natural ventilation are uncertain, unreliable and difficult to control. It may be satisfactory in some cases, but not always.

2.3.2 Natural supply and mechanical exhaust

Because of the unpredictable climatic conditions, ventilations as described in section (2.3.1) may not be a good solution for the ventilation problem. Moreover, security reasons prevent the people from keeping the windows open at night, contributing to poor ventilation. To overcome these difficulties, a system of ventilation for forced exhausting of the polluted/stale air is introduced. The pick up points of this stale air are generally the high moisture areas such as kitchen, utility room, bath room etc. The suction (negative pressure) created by the exhaust fan pulls fresh air through the openings (supply points) of the other rooms in the house and provides a fair circulation of fresh air. By properly locating the pick up and supply points, a very good level of ventilation can be achieved.

Therefore air change is a vital factor for keeping a better environment. The air change required depends on the surroundings and also the nature of the work. A bank or office will require much lesser air changes than a foundry shop or boiler house. A general guideline for the air changes is given in Table 2.1.

Table 2.1 General values of ACH for ventilation purpose

Serial No.	Location	Air changes per hour (ACH)
1	Assembly hall	08 to 12
2	Banks	06 to 10
3	Restaurants	15 to 20
4	Club rooms	08 to 12
5	Factories	10 to 20
6	Hospitals	08 to 12
7	Kitchen-domestic	10 to 20
8	Kitchen-commercial	20 to 30
9	Laboratories	10 to 15
10	Offices	08 to 12
11	Paint shop	30 to 50
12	Toilets-common	15 to 20

To determine the number of exhaust fans required for a room, first we have to find the volume of the room in m^3. This volume is to be multiplied by the value of the ACH from the table. This product will be the displacement capacity of the exhaust fans in an hour. Now we have to look at the technical specification of the fan manufacturers to find out the displacement capacities of the fans. Table 2.2 gives the performance tables of typical exhaust fans.

Table 2.2 Performance table of typical exhaust fans

Fan sweep (mm)	Fan speed (RPM)	Power input (W)	Sound level (dB)	Air displacement (Free air flow condition) CMH
230	1370	40	49	760
300	920	45	46	1270
300	1420	82	FQ	1920
380	920	78	Q	2460
380	1420*	150	NI	4000
450	720	90	Q	3400
450	920	132	FQ	4340
450	1420*	370	FN	6800
600	720	240	FQ	7900
600	920*	500	NI	10,450

*Not recommended to be used with louver shutters

CMH Cubic metre per hour
Q Quiet (50-55 dB)
FQ Fairly quiet (55-60 dB)
NI Normal Industrial noise (60-65 dB)
FN Fairly noisy (Above 65 dB)

EXAMPLE 2.1

A common toilet in a public office measuring 12 m x 4.5 m x 3.0 m is to be provided with mechanical exhausting. Design the capacity of the exhaust fans to be used.

Solution:

Volume of the room = 12 x 4.5 x 3.0
= 162 m^3

From Table 2.1 the required level of ACH is between 15 and 20. Assuming an ACH value of 20

Total air to be displaced in an hour = 162 x 20
= 3240 m^3

From Table 2.2, a 450 mm sweep, 90W, 720 rpm exhaust fan will provide a total air displacement of 3400 cubic metre per hour. The noise level of this fan is quiet low (50–55 dB only) and most suitable for the application.

2.3.3 Mechanical supply and natural exhaust

Most of the conventional ventilating systems use circulating fans to provide the required level of ventilation. Circulating fans include ceiling fans, table fans, wall fans, pedestal fans etc. These fans create an air movement that makes the people inside the building more comfortable. Ceiling fans are considered to be the most effective of these fans, since they effectively circulate the air in a room to create a draft throughout the room.

Ceiling fans are appropriate only if the room height is more than 2.5 m. Ceiling fans provide the best results when mounted at a height of 2.5 to 3.3 m above the floor level and 0.3 to 0.45 m below the ceiling. Ceiling fans shall never be installed so that their blades are less than 0.2 m from the ceiling and the tips of the blades are less than 0.45 m from the nearest wall.

Ceiling fans are specified in terms of the sweep diameter. The standard sweep sizes are 900 mm, 1200 mm and 1500 mm. A 900 mm or 1200 mm fan can cool rooms up to 25 m^2 in size. Higher size fans shall be used for larger rooms. Multiple fans can provide better ventilation in rooms longer than 6 m. Smaller fans are effective on an area of 2 m in diameter. Larger fans are effective up to a diameter of 3.5 m.

A larger fan blade will provide comfortable cooling at a lower velocity than a smaller blade. This is important in office spaces where loose papers and other objects are likely to be disturbed by strong breeze. An expensive fan that works quietly and efficiently will offer longer trouble free service than a cheaper fan. It is also advisable to look for the energy star rating of the fans before buying the fan.

2.3.4 Mechanical supply and mechanical exhaust

When the quality of inside air is very important for the process or function of the building space, controlled flow of air with a level of purity is essential. In such cases as in the case of textile mills, pharmaceutical industry or medical facility, a balanced flow of air is obtained by the use of two fans, one for supply of fresh clean air and another for exhausting the stale air. An elaborate design calculation will be required to avoid back drafting and guaranteeing uniform flow of air. Detailed ducting and careful installation are needed in this case.

For more information on ventilation, reference can be made to the following documents.

IS-3362: Code of practice for natural ventilation for residential buildings.
IS-3103: Code of practice for industrial ventilation.

2.4 DESIGN ASPECTS OF CLIMATE CONTROL

When the desired temperature and humidity cannot be achieved by proper ventilation, air-conditioning is resorted to. Since air-conditioning plants require large quantity of electricity, prior consultation shall be arranged between the electrical consultant and the air-conditioning consultant so as to assess the requirement of electricity for the air-conditioning plant.

The design of air-conditioning systems and its associated controls shall be based on the following criteria:

- The nature of application
- The type of building construction
- External and internal load patterns
- Desired space condition
- Permissible limits of control
- Energy conservation aspects

- Possibility of heat recovery and
- Economic factors.

The designer shall also look into the following aspects of load variations before finalising the design.

- Load variations due to climatic changes as in the case of summer months and winter months.
- Loading pattern of the intermediate season.
- Loading pattern during day and night.
- Loading pattern at weekends.
- Loading pattern when electric supply fails and the system operates on standby generators, etc.

Air conditioning is a specialised job carried out by the air conditioning specialists and a detailed discussion on this topic is outside the scope of the book.

2.5 DESIGN ASPECTS OF VERTICAL TRANSPORTATION

Vertical transportation is an integral aspect of high rise buildings. The lifts and escalators form the major means of vertical transportation. These services are designed and installed by lift/escalator manufacturers. The electrical system designer shall collect the relevant information regarding the requirements of electricity for the lift/escalator from the manufacturer. The following aspects are to be taken into account while deciding the electrical requirements for lifts/escalators.

- Number of lifts, its size or passenger carrying capacity and location in the building
- Number of floors served by the lifts
- Height between floors
- Location of the machine room
- Provision for lighting and ventilation
- Location where electric supply is required and in what quantity
- Quality of electric supply
- Occupant load factor
- Control systems used and
- Operation and maintenance.

For more information reference can be made to the following Indian Standard specifications.

IS-1860: Code of practice for installation, operation and maintenance of passenger and goods lifts.
IS-4666: Specification for passenger and goods lifts.

2.6 DESIGN ASPECTS OF MINOR BUILDING SERVICES

All minor building services like audio systems, fire alarm, electric clock, CC TV systems etc. require the electric supply in a limited quantity at specific locations in the building. Therefore, it is the responsibility of the electrical system designer to assess the requirements of the various minor building services as accurately as possible by prior consultations with the respective agencies who are installing those services. A detailed discussion is not attempted here. However, reference can be made to the following Indian Standard specifications for more information.

Audio systems
IS-1881: Code of practice for installation of indoor amplifying and sound distribution systems.
IS-1882: Code of practice for outdoor installation of public address systems.

Fire alarm systems
National building code, 2005.

Call bell services
IS-8884: Code of practice for installation of electric bells and call systems.

Central clock system
IS-8969: Code of practice for installation and maintenance of impulse and electronic master and slave clock system.

2.7 QUESTIONS

1. Explain the importance of building services in the electrical systems design.
2. How are building services classified, explain briefly?
3. What are the basic design aspects of lighting design?
4. What are the design aspects of ventilation services?
5. An assembly hall measuring 45 m × 15 m × 9 m is to be provided with ventilation of the order of 10 ACH. Design a ventilation system using natural supply and mechanical exhausting system.
6. A domestic kitchen measuring 3.6 m × 2.4 m × 3 m requires mechanical ventilation. Assuming suitable values of ACH, design a ventilation arrangement.
7. What are the basic design aspects of air conditioning systems?
8. Why is it necessary that the electrical system designer must closely liaison with the other agencies like air conditioning, lift, fire fighting, etc., service providers?

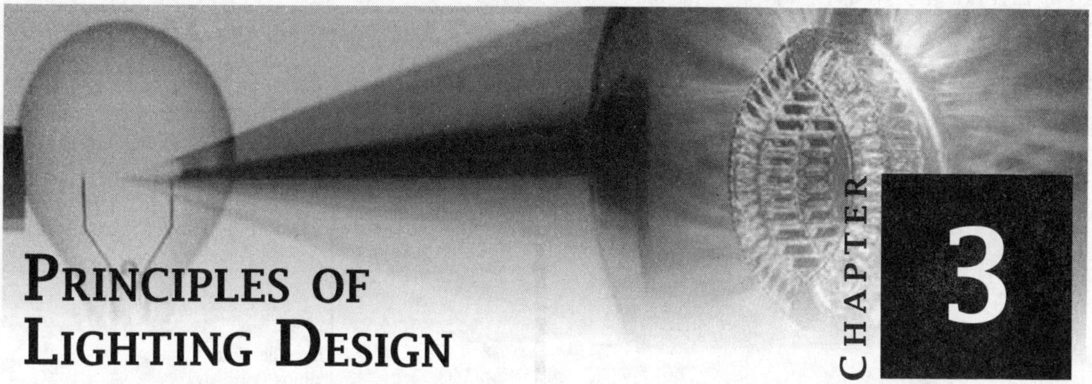

CHAPTER 3

PRINCIPLES OF LIGHTING DESIGN

OBJECTIVES

At the end of this chapter you will

- Understand the operation of various light sources and their characteristics.
- Understand the terms and definitions used in lighting design.
- Be able to design simple lighting schemes.
- Be able to determine the coefficient of utilisation.
- Be able to understand the light loss factor.
- Understand the difference between direct and indirect glare.

The most important aspect of electrical system design for buildings is the design of lighting system. The lighting system must not only provide the desired level of lighting but must also produce visual comfort, proper colour rendition and ambience. Therefore, the designer must have a fairly good knowledge of the light sources and light fixtures that are available in the market. Information regarding the electrical characteristics, photometric details and installation requirements are available from the manufacturers catalogues.

3.1 LIGHT SOURCES

There are six different light sources that are popular today. Incandescent, fluorescent mercury vapour, metal halide, high pressure sodium and low pressure sodium. All lamps except incandescent lamps are gas discharge lamps. Light emitting diode lamps have already entered the lighting arena in a big way. Figure 3.1 shows the development of lamp industry over the years.

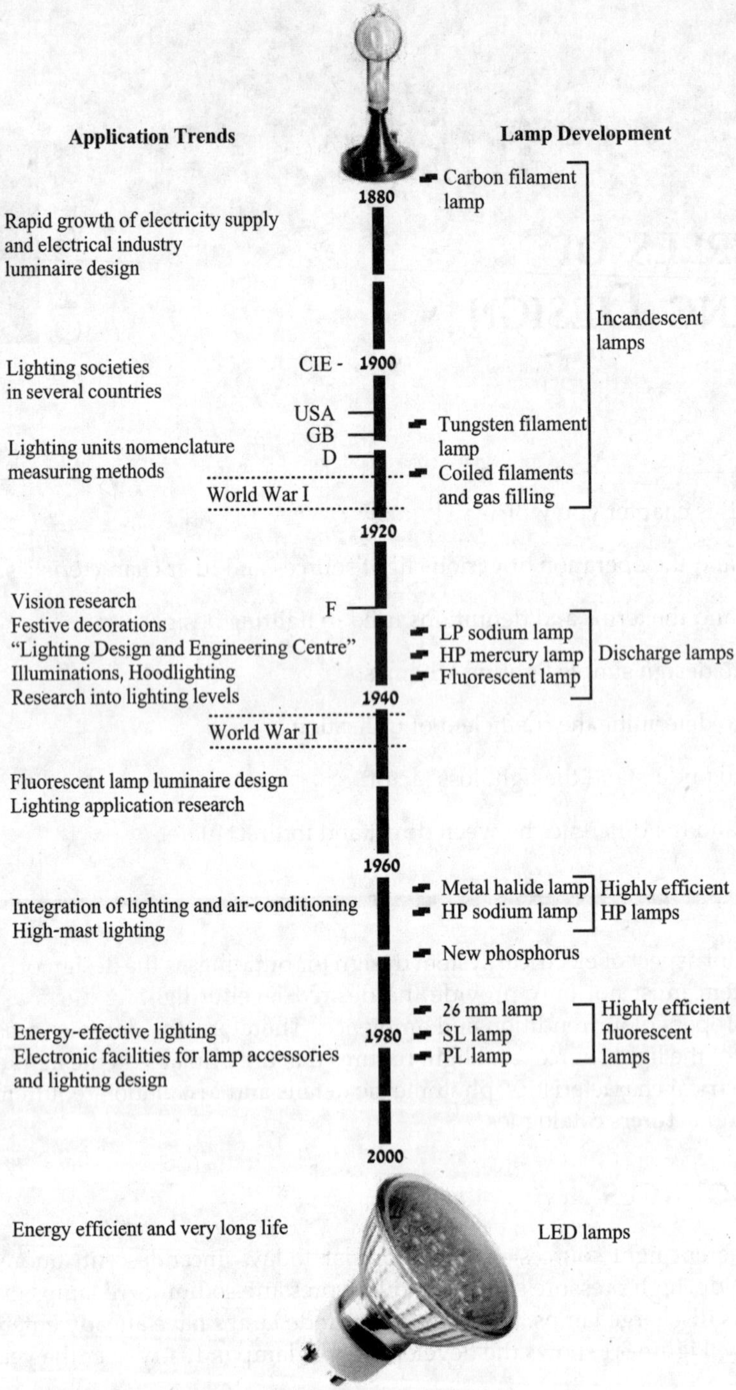

Figure 3.1 Development of lamp industry

3.1.1 Incandescent lamps

Incandescent lamps produce light by electrically heating high resistance tungsten filament to intense brightness. This is one of the most common light sources with the lowest efficacy (lumens per watt) and shortest life. This lamp is still popular due to the simplicity with which it can be used and low price of both lamps and fixtures.

The most common types of incandescent lamp are the "A" or standard "PS" or pear shaped; "R" or reflector "PAR" or parabolic aluminized reflector and the tungsten halogen lamp (quartz). The "R" and "PAR" lamps are generally used for commercial application.

In general the bulbs are clear. Some bulbs are inside frosted to diffuse the concentrated source. These bulbs have a fine coating of silica smoked on the inside of the surface. Over a period of time, the filament evaporates, and to prevent this evaporation, 40 and high wattage lamps are gas filled. A filling of inert gas, such as argon and nitrogen, retards evaporation of the filament, thus allowing higher filament temperature and, hence higher efficacies with less sacrifice of the life. Bulbs under 40 watts are generally evacuated. The life of the lamp and its light output are determined by the filament temperature. The efficiency of incandescent lamps, however, does increase as the lamp wattage increases. In general, incandescent lamps have efficacies of 17 – 23 lumen per watt for up to 500 watts. For higher watts, this range varies from 23 – 33 lumens per watt.

Another type of incandescent lamp is the tungsten halogen lamp, which is also known as quartz lamp. These lamps contain a halogen gas such as iodine or bromine, operate at high temperature and have a smaller envelope. The outer envelope is made of quartz to withstand the very high temperature. At high temperature, the evaporated tungsten associates with a halogen molecule, which prevents it from getting deposited on the bulb wall and instead returns the evaporated tungsten back to the filament. Energy distribution of an incandescent lamp is shown in Figure 3.2.

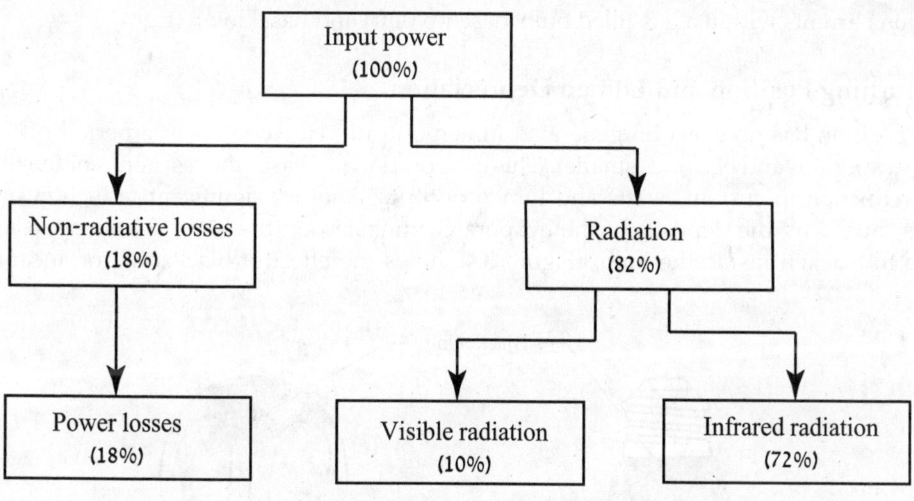

Figure 3.2 Energy distribution of an incandescent lamp

3.1.2 Colour characteristics

Human eye responds differently to various wavelengths of radiation. The brain translates these different wavelengths as colour. A graphic representation of energy emitted by a light source at each wavelength in the spectrum is called a Spacial Power Distribution (SPD) curve. The SPD data are derived through a spectrometer.

Figure 3.3 shows the SPD curve for a typical incandescent lamp. Incandescent lamp source, being a thermal emission type, obeys established physical laws of thermal emission. The curves originate near the ultraviolet range and extend all the way up to the infrared range. Curve A represents a conventional incandescent lamp, which has higher radiant power in the red, orange and yellow regions. This gives incandescent a yellowish-white visible effect on neutral surfaces.

Curve B represents a typical halogen tungsten lamp. This has a slightly higher radiant power in the ultraviolet region and much less in the infrared region. This results in "Whiter" visible effect on neutral surface.

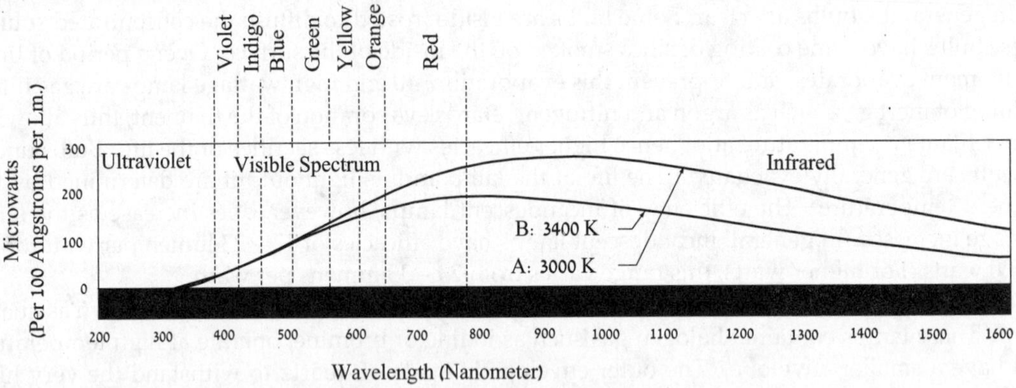

Figure 3.3 SPD of incandescents

Lamp blackening

Convection currents within a gas-filled bulb, base up (left) and base down (right).

3.1.3 Burning Position and Lumen Depreciation

Burning position has no effect on generated lumen output. However, over a period of time, the filament evaporates and becomes smaller. This evaporation increases the resistance of the filaments causing reduction in current, watts and lumen output. Another significant reduction in lumen output is caused by the deposits of the evaporated tungsten on the bulb. These deposits cause the lamp to blacken inside the bulb. Figure 3.4 shows the effect of blackening on incandescent

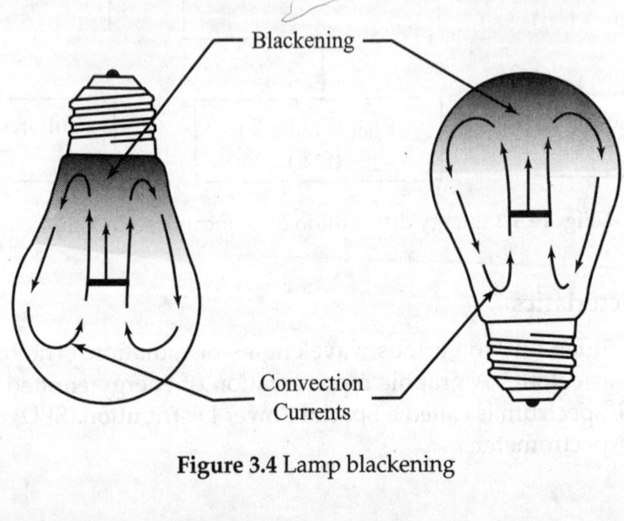

Figure 3.4 Lamp blackening

lamps. In this respect, the burning position is somewhat important. The phenomenon of blackening is minimised in tungsten halogen lamps and thus maintains the initial lumen output.

3.1.4 Energy saving incandescent lamps

The lighting industry, has produced several "energy saving" products to lay emphasis on energy conservation. Two new developments in incandescent lamps are, the krypton filled lamps and the ellipsoidal reflector lamps.

Krypton filled lamps: These lamps use krypton instead of argon inside the bulb. These lamps were primarily designed to replace the existing 40–150 W incandescent lamps. For the same lumen output and rated life, these lamps consume 5–10% less wattage. Lamps are available from 36–143 W to replace 40–150 watt conventional incandescent lamps.

Ellipsoidal reflector lamps: This is a new development in the family of reflector lamps. They are used to replace "A" or "R" lamps in the existing cylindrical down light with black reflecting surface (Figures 3.5a and 3.5b) shows the details of ellipsoidal lamps. When a point source of light is placed at the first focal point of an elliptical reflector, all reflected lights emerge through the second focal point as shown in Figure 3.5a. An ER lamp working on the principles of elliptical reflector produces a pool of light through a narrow opening. This reduces the wastage of light. Figure 3.5b shows the details of an elliptical lamp.

3.1.5 Fluorescent lamps

Fluorescent lamps produce light by creating an arc between two electrodes in atmosphere of low pressure mercury vapour and some inert gas in a glass tube. The mercury vapour produces 253.7 nanometre (nm) vapour energy that strikes the crystals of phosphor and excite them to produce light.

Unlike the incandescent lamp, the fluorescent lamp requires a ballast to strike the electric arc in the tube initially and to maintain that arc. Proper ballast selection is important to optimise light output and lamp life.

3.1.6 Efficacy

The efficacy of most fluorescent lamps is twice that of incandescent. Efficacies ranging from 50 to 70 lumens per watt are common. Efficacy improves as the lamps get longer. The energy distribution of a fluorescent lamp is shown in Figure 3.6.

3.1.7 Ballasts

The principal functions of ballast are:

1. To provide sufficient voltage to start up the lamp
2. To provide power factor correction
3. To provide radio interference suppression.

Energy distribution of a fluorescent lamp. About 22% of total energy is available for visible radiation.

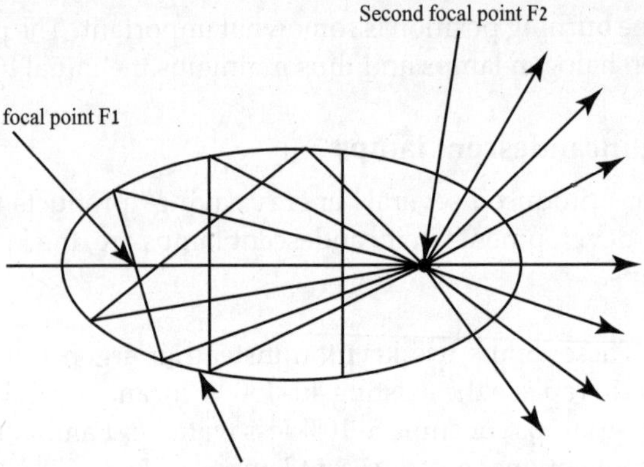

Figure 3.5a Principle of operation of an elliptical reflector

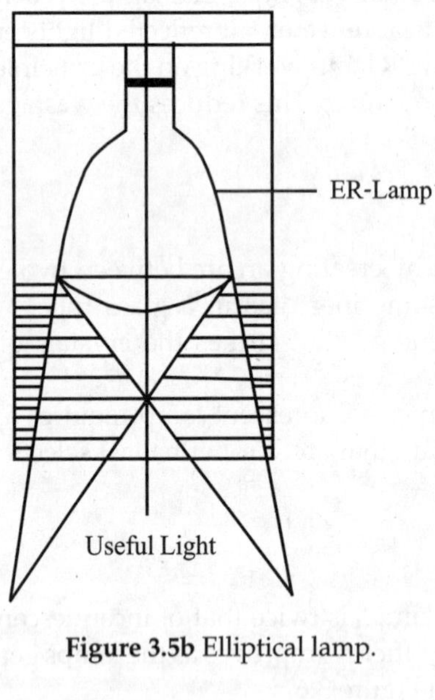

Figure 3.5b Elliptical lamp.

3.1.8 Mortality and lumen depreciation

The life of a fluorescent lamp is significantly dependent on the number of hours it is burnt per start. The most commonly used lamps such as 1200 mm long 40 watt lamps are rated at 20000 hours based on 3 hours per start.

Fluorescent lamps are available in a wide variety of colour, but for most applications the 'cool white', 'warm white' or 'day light' lamps produce acceptable colour. Since fluorescent lamps are linear light sources with relatively low brightness as compared with point sources, they are suited for indoor applications where lighting quality is important and ceiling heights are moderate.

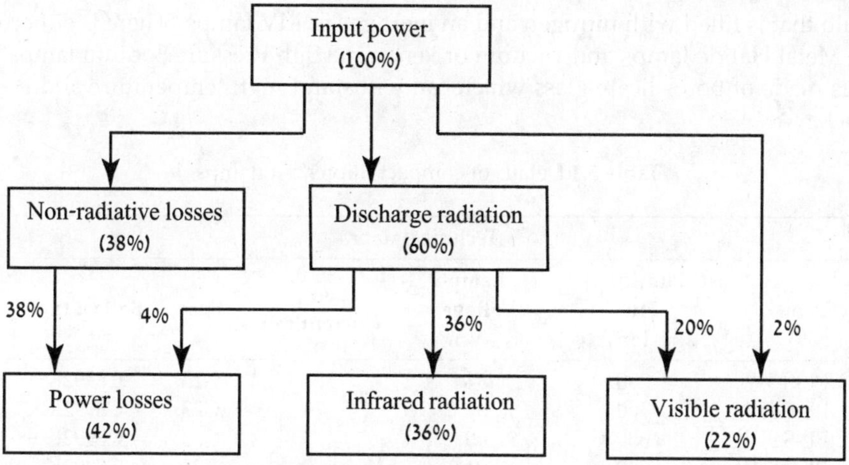

Figure 3.6 Energy distribution of a fluorescent lamp

3.1.9 Energy efficient fluorescent lamps

The compact fluorescent lamps (CFL) were developed after prolonged research as a suitable energy effective alternative to traditional incandescent lamps. Compact fluorescent lamps offer an economic alternative to the incandescent lamps providing the same light output while consuming only 20% of energy. Thus a 9 W CFL can effectively replace a 60 W incandescent lamp saving approximately 80% of precious energy. The other beneficial characteristics are,

1. The life of a CFL is above 7500 hours compared to 1000 burning hours of GLS lamp.
2. The colour characteristics of CFL are almost similar to that of incandescent lamps, thus providing a warm and cozy environment.
3. Being very compact, CFL can ideally suit a small luminaire generally preferred for domestic applications.

3.1.10 Construction and operation

CFL is essentially a gas discharge lamp and operates on the same principle as fluorescent lamp, which requires a ballast and starter for its operation. The tube is coated with improved troposphere fluorescent powder for excellent colour characteristics. In its simplest form, the narrow discharge tubes are bent in U-shape and joined at the tip. The electrodes are connected at the base where the starter is also provided. Two pins project out of the base and are to be connected to a suitable lamp holder. These lamps are ignited at a voltage as low as 140 V. Further researches in the development of CFL have brought about several variants each having its own beneficial characteristics. One such variant is a direct retrofit CFL with a built-in electronic ballast, which can directly be mounted on a conventional GLS lamp holder. Table 3.1 shows the details of different types of CF lamp.

3.1.11 High intensity discharge lamps (HID)

HID is the term used to designate three distinct types of lamp, mercury vapour, metal halide and High pressure sodium. Like fluorescent lamps, they produce light by establishing an arc between two electrodes, however, in HID lamps, the electrodes are only a few centimetres apart, enclosed in opposite ends of a sealed translucent or transparent arc tubes. The arc tube is then enclosed in an outer bulb that is filled with nitrogen and an inert gas in MV lamps. There is either vacuum or nitrogen in Metal Halide lamps and vacuum or xenon in High Pressure Sodium lamps. The outer glass shell is made of borosilicate glass which can withstand high temperature and is resistant to thermal stock.

Table 3.1 Details of compact fluorescent lamps

Technical data					
Type	**Luminous Flux Lm**	**Lamp Voltage V**	**Lamp Current mA**	**Base**	**Ballast type**
PL-S 5W	250	34	180	G23	C8992K
PL-S 7W	400	37	175	G23	C8992K
PL-S 9W	600	60	170	G23	C8992K
PL- S 11W	900	91	160	G23	C8992K

3.1.12 Ballasts

All HID lamps require ballast for their proper functioning. The basic reasons are similar to those of fluorescent lamps. They are categorised as follows.

1. Limits the current: All discharge lamps have negative resistance characteristics. If the lamps are connected to a non-regulated supply, the arc discharge will draw an unlimited amount of current and the lamp will be destroyed. The main function of the ballast is to limit this current flow through the arc tube.
2. Provides a voltage kick to start the arc discharge.
3. Provides power factor correction to offset the coil inductive reactance.
4. Prevents any voltage or current surge caused by the arc discharge from reflecting into the line current.

All HID lamps require several minutes to come to full light output. Also if power is switched off, the arc tube must cool before it can be re-struck again and light produced. Naturally, a few minutes are required for the lamp to switch on again after a power break.

3.1.13 Mercury vapour lamps

Figure 3.7 shows a typical MV lamp. The outer tube is filled with nitrogen and encloses the inner arc tube which contains high purity mercury and argon gas. As the circuit is energised the starting voltage is provided across the starting electrodes and the adjacent main electrode to create an argon arc. This increases the heat and vaporised mercury. The ionised mercury atoms decrease the resistance across the main electrodes and cause the main arc to strike.

Mercury Vapour Lamps are available from 40–1000 watts. Although used extensively in the past, MV lamps are not as popular as other HID lamps today due to relatively low efficacy (24-60 lumens/watt). The colour rendering properties of MV lamps are not as good as those of incandescent lamps. A significant portion of the radiation is in the ultraviolet region resulting in a bluish light. The energy distribution for a 400 watt MV lamp is shown in Figure 3.8.

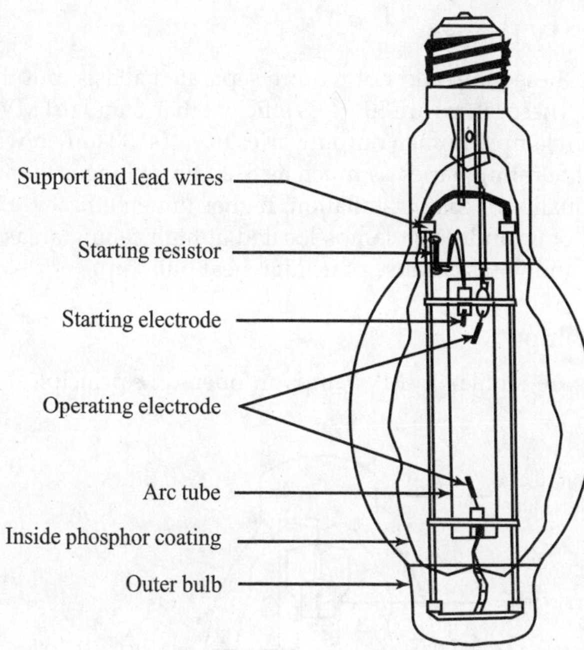

Figure 3.7 Typical construction of a mercury vapour lamp

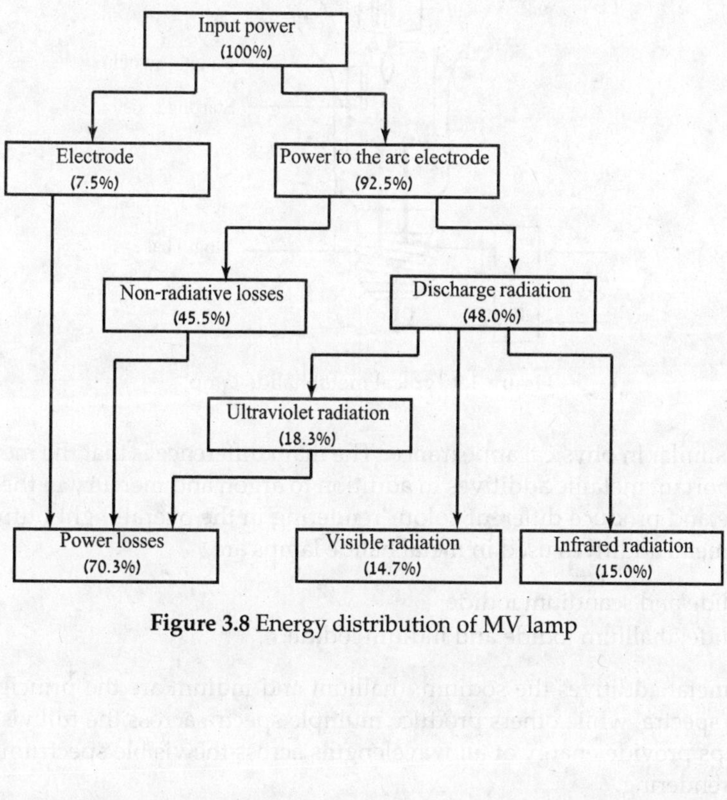

Figure 3.8 Energy distribution of MV lamp

3.1.14 Self ballasted MV lamps

As their names indicate, these lamps do not require separate ballasts, but the ballast circuit is an integral part of the lamp. These lamps are 50% less efficient than standard MV lamps, but 50% more efficient than incandescent lamps. Lumen output varies from 14–30 lumen/watt. Depending on the type and wattage, their life rating varies as much as 8–20 times longer than that for incandescent lamp. Their combined quality of easy installation, higher lumen efficacy, and longer life make it suitable for replacement of incandescent lamps located at high ceiling areas. These lamps should not be used in enclosed luminaires because of its high heat build up.

3.1.15 Metal halide lamps

The metal halide lamps are similar to MV lamps in operating principle. Figure 3.9 shows the constructional details.

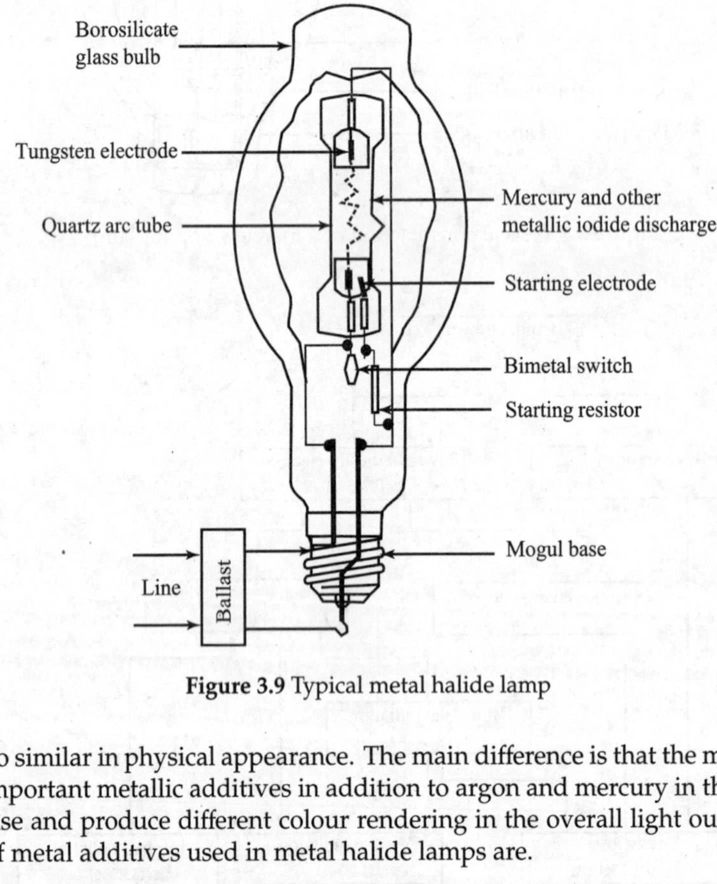

Figure 3.9 Typical metal halide lamp

They are also similar in physical appearance. The main difference is that the metal halide lamps contain some important metallic additives in addition to argon and mercury in the arc tube, which partially vaporise and produce different colour rendering in the overall light output. Two typical combinations of metal additives used in metal halide lamps are.

1. Sodium iodide and scandium iodide
2. Sodium iodide, thallium iodide and indium iodide.

Of the different metal additives the sodium, thallium and indium are the principal producers of dominant colour spectra, while others produce multiple spectra across the full visible region. The metal halide lamps provide energy at all wavelengths across the visible spectrum, creating a well balanced colour rendering.

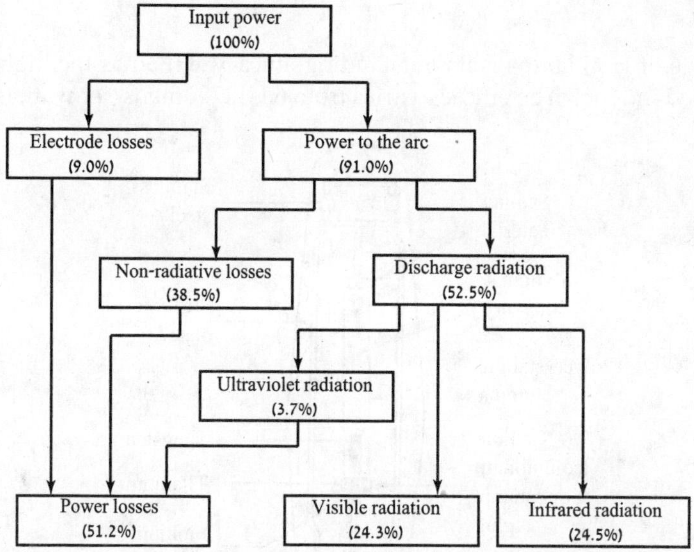

Figure 3.10 Energy distribution of MH lamp

3.1.16 Efficacy

The biggest advantages of metal halide lamps over MV lamps of the same wattage are their substantially higher efficacy. The efficacy varies from 66–100 lumen/watt compared to 24–60 lumen per watt of MV lamps. Their substantially higher efficacy with better colour rendering makes these sources ideally suitable for high ceiling areas where a lot of light with good colour is required.

3.1.17 Burning position

The metal halide lamps are extremely sensitive to burning positions. For the proper functioning of the bimetal switch, each vertically operated lamp has separate designs, one to burn with base up and the other to burn with base down. The lumen output of either is identical as long as they are operated in the vertical position with a tolerance of ±15° from vertical axis. When operated horizontally, these lamps produce significantly low output and also change its colour. Ballasts designed specially for metal halide lamps should be used. The energy distribution of a 400 W metal halide is shown in Figure 3.10.

3.1.18 High pressure sodium vapour lamp (HPS)

The physical appearance, construction, operation and radiation is somewhat different for HPS lamps from others discussed so far. Figure 3.11 shows typical construction of an HPS lamp. The arc tube is made of a ceramic translucent aluminum oxide, which can withstand temperature as high as 1300 °C. The outer envelope is made of a borosilicate glass that can withstand as high as 400 °C.

These lamps produce energy at all wave lengths; however, the major portion of energy is concentrated in the Yellow Orange part of the spectrum. Because of this, the light produced by this lamp is a golden white colour. Figure 3.12 shows the SPD of HPS lamps.

The HPS lamps do not have a starting electrode. The starting is done with the help of a special circuitry in the ballast; a high voltage with high frequency pulse is used to ionise the xenon starting gas. As the lamp warms up, it goes through different colour radiation stages. At first when xenon and mercury are ionised, a bluish white glow takes place, next comes the monochromatic yellow of the sodium at low pressure. Finally a broad spectrum appears to its fullest brightness.

3.1.19 Efficacy

The main advantage of HPS lamps is its outstanding efficacy. HPS has the highest efficacy of all lamps normally used indoors. The efficacy varies from 60–127 lumens per watt.

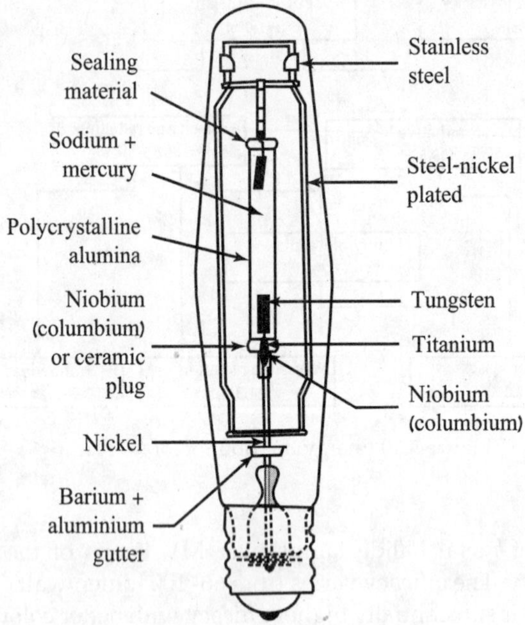

Figure 3.11 Typical high pressure SV lamp

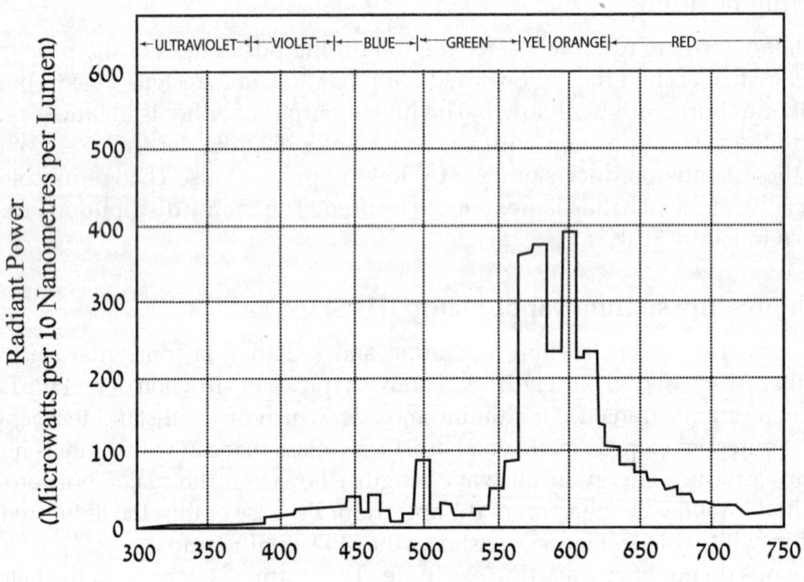

Figure 3.12 SPD of HP SV lamp

HPS lamps are available in 35–1000 watt size. Another attraction of these lamps is that the life of these lamps is around 24000 hours. With moderate colour rendition, outstanding lumens efficacy and long life, these lamps are the best energy savers in the family of HID lamps. Although HPS lamp first found its practical use in outdoor lighting, it now is a readily accepted light source indoors in industrial plants. It is also being used in many commercial and institutional applications as well. The energy distribution of a 400 WHPS lamps is shown in Figure 3.13.

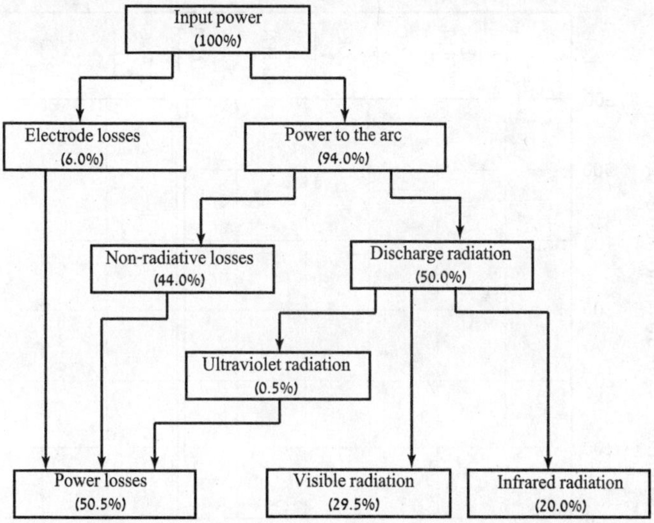

Figure 3.13 Energy distribution of HP SV lamp

3.1.20 Low pressure sodium lamp (LPS)

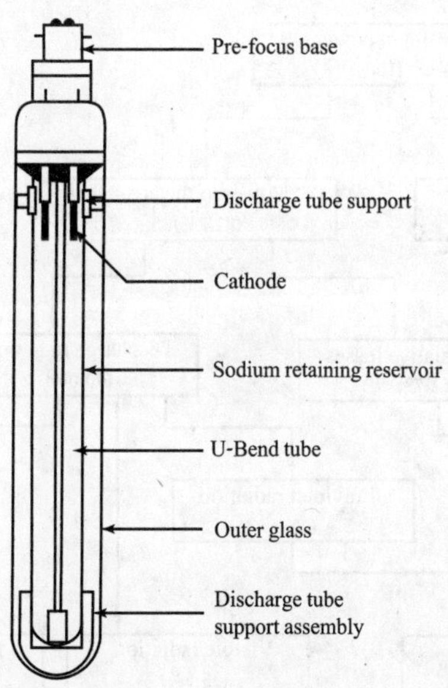

Figure 3.14 Typical low pressure SV lamp

The Low Pressure Sodium Lamp is the most efficient light source presently available providing up to 183 lumens per watt. The light producing element in an LPS is a U-shaped arc tube containing sodium vapour. The outer jacket is made of borate glass. Figure 3.14 shows the construction details of an LPS lamp. The use of these lamps indoors is severely restricted, because, it has a monochromatic (yellow) light output. Consequently, most colour illuminated by this source appears as tones of grey. The SPD of a typical LPS lamp is shown in Figure 3.15.

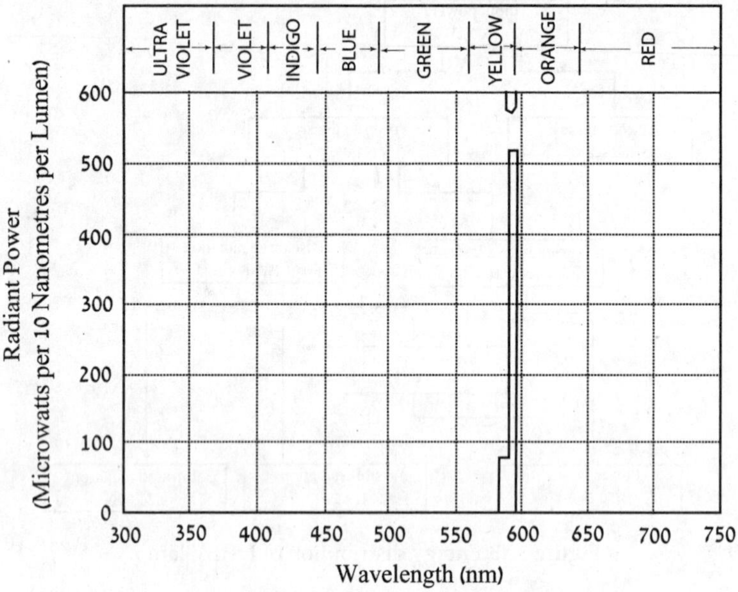

Figure 3.15 SPD of LP SV lamp

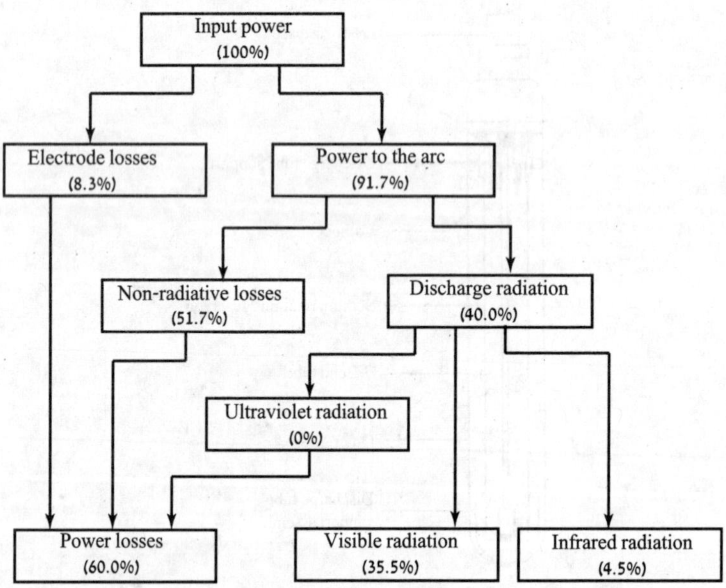

Figure 3.16 Energy distribution diagram of LP sodium lamp

The greatest advantages of LPS lamps are its outstanding lamp lumen efficacy. Lamp efficacy is 135 lumen per watt for lamps rated 35 watts, and goes up to 183 lumens per watt for lamps rated 180 W. They also have long and reliable life of about 18000 hours. Figure 3.16 shows the energy distribution curve of a typical LPS lamp. LPS lamps are available in the range from 18–180 watts. Ballasts designed specially for LPS must be used. The primary use of these lamps is in street lighting as well as outdoor area and security lighting.

3.1.21 Light Emitting Diodes (LED)

Light Emitting Diodes or **LEDs** are among the most widely used of all the different types of semiconductor diodes available today. They are the most visible type of diodes that emit a fairly narrow bandwidth of either visible light at different coloured wavelengths, invisible infra-red light for remote control applications or laser type light when a forward current is passed through them. A "**Light Emitting Diode**" or LED as it is more commonly called, is basically just a specialized type of PN junction diode, made from a very thin layer of fairly heavily doped semiconductor material. When the diode is forward biased, electrons from the semiconductors conduction band recombine with holes from the valence band releasing sufficient energy to produce photons which emit a monochromatic (single colour) of light. Because of this thin layer a reasonable number of these photons can leave the junction and radiate away producing a coloured light output. Thus, we can say that when operated in a forward biased direction **Light Emitting Diodes** are semiconductor devices that convert electrical energy into light energy. Figure 3.17 shows the constructional details of an LED.

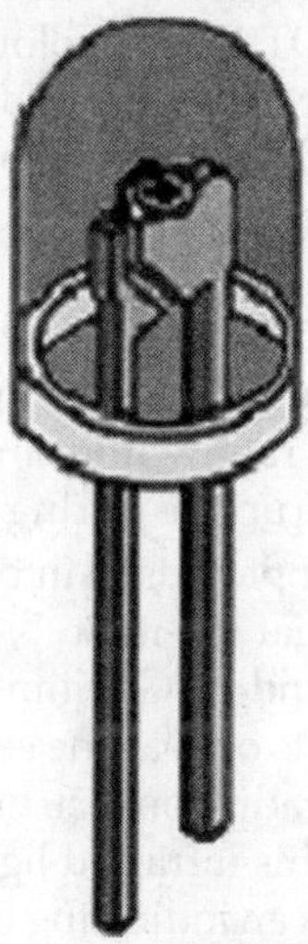

Figure 3.17 A light emitting diode

Construction of a light emitting diode is very different from that of a normal signal diode. The PN junction of an LED is surrounded by a transparent, hard plastic epoxy resin hemispherical shaped shell or body which protects the LED from both vibration and shock. Surprisingly, an LED junction does not actually emit that much light so the epoxy resin body is constructed in such a way that the photons of light emitted by the junction are reflected away from the surrounding substrate base to which the diode is attached and are focused upwards through the dome shaped top of the LED, which itself acts like a lens concentrating the amount of light. That is why the emitted light appears to be brightest at the top of the LED. However, not all LEDs are made with a hemispherical shaped dome for their epoxy shell. Some LEDs have a rectangular or cylindrical shaped construction that has a flat surface on top. Also, nearly all LEDs have their cathode, (K) terminal identified by either a notch or flat spot on the body, or by one of the leads being shorter than the other, (the Anode, A).

Unlike normal incandescent lamps which generate large amounts of heat when illuminated, the light emitting diode produces a "cold" form of generation of light which leads to high efficiencies than the normal "light bulb" because most of the generated energy radiates away within the visible spectrum. Because LEDs are solid-state devices, they can be extremely small and durable and provide much longer lamp life (up to about 50000 hours) than normal light sources.

Light Emitting Diode Colours

So how does a light emitting diode get its colour? Unlike normal signal diodes which are made for detection or power rectification, and which are made from either Germanium or Silicon semiconductor materials, **Light Emitting Diodes** are made from exotic semiconductor compounds such as Gallium Arsenide (GaAs), Gallium Phosphide (GaP), Gallium Arsenide Phosphide (GaAsP), Silicon Carbide (SiC) or Gallium Indium Nitride (GaInN) all mixed together at different ratios to produce a distinct wavelength of colour. Different LED compounds emit light in specific regions of the visible light spectrum and therefore produce different intensity levels. The exact choice of the semiconductor material used will determine the overall wavelength of the photon light emissions and therefore the resulting colour of the light emitted.

Thus, the actual colour of a light emitting diode is determined by the wavelength of the light emitted, which in turn is determined by the actual semiconductor compound used in forming the PN junction during manufacture and NOT by the colouring of the LEDs plastic body although these are slightly coloured to both enhance the light and indicate its colour when its not be used. Light emitting diodes are available in a wide range of colours with the most common being RED, AMBER, YELLOW and GREEN and are thus widely used as visual indicators and as moving light displays.

Recently blue and white coloured LEDs are also available but these tend to be much more expensive than the normal standard colours due to the production costs of mixing together two or more complementary colours at an exact ratio within the semiconductor compound and also by injecting nitrogen atoms into the crystal structure during the doping process.

Table 3.2 shows that the main P-type dopant used in the manufacture of **Light Emitting Diodes** is Gallium (Ga, atomic number 31) and that the main N-type dopant used is Arsenic (As, atomic number 31) giving the resulting compound of Gallium Arsenide (GaAs) crystal structure. The problem with using Gallium Arsenide on its own as the semiconductor compound is that it radiates large amounts of low brightness infra-red radiation (850 nm–940 nm approx.) fromits junction when a forward current is flowing through it. This infra-red light is ok for television remote controls but not very useful if we want to use the LED as an indicating light. But by adding Phosphorus (P, atomic number 15), as a third dopant the overall wavelength of the emitted radiation is reduced to below 680 nm giving visible red light to the human eye. Further refinements in the doping process of the PN junction have resulted in a range of colours spanning the spectrum of visible light as we have seen above as well as infra-red and ultra-violet wavelengths.

By mixing together a variety of semiconductor, metal and gas compounds the following list of LEDs can be produced.

- Gallium Arsenide (GaAs)—infra-red
- Gallium Arsenide Phosphide (GaAsP)—red to infra-red, orange

Table 3.2 Typical characteristics of light-emitting diodes

LED Characteristics			
Semiconductor Material	**Wavelength**	**Colour**	**VF @ 20 mA**
GaAs	850–940 nm	Infra-Red	1.2 V
GaAsP	630–660 nm	Red	1.8 V
GaAsP	605–620 nm	Amber	2.0 V
GaAsP:N	585–595 nm	Yellow	2.2 V
AlGaP	550–570 nm	Green	3.5 V
SiC	430–505 nm	Blue	3.6 V
GaInN	450 nm	White	4.0 V

- Aluminium Gallium Arsenide Phosphide (AlGaAsP)—high-brightness red, orange-red, orange, and yellow
- Gallium Phosphide (GaP)—red, yellow and green
- Aluminium Gallium Phosphide (AlGaP)—green
- Gallium Nitride (GaN)—green, emerald green
- Gallium Indium Nitride (GaInN)—near ultraviolet, bluish-green and blue
- Silicon Carbide (SiC)—blue as a substrate
- Zinc Selenide (ZnSe)—blue
- Aluminium Gallium Nitride (AlGaN)—ultraviolet

Like conventional PN junction diodes, LEDs are current-dependent devices with its forward voltage drop VF, depending on the semiconductor compound (its light colour) and on the forward biased LED current. The point where conduction begins and light is produced is about 1.2V for a standard red LED to about 3.6V for a blue LED. The exact voltage drop will of course depend on the manufacturer because of the different dopant materials and wavelengths used. The voltage drop across the LED at a particular current value, for example 20 mA, will also depend on the initial conduction VF point. As an LED is effectively a diode, its forward current to voltage characteristics curves can be plotted for each diode colour as shown in Fig. 3.18.

Before a light emitting diode can "emit" any form of light it needs a current to flow through it, as it is a current dependent device with their light output intensity being directly proportional to the forward current flowing through the LED. As the LED is to be connected in a forward bias condition across a power supply it should be current limited using a series resistor to protect it from excessive current flow. Never connect an LED directly to a battery or power supply as it will be destroyed almost instantly because too much current will pass through and burn it out.

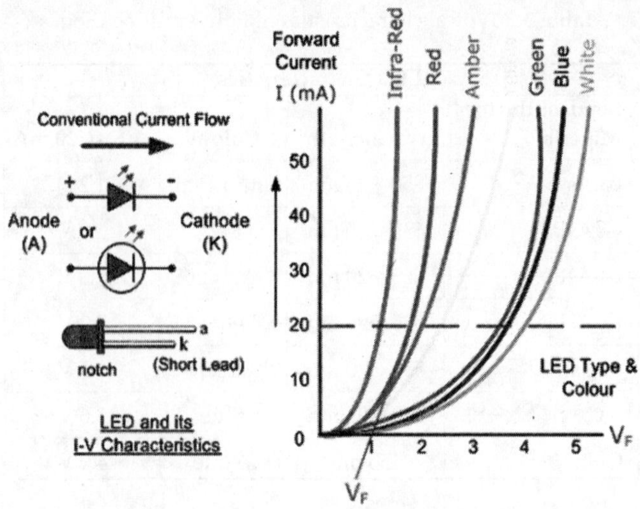

Figure 3.18 Light Emitting Diode (LED) Schematic symbol and its I-V Characteristic Curves showing the different colours available

Each LED has its own forward voltage drop across the PN junction and this parameter which is determined by the semiconductor material used, is the forward voltage drop for a specified amount of forward conduction current, typically for a forward current of 20 mA. In most cases LEDs are operated from a low voltage DC supply, with a series resistor, RS used to limit the forward current to a safe value from say 5 mA for a simple LED indicator to 30 mA or more where a high brightness light output is needed.

LEDs for lighting

A direct comparison of LEDs to other prevailing lighting technologies, such as incandescent and fluorescent, indicates the energy savings that can be realized. Incandescent light uses the maximum energy, fluorescent is second, and LEDs are the most efficient of the three. Unlike incandescent lamps, the electronics used to power an LED lighting system are complex. The first requirement of an LED based lighting scheme is Direct Current (DC) source, since LEDs are direct current devices. The second requirement is a constant-current DC driver. Standard voltage-based design techniques for common light-emitting diodes cannot be applied when designing for LED lighting modules. When an LED lighting module begins to conduct, they cease to act like standard light-emitting diodes: the forward voltage across the device begins to fall and current through the device increases correspondingly, depending on the device temperature. The higher the device temperature, the chances are that the device will face a thermal run away situation. Thus there is a chance for the failure of the device if the current through the LED module is not limited properly. Asimple resistor in series with it is not sufficient, as small changes in applied voltage will produce larger swings in current and the devices brightness. LEDs for lighting require careful current management, because series currents that pass through the devices can be affected by temperature and applied voltage. This becomes important when driving large arrays of LED lighting devices. Although good design generally incorporates the use of a current-limiting resistor, this resistor is used more as a fuse in the event of a mechanical short in the driven array of LEDs. Using current-based driving takes the pressure off the surrounding circuitry to provide precise voltages, and only requires that the minimum voltage along with the proper amount of drive current be available. The current driver also provides a measure of reliability in systems that will be exposed to extreme conditions. Thus, light output will be consistent, no matter what the ambient temperature is.

Hence, an LED driver is required in an LED lighting system, similar to the ballast in a fluorescent lighting system. Since the electric current passing through the LED must flow in one direction, Direct Current (DC) power must be used. A small fluctuation on the input forward voltage will cause wide variations in the current passing through the diode and therefore the DC power source should be either a constant current source or have a current limiting device.

Another factor to be considered when discussing the use of LEDs for lighting is the heat management. Proper thermal management is extremely important when designing with LEDs, which are more sensitive to heat than some other types of devices. A recent technology used in heat management is the use of a wide ceramic substrate to achieve better heat dissipation, as well as high reliability. The ceramic substrate is an excellent way to get the heat out of the lighting LED, as it offers superior heat conduction over a greater area. Therefore, it is far more effective in dissipating the heat generated into the heatsink. Some other solutions rely upon heat conduction through the mounting lugs to handle the heat generated within the LED. The biggest challenge to any LED lighting design is the heatsink. However, doing this effectively means achieving high longevity of the device at its rated output.

Attention must be paid to the drive electronics for both fluorescent- and LED-based lighting when designing a dimmable luminaire. Fluorescent bulbs require dimmable ballast, and this ballast must quench the luminaire when the line voltage to it falls below a certain point and the arc can no longer be sustained. Otherwise, the life of the lamp will be severely shortened. Likewise, the fluorescent ballast must be of a special type, able to withstand the chopped AC, electrical noise, and voltage spikes created by inexpensive SCR dimmers. The complexity of a dimmable driver increases exponentially as the driver must accept what is now a highly distorted AC line waveform, turn it into useable current, then power and dim its load accordingly with the input. This must be achieved without throwing a greater load upon the input line, since as voltage goes down, current must go up to maintain the same output. If the luminaire will not be dimmed, then a relatively simple constant-current driver can be used. If the fixture will be dimmed, then the driver must be designed for dimming. Several solutions and technologies for dimming LED modules currently exist, and others are being introduced regularly.

Dimming a single LED is normally achieved by placing a resistor or variable resistor (potentiometer) in series with it; but this carries the risk of inconsistent light output, especially if the supply voltage to the series resistor is allowed to vary. Light output can also vary with ambient temperature while using this method. A current regulator circuit will be a better solution. Using an in-line resistor does not work as well for multiple LED configurations, such as series-parallel arrays. A better method in this case begins with the use of a current driver. A current driver circuit can keep the array from thermal runaway if driving it near its maximum-rated level, since the forward voltage tends to drop with increased heat in these types of devices. Dimming the array can now be accomplished through the use of Pulse Width Modulation (PWM). The PWM method can control the driver and therefore operate the array of LEDs with the recommended forward current. Dimming is achieved by turning the LEDs on and off at high frequency. PWM actually varies the duty cycle of the on and off portion of the current that feeds the LEDs. The frequency is fast enough that the human eye cannot see it; thus, dimming is efficiently achieved. LEDs are quite robust and can handle dimming very well, whereas dimming can be life-shortening to fluorescent lights. In low temperatures, the LEDs light instantly, while fluorescents require a definite time to come to full brightness.

LED Advantages

A dramatic example of LED superiority over that of gas-discharge lighting is in that of industrial large-area lighting; for instance, landscape, street, and lot lighting. LEDs are more efficient than gas-discharge lighting, provide a more pleasant light, and last longer. Also, because the light produced by LEDs does not contain any UV component, LEDs do not attract insects. This is a major advantage in exterior lighting.

LED bulbs offer a number of significant advantages over other lighting types. One is longer operating life; up to 50,000 hours to half brightness with LEDs. This is 60% longer life than an average long-life sodium-vapor commercial luminaire.

As previously mentioned, LEDs also offer increased durability because they are solid state devices, and LEDs also offer more design flexibility given their small footprint and freedom of placement. For instance, certain types of arch lighting and other creative lighting solutions that mount lights on curved surfaces are newly enabled by LEDs.

The growing popularity of LEDs is also linked to their benefit to the environment. They require little energy to operate, and lack hazardous materials such as Mercury (Hg). It is likely that the adoption of LEDs into lighting designs will continue to increase due to their environmental friendliness, driven by both customer awareness and increasing legislation related to energy conservation.

3.1.22 Photometric analysis

Photometric report

A good lighting design is dependent on the proper selection of luminaires to produce light of required quantity and quality at the right location. Photometrics is the state of art that analyses the luminaires from their operational standpoint and aids the lighting designer in making a proper selection. In general, each type of luminaire manufactured by a company may be totally different from another type of luminaire using the same lamp. The photometrics of each type of luminaire are available from the manufacturer, and are prepared in the format recommended by the Illumination Engineering Society (IES). A typical photometric report of indoor type luminaire includes the following:

1. Luminaires identification and general description
2. Candlepower distribution curve
3. Candlepower summaries, average and zonal
4. Zonal lumens and percentage
5. Luminaires efficiency
6. Maximum and average brightness
7. CIE-IES luminaires classification
8. Shielding angle
9. S/M H ratio/spacing criterion
10. Coefficient of utilisation.

A detailed discussion of the photometric report is outside the scope of this book.

3.2 LIGHTING CALCULATIONS

Lighting design is the process of selecting the right kind of luminaires to produce light of the required quantity and quality at the right location. All light designs are related to two major considerations.

1. Quantity of lighting
2. Quality of lighting.

While the quantity requirement is an unavoidable one, quality also plays an equally important role. There are two basic design techniques to determine quantity of illuminance, the point by point method and the lumen method.

3.3 POINT BY POINT METHOD

This method permits determination of illumination level due to one or more source upon a specific point within an area. The inverse-square law is made use of in this, which states that the illumination at a point on a surface perpendicular to the light ray is equal to the luminous intensity of the source at that point, divided by the square of the distance between the source and the point of calculation.

$$\text{Illuminance } E = \frac{I}{D^2}$$

Where,

I = luminous intensity in Candela and
D = Distance in metre between the source and the point of calculation.

If the surface is not perpendicular to the light rays, suitable trigonometric function must be used. Figure 3.19 shows the details of calculation.

1. Source directly above the point.

$$\text{Illuminance } E = \frac{I}{D^2}$$

2. When light arrives at the point on a horizontal surface at an angle.

$$\text{Illuminance } E_h = \frac{I_\theta \cos\theta}{D^2}$$

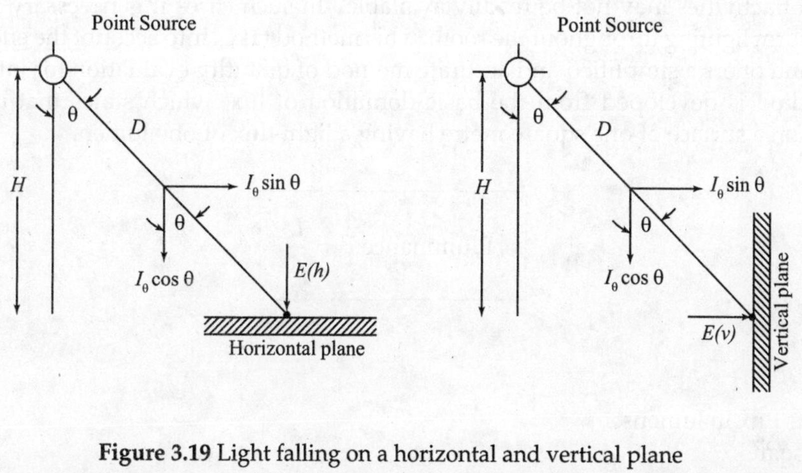

Figure 3.19 Light falling on a horizontal and vertical plane

EXAMPLE 3.1

A point source produces 3000 cd in the direction of interest. The angle of incidence with respect to the vertical is 30°. Determine the illuminance at a point 5 m away from the light source.

Solution:

$$\text{The illuminance } E = \frac{I_\theta \cos\theta}{D^2} = \frac{3000}{5^2}\cos 30 = 103.92 \text{ lux}$$

Although this method is very simple, a number of assumptions made in the calculation limit its application.

1. This formula is true only for point sources
2. For luminaires with asymmetrical candle power distribution, the CP distribution curve of the required vertical angle must be used
3. Application with rows of fluorescent lamps (linear source) will be unsuitable
4. Application of this method for interior lighting with number of luminaires will necessitate the use of computers, since contribution of luminaire at each location must be analysed for all points in the room
5. This method does not take into account surface inter-reflectance.

Consequently, this method is most suited for large areas, especially outdoor lighting where a few luminaires are involved and where no surface reflectance are to be accounted. It can also be used effectively in interior where task lighting is to be calculated on a surface lighted by a few luminaires and where the room has negligible surface reflectance.

3.4 AVERAGE LUMEN METHOD

This method offers a much more simplified way of calculating an average uniform illuminance level on a plane in interiors. In many applications complete information regarding the task location or the type of activities may not be readily available. In such cases it is necessary to provide an average uniform lighting throughout the room. This method takes into account the effects of surface reflectance and offers a simplified and accurate method of quantity evaluation for interiors.

This method is developed from the basic definition of lux, which states that one lux is the illuminance on a surface of one square metre having a light flux of one lumen.

$$\text{Illuminance } E = \frac{L}{A}$$

Where,

L = luminous flux in lumens.
A = Area in sq.m.

In reality, all the luminous flux generated by the lamps will not fall on the work plane. A number of factors like, luminaire candle power distribution, efficiency, room size and shape, surface reflectance and luminaire height, will affect the total number of lumens reaching the work plane. Thus the formula is to be multiplied by a coefficient of utilisation (CU) to take into consideration all the above factors. Therefore,

$$E = \frac{L \times CU}{A}$$

This, however, is the initial illuminance level. The luminaire, lamp and even the surface will accumulate dust over a period of time and the lamp lumen output will depreciate with time. To obtain the maintained illuminance level, the formula now must be multiplied by a 'light loss factor' (LLF) to account for the depreciation in light output. Hence,

$$E = \frac{L \times CU \times LLF}{A}$$

where, L-represents the total number of lumens produced by all luminaires in a room. The formula can be modified as,

$$E = \frac{L_n \times N \times CU \times LLF}{A}$$

N = Number of luminaire
L_n = Initial lumen output per luminiare.

$$N = \frac{A \times E}{L_n \times CU \times LLF}$$

From the above, it is obvious that for a given level of illuminance and area, the only means of reducing the number of luminaires is by using the highest values of L_n, CU and LLF. The lower the number of luminaires, the less the power consumption.

A lighting designer must, therefore look at all aspects of luminaire selection to determine which will offer the highest values of three critical factors viz: L_n, CU and LLF.

3.5 LUMEN

The only way of obtaining maximum lumen is the proper selection of light source. The largest lumen producing source appears to be the immediate choice. But this should not be the only criterion. Colour rendering, lumen maintenance, life and cost are some of the other features that also have to be considered.

The surface reflectance and room proportions will have marked effect on CU values. Ceiling reflectance has the most significant effect on CU values with luminaires that produce light upwards, indirect and direct-indirect type of luminaires fall in this category. Wall reflectance has significant effect on almost all luminaires, and particularly for luminaires with wide spread distribution. Floor reflectance has least effect on CU values, since most tasks are located above work plane. Light reflecting from floor will have to rebound from the ceiling to reach the work plane. Most floors are intentionally made dark because of maintenance problem, and hence offer low reflectance.

However, there are certain areas where highly reflecting floors are useful. When the task is located below the work plane, as in the case of aircraft repair work, a high reflecting floor will enhance the light available at the task location. Every lighting designer must see that maximum possible value of CU is obtained so that maximum useful light is available for the same input power and hence contribute to energy saving.

3.6 LIGHT LOSS FACTOR (LLF)

Periodic schedule and appropriate maintenance is very essential in maintaining the initial lighting level. An LLF would not be required if light output remains constant throughout the useful life. However, inherent light loss characteristics of the luminaire and unfavourable ambient conditions will force the net output to deteriorate with time. An LLF is introduced in lighting calculations to make up for expected loss of light in the lighting system. Figure 3.20 shows the effect of light loss.

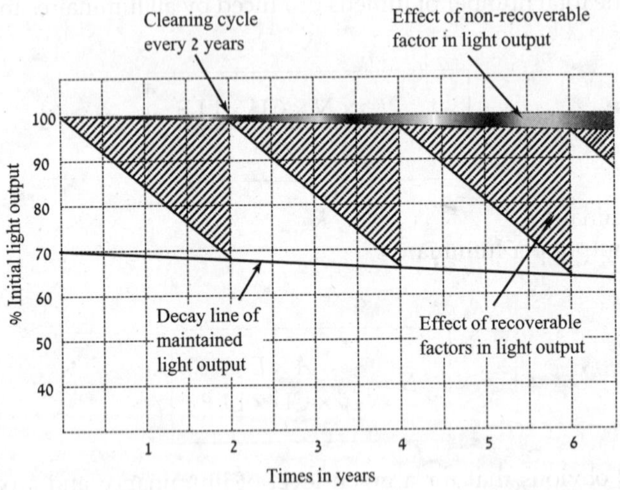

Figure 3.20 Effect of light loss over a period of time

Inclusion of an LLF will cause a higher initial illuminance level that drops with time and rises to initial value, after scheduled maintenance. But even with proper maintenance, it will be difficult to get back to initial level of illumination; because of the inherent light loss. The shaded portion shows the wasted energy, which increases with longer maintenance intervals. Light loss cannot be avoided, but a substantial portion of wasted energy can be saved with shorter maintenance intervals. The various factors that contribute to light loss are of two types.

1. Recoverable
2. Non recoverable.

The recoverable factors include:

1. Luminaire dirt depreciation (LDD)
2. Room surface dirt depreciation (RSDD)
3. Lamp lumen depreciation (LLD) and
4. Lamp burn out (LBO).

3.6.1 Luminaire dirt depreciation (LDD)

The greatest loss of light output is due to the dirt accumulation on lamps and luminaire reflecting surface. Proper selection of maintenance schedule can greatly improve this factor. In general the following guidelines may be followed for maintenance schedule.

Air-conditioned spaces	: Once in two years
Non A/C offices, School, etc.	: Once in a year
Industrial areas	: 3-6 times a year
Food preparation area	: every week.

3.6.2 Room surface dirt depreciation (RSDD)

This factor takes into account the dirt or dust accumulation on surfaces. A proper schedule for cleaning the reflecting surface must be followed to maintain the reflectance.

3.6.3 Lamp lumen depreciation and lamp burn out (LLD and LBO)

Lamp lumen depreciation is an inherent characteristic of all re-lamps. Hence, unless a suitable lamping policy is followed, the illumination level may be adversely affected. Two types of re-lamping method used are

1. Spot re-lamping
2. Group re-lamping.

Spot re-lamping refers to changing of lamps as and when a lamp burns out, whereas group re-lamping is the process of replacing all the lamps in an installation after the useful life period of the lamps irrespective of the fact whether the lamps are in working condition or not.

3.6.4 Non-recoverable factors

These include:

1. Luminaire ambient temperature (LAT)
2. Voltage variation (VV)
3. Ballast factor (BF) and
4. Luminaire surface depreciation (LSD)

These factors present those conditions of a lighting system that reduce the light output, where nothing in terms of periodic maintenance can recover the losses.

3.6.5 Luminaire ambient temperature (LAT)

A variation in the ambient temperature does not have much effect on the incandescent and HID lamps. However, the fluorescents are affected by a change in ambient temperature. Fluorescents produce a peak output at about 25 °C.

3.6.6 Voltage variation factor (VV)

For incandescent lamps a variation of 1% voltage may cause as much as 3% variation in light output. For HID and fluorescents also variation in voltage affects their output.

3.6.7 Ballast Factor (BF)

Ballast factor is defined as the ratio of light output by a commercial ballast to that by reference ballast. If the ballast used in the luminaire is different from that specified by the manufacturer, light output will differ.

3.6.8 Luminaire surface depreciation factor (LSD)

Changes in the various components used in the manufacture of luminaires can cause reduction in light output. For example, due to aging, the painted or polished surface will have reduced reflectance. Similarly, the plastic louvers tend to turn yellow, thereby reducing the light output.

All four non-recoverable factors will depreciate the output permanently and nothing can be done to recover them.

3.7 QUALITY OF LIGHTING

Quality assessment in lighting is related to measurement of glare, colour appearance and modelling. Glare can be broadly divided into two types: direct glare and indirect glare. Direct glare comes into picture with all tasks with heads up position. On the other hand, indirect glare is associated with tasks that involve heads down position. Figure 3.21 shows the two types of glare normally encountered.

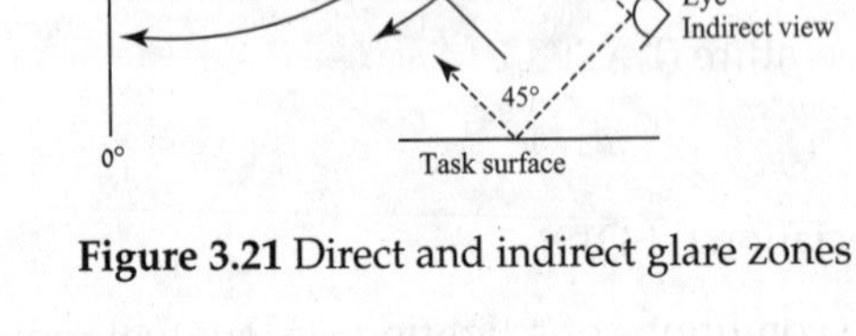

Figure 3.21 Direct and indirect glare zones

3.7.1 Direct glare

Direct glare mainly results from two types of source. This first is an excessively high luminous source staring into persons eyes so as reduce his ability to see the task. This is known as "disability glare". An example of this type of glare is looking straight at the high beam of an oncoming car. The other type of direct glare is due to peripheral stray lights that are not necessarily aimed at the eye. This is known as "discomfort glare".

3.7.2 Indirect glare

Indirect glare is the result of mirror effect of the tasks that reflect the light source from tasks to the eyes of the viewer. This results in a loss of visibility and visual performance. This is most easily visualised if a mirror is placed at the task surface and image of light fixture is seen at the normal viewing angle. This form of indirect glare is called 'veiling reflection' because its effects are similar to those that would result if a thin veil were placed between the workers eye and the task surface. Veiling reflection reduces the contract between the foreground and background. This means that room lighting may hinder the visibility rather than improve it. A ray of light arriving from an angle and direction such that visibility is poor is termed a "Poor ray". In contrast, a ray arriving at an angle and direction that improves visibility is a "good ray". Figures 3.22 and 3.23 show the grading of light rays with respect to the direction and angle of arrival of light rays.

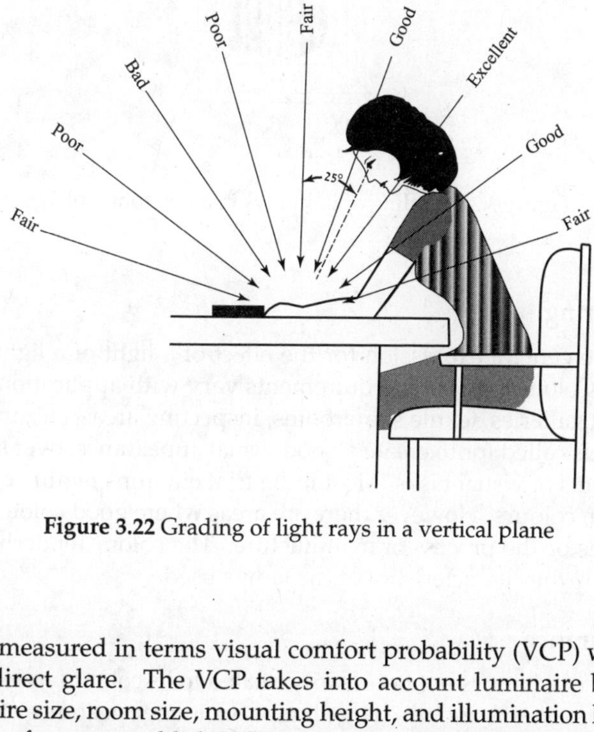

Figure 3.22 Grading of light rays in a vertical plane

The direct glare is measured in terms visual comfort probability (VCP) which in essence is the discomfort factor of direct glare. The VCP takes into account luminaire brightness at different angles of view, luminaire size, room size, mounting height, and illumination level and room surface reflectance. Most manufacturers publish VCP tables for their luminaires. A VCP value of 70 or higher provides an acceptable brightness for an office situation.

The indirect glare is measured in terms of "equivalent sphere illumination" (ESI). In order to provide a good quality design, it is necessary that VCP and ESI values be as high as possible. This poses some problems, since their principles of operation oppose each other. The direct glare zone is between 45°and 90°. For a high VCP value, it is necessary that the luminance associated with this zone be as low as possible and most of the light arrive at 0–45°angle.

The indirect glare zone, on the other hand, is between 0 and 45 degrees. For a high value of ESI, it is necessary that the luminance associated with this zone be as low as possible and that the light emerges between 45 and 90°. A good quality lighting design will have to evaluate the importance of each in the application and create a design that will be most suitable for the purpose.

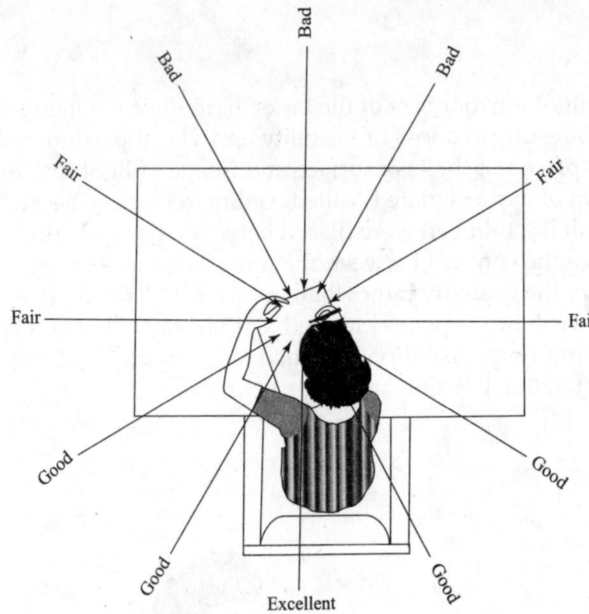

Figure 3.23 Grading of light rays in a horizontal plane

3.7.3 Colour rendering

Colour rendering is the general expression for the effect of a light of a light source on the colour appearance of object. Colour rendering requirements vary with applications. It is very critical in areas like hospitals, art galleries, textile showrooms, inspecting areas, etc., in office environment; a good colour rendering is called for to create a good visual appearance, over and above the purpose meeting the requirements of visual tasks. Most industrial environs require only recognition of and discrimination between colours. However there are areas where good colour rendering is required because of the products or the process of manufacture. The colour rendering ability of a lighting system is determined by the characteristics of the lamps used.

3.7.4 Colour appearance

The colour appearance of a lamp is not necessarily related to its colour rendering properties. This is true particularly for fluorescents and discharge lamps. Colour appearance of lamps can be assessed from their correlated colour temperature (CCT).

The correlated colour temperature of light source is the absolute temperature of a black body whose chromaticity most nearly resembles that of the light source. A completely radiating source would be red at a temperature of 800–900 K, yellow at 3000 K, white (natural) at about 5000 K and cool pale bluish at 8000–10000 K.

3.8 DESIGN PROCEDURES

The best option available for lighting design is the use of standard lighting software. If no such facility is available, any one of the following techniques can be used for the design.

- Short hand method

- Long hand method
- Simplified method.

3.8.1 Short hand method

In this method it is assumed that the average maintained quantity of lumens arriving on the work plane will be half the quantity of the total new lamp lumens. This method assumes "normal" sized rooms, "normal" surface reflectance and "normal" dirt conditions. A normal sized room is the one in which the mounting height (distance from the luminaire bottom to the work plane) is less than half the smallest room dimension. It is also assumed that conventional fixtures are used. A high bay fixture is used when the mounting height is over 6 m. Otherwise low bay fixtures are recommended. The total number of fixtures (luminaires) to be used is then calculated as follows:

Step-1: Obtain the required level of illumination in lux.

Step-2: Choose a particular lamp and obtain the lamp lumens from the lamp data.

Step-3: Delivered maintained lamp lumen (DMLL) = 0.5 × Rated lamp lumens.

Step-4: Number of fixtures (luminaires) = $\dfrac{\text{Room area } \times \text{Lux}}{\text{DMLL}}$

Select the nearest round figure.

Step-5: Area per fixture = $\dfrac{\text{Room area}}{\text{Number of fixtures}}$

Step-6: Spacing between fixtures = $\sqrt{\text{Area per fixture}}$

Spacing should not be more than the mounting height (MH) for high bay fixtures and 1.5 times the mounting height for low bay fixtures. Check for conformity. If the spacing is too large, go back to step-2 and start again with a lower lamp rating.

3.8.2 Long hand method

This is a more accurate and effective method of estimating the number of light fixtures (luminaires) and their spacing, because it takes into account the differences in photometric performance caused by room geometry and system lumen depreciation. The core idea of this method is the determination of coefficient of utilisation (CU) on the concept that the area to be lighted has three cavities or spaces that have effective reflectances with respect to each other and the work plane. The given space to be lighted is divided into three cavities that is shown in Figure 3.24. The space between the ceiling and the bottom of the luminaire is known as ceiling cavity (CC) and the distance is h_{cc}, or the height of ceiling cavity. The space between the work plane and the bottom of luminaire is room cavity (RC) and the distance is h_{rc} or the height of room cavity. The space between the work plane and the floor is Floor Cavity and the height is h_{fc} or the height of floor cavity. The room cavities are shown in Figure 3.24.

The cavity ratios represent the geometric properties of the ceiling, room and floor cavities and can be found using the formula,

$$\text{Cavity ratio} = \frac{2.5 \times \text{height of cavity } (h) \times \text{perimeter}}{\text{Area}}$$

Where, $h = h_{cc}$, for ceiling cavity ratio (CCR)
$= h_{rc}$, for room cavity ratio (RCR)
$= h_{fc}$ for floor cavity ratio (FCR).

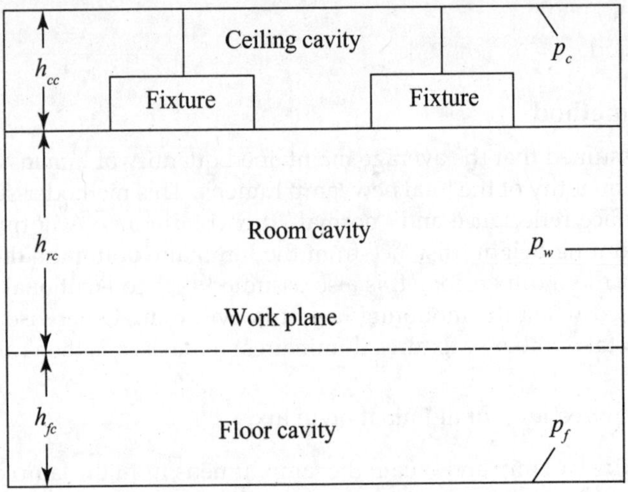

Figure 3.24 Details of room cavities

Step-1: Study the site plan and elevation of the installation. Determine the lighting needs in consultation with standards and the end user.

Step-2: Select the light sources and luminaire, appropriate to the installation geometry, nature of visual task and energy criteria.

Step-3: Calculate the ceiling, room and floor cavity ratios and using Table 3.3 find out the effective ceiling and floor cavity reflectance.

Step-4: Use the given chart for coefficient of utilisation, the effective ceiling and floor reflectance's and RCR find out the value of effective coefficient of utilisation.

Step-5: Select the light loss factor (LLF), this factor was previously referred to as maintenance factor (MF). Generally accepted values are,

A/C rooms, clean rooms, etc.	0.8
Industrial environment	0.7
Dusty areas	0.6.

Step-6: Use the formula to arrive at the quantity of luminaires (fixtures)

$$\text{Number of luminaires} = \frac{\text{Area} \times \text{Illumination}}{\text{CU} \times \text{LLF} \times \text{Lumen output of luminaire}}$$

Step-7: Arrange the luminaires symmetrically giving due considerations to spacing criterion to achieve uniformity of illumination.

$$\text{Area per luminaire (fixture)} = \frac{\text{Room area}}{\text{Number of fixtures}}$$

$$\text{Therefore fixture spacing} = \sqrt{\text{Area per fixture}}$$

Spacing should not be more than the mounting height (MH) for high bay fixtures and 1.5 times the mounting height for low bay fixtures. Check for conformity. If the spacing is too large, go back to step-2 and start again with a lower lamp rating.

3.8.3 Simplified design procedure

A simplified design procedure for lighting design by lumen method for a simple interior is given below (assumption: COU chart based on room index is available).

Step-1: Study the site plan and elevation of the installation. Determine the lighting needs in consultation with standards and the end user.
Step-2: Select the light sources and luminaire, appropriate to the installation geometry, nature of visual task and energy criteria.
Step-3: Calculate the room index (RI) using the formula,

$$\text{Room Index (RI)} = \frac{\text{Length} \times \text{Breadth}}{(\text{Length+breadth}) \times \text{Mounting height}}$$

Step-4: Based on this, coefficient of utilisation value can be obtained from the COU table for the selected luminaire. These values are generally based on the RI and room surface reflectance which are generally 70% for light coloured, 50% for average and 30% for dark walls and 50%, 30% and 10% for corresponding ceilings.
Step-5: Select the light loss factor (LLF), this factor was previously referred to as maintenance factor (MF). Generally accepted values are,

A/C rooms, clean rooms, etc.	0.8
Industrial environment	0.7
Dusty areas	0.6

Step-6: Use the formula to arrive at the quantity of luminaires

$$\text{Number of luminaires} = \frac{\text{Area} \times \text{Illumination}}{\text{CU} \times \text{LLF} \times \text{Lumen output of luminaire}}$$

Step-7: Arrange the luminaires symmetrically giving due considerations to spacing criterion to achieve uniformity of illumination.

$$\text{Area per luminaire (fixture)} = \frac{\text{Room area}}{\text{Number of fixtures}}$$

$$\text{Therefore fixture spacing} = \sqrt{\text{Area per fixture}}$$

If this spacing is too wide, repeat the process with a fixture with lower wattage of lamps. Normally spacing should not be more than the mounting height for high bay fixtures and 1.5 times the mounting height for low bay fixtures.

3.9 ARRANGEMENT OF FIXTURES

Fixtures shall be arranged from the middle of the room to the outside. A square array is the best option but rectangular array also will work satisfactorily as long as the spacing to mounting height ratio is not violated. With an odd quantity of fixture rows or columns, there will be a fixture line on the centre line. With even quantities of rows or columns, the locations are half the spacing of the centre line. The closest fixture to a wall should be one half the spacing or less.

EXAMPLE 3.2

A classroom measuring 6.5 m × 8 m is to be provided with an illumination level of 300 lux. The height of the ceiling is 4.5 m. The height of the ceiling is 4.5 m. The height of the working plane is 1.0 m above floor level. The ceiling/wall/floor reflectances are 70/50/20.

Design a lighting scheme for the classroom using general purpose 2 × 40 W fluorescent fixtures whose coefficient of utilisation chart is given below. Assume that the luminaires are suspended from the ceiling at 1.0 m below the ceiling level. The light loss factor (LLF) may be taken as 0.70. Spacing of lamps shall not exceed the mounting height. Initial lamp lumens = 4000.

Table of coefficient of utilisation
(20% effective floor cavity reflectance)

Effective ceiling cavity reflectance	**80%**			**50%**		
Wall reflectance	**50**	**30**	**10**	**50**	**30**	**10**
RCR						
10	0.33	0.26	0.22	0.31	0.26	0.22
9	0.43	0.35	0.27	0.40	0.35	0.29
8	0.58	0.42	0.35	0.48	0.42	0.36
7	0.58	0.50	0.42	0.55	0.48	0.42
6	0.64	0.57	0.49	0.61	0.54	0.47
5	0.72	0.65	0.59	0.65	0.60	0.56
4	0.77	0.71	0.64	0.71	0.65	0.60
3	0.82	0.76	0.70	0.74	0.69	0.63
2	0.87	0.82	0.77	0.78	0.74	0.70
1	0.91	0.87	0.83	0.81	0.78	0.75

Solution:

Step-1: Ceiling cavity height h_{cc} = 1.0 m
Floor cavity height h_{fc} = 1.0 m
Room cavity height h_{rc} = 2.5 m

Step-2: Ceiling cavity ratio (CCR) $= 2.5\, h_{cc} \times \dfrac{\text{perimeter}}{\text{area}}$

$$= 2.5 \times 1.0 \times \frac{(6.5 + 8)2}{6.5 \times 8}$$
$$= 1.40$$

Similarly,

Floor cavity ratio (FCR) = 1.40

Room cavity ratio (RCR) $= 2.5 \times 2.5 \times \dfrac{(6.5 + 8)2}{6.5 \times 8}$

$= 3.48$ (say 3.50)

Step-3: By referring to Table 3.3

Corresponding to ceiling/wall reflectance of 70%/50% and CCR of 1.4, the effective ceiling cavity reflectance is 55%. Similarly, with a floor/wall reflectance of 20%/50% and FCR of 1.4, the effective floor reflectance is 18% which can be rounded to 20%.

Step-4: From the Coefficient of Utilisation table of the light fixture, corresponding to a RCR of 4 and 3 we get the CU values as shown below.

RCR	CU value 80%/50%/20%	CU value 50%/50%/20%
4	0.77	0.71
3	0.82	0.74

We have to calculate the required CU value for an effective ceiling reflectance of 55% and an RCR = 3.5. For RCR = 3.0, the CU value for 55% effective ceiling reflectance can be calculated as,

$$CU = \frac{0.82 - 0.74}{30} \times 5 + 0.74$$
$$= 0.753$$

Similarly, for RCR = 4.0 and effective ceiling reflectance = 55%,

$$CU = \frac{0.77 - 0.71}{30} \times 5 + 0.71$$
$$= 0.720$$

Therefore, we have the CU values as,

RCR	Ceiling/wall/floor reflectance 55%/50%/20%
4	0.72
3	0.753

Hence, the value of CU for an RCR = 3.5 can be interpolated from the above table as,

$$CU = 0.753 - (0.753 - 0.720)\ 0.5$$
$$= 0.7365$$

Step-5: The number of luminaires required $= \dfrac{\text{Area} \times \text{Lux}}{\text{CU} \times \text{LLF} \times \text{Lumens}}$

$$= \frac{6.5 \times 8 \times 300}{0.7365 \times 0.7 \times 4000}$$

$$= 7.564 \text{ (say 8 nos.)}$$

$$\text{Area per luminaire (fixture)} = \frac{\text{Room area}}{\text{Number of fixtures}}$$

$$= \frac{6.5 \times 8}{8} = 6.5 \text{ m}^2$$

Therefore,

$$\text{Spacing between fixtures} = \sqrt{6.5} = 2.55 \text{ m}$$

For low bay fixtures (mounting height below 6 m), spacing can be upto 1.5 times the mounting height. In this case, mounting height is equal to 2.5 m and therefore the spacing should be below 1.5 ×2.5 = 3.75 m. Hence the design is satisfactory.

A typical layout of the lamps is shown below.

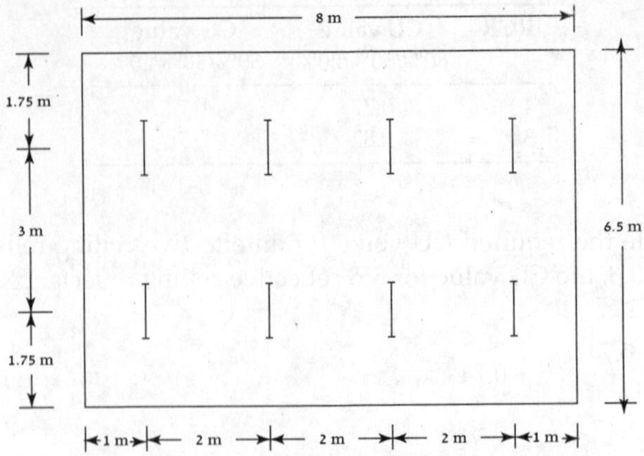

Figure 3.25

EXAMPLE 3.3

An industrial workshop measuring 12 m × 6 m is to be provided with an illumination level of 200 lux. Design the lighting layout giving due consideration for spacing. The following data pertains to the work shop.

1. Wall reflectance = 50%
2. Ceiling reflectance = 70%
3. Floor reflectance = 10%
4. Height of floor cavity = 1.0 m
5. Height of ceiling cavity = 2.5 m
6. Height of room cavity = 4.0 m

Use 2 × 40 W industrial fluorescent fixture with vitreous enamel reflectors whose coefficient of utilisation chart is given below.

Lamp rating = 2×40 W
Initial lamp lumens = 4500
Maximum ratio of spacing to mounting height = 1.5

Coefficient of utilisation chart

Room reflectance			Room index									
Floor	**Ceiling**	**Wall**	**0.6**	**0.8**	**1.0**	**1.25**	**1.50**	**2.0**	**2.5**	**3.0**	**4.0**	**5.0**
		50	0.36	0.46	0.51	0.57	0.61	0.66	0.70	0.72	0.75	0.77
10	70	30	0.51	0.40	0.47	0.52	0.56	0.62	0.66	0.68	0.72	0.74
		10	0.28	0.36	0.42	0.48	0.52	0.58	0.62	0.65	0.70	0.72
10	50	50	0.35	0.45	0.50	0.56	0.60	0.64	0.68	0.70	0.73	0.75
		30	0.31	0.40	0.46	0.52	0.55	0.61	0.64	0.67	0.71	0.73
		10	0.28	0.36	0.42	0.48	0.52	0.58	0.62	0.64	0.68	0.71
10	30	30	0.31	0.40	0.45	0.51	0.54	0.59	0.64	0.66	0.69	0.71
		10	0.28	0.36	0.42	0.47	0.51	0.57	0.61	0.63	0.68	0.70

Solution

Step-1 and Step-2 are not required as simplified design procedure is used.

Step-3:

$$\text{Room index} - \text{RI} = \frac{\text{length} \times \text{breadth}}{(\text{length+breadth}) \text{ mounting height}}$$

$$= \frac{6 \times 12}{(6+12)4}$$

$$= 1.0$$

Step-4: From the Table for coefficient of utilisation, corresponding a value of RI = 1.0 and ceiling/wall/floor reflectance's of 70%/50%/10%, the value of coefficient of utilisation (CU) can be obtained as, CU = 0.51.

Step-5: Since the lighting is for an industrial area, the light loss factor may be assumed to be 70%. The light loss factor (LLF) = 0.70

Step-6:

$$\text{Number of luminaires required} = \frac{\text{Area} \times \text{lux}}{\text{CU} \times \text{LLF} \times \text{lumens}}$$

$$= \frac{6 \times 12 \times 200}{0.51 \times 0.70 \times 4500}$$

$$= 8.963 \text{ say 9 numbers}$$

$$\text{Area per fixture (Luminaire)} = \frac{6 \times 12}{9} = 8 \text{ m}^2$$

$$\text{Therefore, spacing between fixtures} = \sqrt{8}$$

$$= 2.828 \text{ m say 3 m.}$$

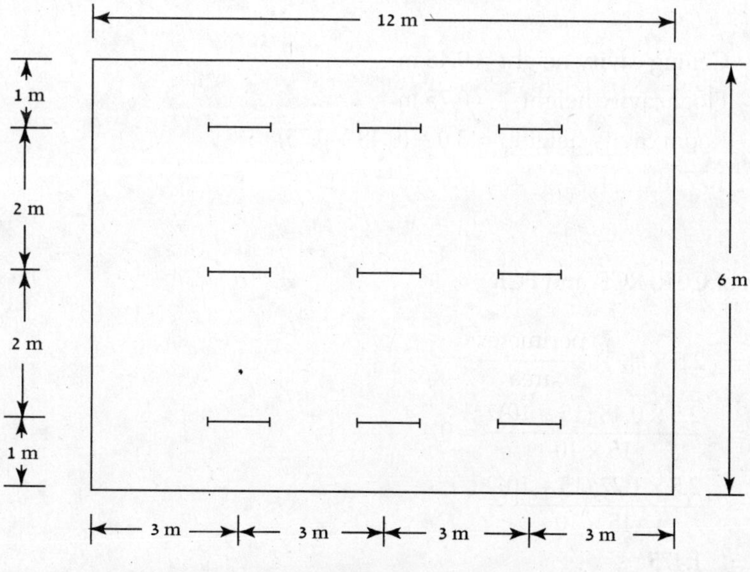

Figure 3.26

In this case the spacing can be 1.5 times the mounting height. i.e., 1.5 × 3.0 = 4.5 m. Hence the design is satisfactory. Typical layout of the lamps is shown below.

EXAMPLE 3.4

It is proposed to provide lighting to a room of size 15 m long, 10 m wide and 3 m high. The work plane may be considered to be 0.75 m above floor level. The ceiling/wall/floor reflectance are 80%/50% /20%. Use 2 × 40 W luminaires with acrylic prismatic lens. Find the values of coefficient of utilisation if, 1) The luminaires are suspended 0.48 m below the ceiling and 2) The luminaires are ceiling mounted. The CU table for the luminaire is given below.

Table of Coefficient of Utilisation

Effective floor cavity reflectance = 20%																		
RC	**80**				**50**							**30**			**10**			**0**
RW	70	50	30	10	70	50	30	10	50	30	10	50	30	10	50	30	10	0
RCR																		
1	72	69	67	65	70	68	66	64	65	63	62	63	61	60	60	59	58	57
2	67	62	58	55	65	61	58	55	59	56	53	57	54	52	55	53	51	50
3	62	56	51	48	60	55	51	47	53	49	47	51	48	46	50	47	45	44
4	57	50	45	42	56	49	45	41	48	44	41	46	43	40	45	42	40	38
5	53	45	40	36	51	44	39	36	43	39	35	42	38	35	41	37	35	33
6	49	41	36	32	48	40	35	32	39	35	31	38	34	31	37	34	31	30
7	45	37	32	28	44	36	31	28	35	31	28	35	30	27	34	30	27	26
8	42	33	28	24	41	33	28	24	32	27	24	31	27	24	30	27	24	23
9	38	30	25	21	37	29	25	21	29	24	21	28	24	21	27	24	21	20
10	36	27	22	19	35	27	22	19	26	22	19	26	22	19	25	21	19	17

Solution:

Case-1: Luminaire suspended 0.48 m below the ceiling.

Step-1:

h_{cc} = Ceiling cavity height = 0.48 m

h_{fc} = Floor cavity height = 0.75 m

h_{rc} = Room cavity height = 3.0 – (0.48 + 0.75)

= 1.77 m

Step-2:

Determination of CCR, RCR and FCR

$$\text{CCR} = 2.5 \times h_{cc} \times \frac{\text{perimeter}}{\text{area}}$$

$$= \frac{2.5 \times 0.48\,(15 + 10)2}{15 \times 10} = 0.4$$

$$\text{RCR} = \frac{2.5 \times 1.77\,(15 + 10)2}{15 \times 10}$$

$$= 1.475$$

$$FCR = \frac{2.5 \times 0.75\,(15 + 10)2}{15 \times 10}$$
$$= 0.625$$

Step-3: Determination of effective reflectance's of ceiling and floor cavities. Referring to the Table 3.3. Corresponding to a ceiling reflectance = 80%, wall reflectance = 50% and CCR = 0.4, the effective ceiling cavity reflectance is found to be 74%.

Similarly, for a floor reflectance of 20%, wall reflectance of 50% and FCR = 0.625, the effective floor reflectance is given as 19%, which can be rounded to 20%.

Step-4: Now we have to find the coefficient of utilisation corresponding to an effective ceiling reflectance of 74%, wall reflectance of 50% and floor reflectance of 20%. From the given table of CU, corresponding to a ceiling/wall/floor reflectance of 80/50/20 and 70/50/20 percent, the coefficient of utilisation for an RCR value of 1 and 2 are given as

RCR	80/50/20%	70/50/20%
1	0.69	0.68
2	0.62	0.61

For RCR = 1, the value of CU for an effective ceiling/wall/floor reflectance of 74/50/20% can be interpolated as

$$CU = \frac{(0.69 - 0.68)}{10} \times 4 + 0.68$$
$$= 0.684$$

Similarly, for RCR = 2 and an effective ceiling/wall/floor reflectance of 74/50/20%, the value of CU is

$$CU = \frac{(0.69 - 0.68)}{10} \times 4 + 0.61$$
$$= 0.614$$

Now, we have the CU values for RCR = 1 and 2 as

RCR	74/50/20
1	0.684
2	0.614

Therefore, the interpolated value of CU for an RCR = 1.475 is given by,

$$CU = 0.684 - (0.684 - 0.614)0.475$$
$$= 0.6508$$

Case-2: The luminaire mounted on the ceiling directly.

Step-1:

$$h_{cc} = 0.0 \text{ m}$$
$$h_{rc} = 3.0 - 0.75$$
$$= 2.25 \text{ m}$$
$$h_{fc} = 0.75 \text{ m}$$

Step-2:

$$\text{CCR} = \frac{2.5 \times h_{cc} \times \text{perimeter}}{\text{area}}$$

$$= 0.0$$

since $h_{cc} = 0.0$

$$\text{RCR} = \frac{2.5 \times 2.25 \times 2\,(15 + 10)}{15 \times 10}$$

$$= 1.875$$

$$\text{FCR} = \frac{2.5 \times 0.75 \times 2\,(15 + 10)}{15 \times 10}$$

$$= 0.625$$

Step-3: Referring to Table 3.3, with a ceiling reflectance of 80%, wall reflectance of 50% and CCR = 0.0, the effective ceiling cavity reflectance = 80%.

Similarly, with a floor reflectance of 20%, wall reflectance of 50% and FCR = 0.625, the effective floor cavity reflectance = 19% which can be rounded to 20%.

Step-4:
From the Table for coefficient of utilisation, corresponding to ceiling/wall/floor reflectance of 80/50/20%, the values of CU for RCR = 1 and 2 are given below.

RCR	**80/50/20%**
1	0.69
2	0.62

The value of CU corresponding to an RCR = 1.875 can be interpolated from the above table as

$$\text{CU} = 0.69 - (0.69 - 0.62)\,0.875$$

$$= 0.629$$

From the above results it can be inferred that when the luminaire is suspended below the ceiling the value of CU is higher compared to a situation where the luminaire is fixed on the ceiling.

Table 3.3 Per Cent Effective Ceiling or Flore Reflectance for Various Reflectance Combinations (Reprinted from the IES Lighting Handbook, 1981, Reference Volume)

EFFECTIVE CAVITY REFLECTANCE

BASE REFLECTANCE PER CENT	90									
WALL REFLECTANCE PER CENT	90	80	70	60	50	40	30	20	10	0
ROOM CAVITY RATIO										
0.2	89	88	88	87	85	85	85	84	84	82
0.4	88	87	86	85	84	83	81	80	79	76
0.6	87	86	84	82	80	79	77	76	74	73
0.8	87	85	82	80	77	75	73	71	69	67
1.0	86	83	80	77	75	72	69	66	64	62
1.2	85	82	78	75	72	69	65	63	60	57
1.4	85	80	77	73	69	65	62	59	57	52
1.6	84	78	75	71	67	63	59	56	53	50
1.8	83	78	73	69	64	60	56	53	50	48
2.0	83	77	72	67	62	56	53	50	47	43
2.2	82	76	70	65	59	54	50	47	44	40
2.4	82	75	68	64	58	53	48	45	41	37
2.6	81	74	67	62	56	51	46	42	38	35
2.8	81	73	66	60	54	49	44	40	36	34
3.0	80	72	64	58	52	47	42	38	34	30
3.2	78	71	63	56	50	45	40	36	32	28
3.4	78	70	62	54	48	43	38	34	30	27
3.6	78	69	61	53	47	42	36	32	28	25
3.8	78	69	60	51	45	40	35	31	27	23
4.0	77	69	58	51	44	39	33	29	25	22
4.2	77	62	57	50	43	37	32	28	24	21
4.4	76	61	56	49	42	36	31	27	23	20
4.6	76	60	55	47	40	35	30	26	22	19
4.8	75	59	54	46	38	34	28	25	21	18
5.0	75	59	53	45	38	33	28	24	20	16
6.0	73	61	48	41	34	28	24	20	16	11
7.0	70	58	45	38	30	27	21	18	14	8
8.0	68	55	42	35	27	23	18	15	12	6
9.0	66	52	38	31	25	21	16	14	11	5
10.0	65	51	36	29	22	19	15	11	9	4

BASE REFLECTANCE PER CENT	80									
WALL REFLECTANCE PER CENT	90	80	70	60	50	40	30	20	10	0
ROOM CAVITY RATIO										
0.2	79	78	78	77	77	76	76	75	74	72
0.4	79	77	76	75	74	73	72	71	70	68
0.6	78	76	75	73	71	70	68	66	65	63
0.8	78	75	73	71	69	97	65	63	61	57
1.0	77	74	72	69	67	65	62	60	57	55
1.2	76	73	70	67	64	61	58	55	53	51
1.4	76	72	68	65	62	59	55	53	50	48
1.6	75	71	67	63	60	57	53	50	47	44
1.8	75	70	66	62	58	54	50	47	44	41
2.0	74	69	64	60	56	52	48	45	41	38
2.2	74	68	63	58	54	49	45	42	38	35
2.4	73	67	61	56	52	47	43	40	36	33
2.6	73	66	60	55	50	45	41	38	34	31
2.8	73	65	59	53	48	43	39	36	32	29
3.0	72	65	58	52	47	42	37	34	30	27
3.2	72	65	37	51	45	40	35	33	28	25
3.4	71	64	56	49	44	38	34	32	27	24
3.6	71	63	54	48	43	38	32	30	25	23
3.8	70	62	53	47	41	36	31	28	24	22
4.0	70	61	53	46	40	35	30	26	22	20
4.2	69	60	52	45	39	34	29	25	21	18
4.4	69	60	51	44	38	33	28	24	20	17
4.6	69	59	50	43	37	32	27	23	19	15
4.8	68	58	49	42	36	31	26	22	18	14
5.0	68	58	48	41	35	30	25	21	18	14
6.0	66	55	44	38	31	27	22	19	15	10
7.0	64	53	41	35	28	24	19	16	12	7
8.0	62	50	38	32	25	21	17	14	11	5
9.0	61	48	36	30	23	19	15	13	10	4
10.0	59	46	33	27	21	18	14	11	8	3

BASE REFLECTANCE PER CENT	70									
WALL REFLECTANCE PER CENT	90	80	70	60	50	40	30	20	10	0
ROOM CAVITY RATIO										
0.2	70	69	68	68	67	67	66	66	65	64
0.4	69	68	67	66	65	64	63	62	61	58
0.6	69	67	65	64	63	61	59	58	57	54
0.8	68	66	64	62	60	58	56	55	53	50
1.0	68	65	62	60	58	55	53	52	50	47
1.2	67	64	61	59	57	54	50	48	46	44
1.4	67	63	60	58	55	51	47	45	44	41
1.6	67	62	59	56	53	47	45	43	41	38
1.8	66	61	58	54	51	46	42	40	38	35
2.0	66	60	56	52	49	45	40	38	36	33
2.2	65	60	55	51	48	43	38	36	34	32
2.4	65	60	54	50	46	41	37	35	32	30
2.6	65	59	54	49	45	40	35	33	30	28
2.8	65	59	53	48	43	38	33	30	28	26
3.0	66	58	52	47	42	37	32	29	27	24
3.2	64	58	51	46	40	36	31	28	25	23
3.4	64	57	50	45	39	35	29	27	24	22
3.6	63	56	49	44	38	33	28	25	22	20
3.8	63	56	49	43	37	32	27	24	21	19
4.0	63	55	48	42	36	31	26	23	20	17
4.2	62	55	47	41	35	30	25	22	19	16
4.4	62	54	46	40	34	29	24	21	18	15
4.6	62	53	45	39	33	28	24	21	17	14
4.8	62	53	45	38	32	27	23	20	16	13
5.0	61	52	44	36	31	26	22	19	16	12
6.0	60	51	41	35	28	24	19	16	13	9
7.0	58	48	38	32	26	22	17	14	11	6
8.0	57	46	35	29	23	19	15	13	10	5
9.0	56	45	33	27	21	18	14	12	9	4
10.0	55	43	31	25	19	16	12	10	8	3

BASE REFLECTANCE PER CENT	60									
WALL REFLECTANCE PER CENT	90	80	70	60	50	40	30	20	10	0
ROOM CAVITY RATIO										
0.2	60	59	59	59	58	57	56	56	55	53
0.4	60	59	59	58	57	55	54	53	52	50
0.6	60	58	57	56	55	53	51	51	50	46
0.8	59	57	56	55	54	51	48	47	46	43
1.0	59	57	55	53	51	48	45	44	43	41
1.2	59	56	54	51	49	46	44	42	40	38
1.4	59	56	53	49	47	44	41	39	38	36
1.6	59	55	52	48	45	42	39	37	35	33
1.8	58	55	51	47	44	40	37	35	33	31
2.0	58	54	50	46	43	39	35	33	31	29
2.2	58	53	49	45	42	37	34	31	29	28
2.4	58	53	48	44	41	36	32	30	27	26
2.6	58	53	48	43	39	35	31	28	26	24
2.8	58	53	47	43	38	34	29	27	24	22
3.0	57	52	46	42	37	32	28	25	23	20
3.2	57	51	45	41	36	31	27	23	22	18
3.4	57	51	45	40	35	30	26	23	20	17
3.6	57	50	44	39	34	29	25	22	19	16
3.8	57	50	43	38	33	29	24	21	19	15
4.0	57	49	42	37	32	28	23	20	18	14
4.2	56	49	42	37	32	27	22	19	17	14
4.4	56	49	42	36	31	27	22	19	16	13
4.6	56	49	41	35	30	26	21	18	16	13
4.8	56	48	41	34	29	25	21	18	15	12
5.0	56	48	40	34	28	24	20	17	14	11
6.0	55	45	37	31	25	21	17	14	11	7
7.0	54	43	35	30	24	20	15	12	9	5
8.0	53	42	33	28	22	18	14	11	8	4
9.0	52	40	31	26	20	16	12	10	7	3
10.0	51	39	29	24	18	15	11	9	7	2

BASE REFLECTANCE PER CENT	50									
WALL REFLECTANCE PER CENT	90	80	70	60	50	40	30	20	10	0
ROOM CAVITY RATIO										
0.2	50	50	49	49	48	48	47	46	46	44
0.4	50	49	48	48	47	46	45	45	44	42
0.6	50	48	47	46	45	44	43	42	41	48
0.8	50	48	47	45	44	42	40	38	38	36
1.0	50	48	46	44	43	40	38	37	36	34
1.2	50	47	45	43	41	39	36	35	34	29
1.4	50	47	45	42	40	38	35	34	32	27
1.6	50	47	44	41	39	36	33	32	30	26
1.8	50	46	43	40	38	35	31	30	28	25
2.0	50	46	43	40	37	34	30	28	26	24
2.2	50	46	42	38	36	33	29	27	24	22
2.4	50	46	42	37	35	31	27	25	23	21
2.6	50	46	41	37	34	30	26	23	21	20
2.8	50	46	41	36	33	29	25	22	20	19
3.0	50	45	40	36	32	25	24	21	19	17
3.2	50	44	39	35	31	27	23	20	18	16
3.4	50	44	39	35	30	26	22	19	17	15
3.6	50	44	39	34	29	25	21	18	16	14
3.8	50	44	38	34	29	25	21	17	15	13
4.0	50	44	38	33	28	24	20	17	15	12
4.2	50	43	37	32	28	24	20	17	14	12
4.4	50	43	37	32	27	23	19	16	13	11
4.6	50	43	36	31	26	22	18	15	13	10
4.8	50	43	36	31	26	22	18	15	12	9
5.0	50	42	35	30	25	21	17	14	12	9
6.0	50	42	34	29	23	19	15	13	10	6
7.0	49	41	32	27	21	18	14	11	8	5
8.0	49	40	30	25	19	16	12	10	7	3
9.0	48	39	29	24	18	15	11	9	7	3
10.0	47	37	27	22	17	14	10	8	6	2

Continued...

BASE REFLECTANCE PER CENT	40										30										20										10										0									
WALL REFLECTANCE PER CENT	90	80	70	60	50	40	30	20	10	0	90	80	70	60	50	40	30	20	10	0	90	80	70	60	50	40	30	20	10	0	90	80	70	60	50	40	30	20	10	0	90	80	70	60	50	40	30	20	10	0
ROOM CAVITY RATIO																																																		
0.2	40	40	39	39	39	38	38	37	36	36	31	31	30	30	29	29	29	28	28	27	21	20	20	20	20	20	19	19	19	17	11	11	11	10	10	10	10	9	9	9	2	2	2	1	1	1	1	0	0	0
0.4	41	40	39	39	38	37	36	35	34	34	31	31	30	30	29	28	28	27	26	25	22	21	20	20	20	19	19	18	18	16	12	11	11	11	11	10	10	9	9	8	4	3	3	2	2	2	1	1	0	0
0.6	41	40	39	38	37	36	34	33	32	31	32	31	30	29	28	27	26	26	25	23	23	21	21	20	19	19	18	18	17	15	13	13	12	11	11	10	10	9	8	8	5	5	4	3	3	2	2	1	1	0
0.8	41	40	38	37	36	35	33	32	31	29	32	31	30	29	28	26	25	25	23	22	24	22	21	20	19	19	18	17	16	14	15	14	13	12	11	10	10	9	8	7	7	6	5	4	4	3	2	2	1	0
1.0	42	42	38	37	35	33	32	31	29	27	33	32	30	29	27	25	24	23	22	20	25	23	22	20	19	18	17	16	15	13	16	14	13	12	12	11	10	9	8	7	8	7	6	5	4	3	2	2	1	0
1.2	42	40	38	36	34	32	30	29	27	25	33	32	30	28	27	25	23	22	21	19	25	23	22	20	19	17	17	16	14	12	17	15	14	13	12	11	10	9	7	6	10	8	7	6	5	4	3	2	1	0
1.4	42	39	37	35	33	31	29	27	25	23	34	32	30	28	26	24	22	21	19	18	26	24	22	20	18	17	16	15	13	12	18	16	14	13	12	11	10	9	7	6	11	9	8	7	6	4	3	2	1	0
1.6	42	39	37	35	32	30	27	25	23	22	34	33	29	27	25	23	22	20	18	17	26	24	22	20	18	17	16	15	13	11	19	17	15	14	12	11	9	8	7	6	12	10	9	7	6	5	3	2	1	0
1.8	42	39	36	34	31	29	26	24	22	21	35	33	29	27	25	23	21	19	17	16	27	25	23	20	18	17	15	14	12	10	19	17	15	14	13	11	9	8	6	5	13	11	9	8	7	5	4	3	1	0
2.0	42	39	36	34	31	28	25	23	21	19	35	33	29	26	24	22	20	18	16	14	28	25	23	20	18	16	15	13	11	9	20	18	16	14	13	11	9	8	6	5	14	12	10	9	7	5	4	3	1	0
2.2	42	39	36	33	30	27	24	22	19	18	36	32	29	26	24	22	19	17	15	13	28	25	23	20	18	16	14	12	10	9	21	19	16	14	13	11	9	7	6	5	15	13	11	9	7	6	4	3	1	0
2.4	43	39	35	33	29	27	24	21	18	17	36	32	29	26	24	22	19	16	14	12	29	26	23	20	18	16	14	12	10	8	22	19	17	15	13	11	9	7	6	5	16	13	11	9	8	6	4	3	1	0
2.6	43	39	35	32	29	26	23	20	17	15	36	32	29	25	23	21	18	16	14	12	29	26	23	20	18	16	14	11	9	8	23	20	17	15	13	11	9	7	6	4	17	14	12	10	8	6	5	3	2	0
2.8	43	39	35	32	28	25	22	19	16	14	37	33	29	25	23	21	17	15	13	11	30	27	23	20	18	15	13	11	9	7	23	20	18	16	13	11	9	7	5	3	17	15	13	10	8	7	5	3	2	0
3.0	43	39	35	31	27	24	21	18	16	13	37	33	29	25	22	20	17	15	12	10	30	27	23	20	17	15	13	11	9	7	24	21	18	16	13	11	9	7	5	3	18	16	13	11	9	7	5	3	2	0
3.2	43	39	35	31	27	23	20	17	15	13	37	33	29	25	22	19	16	14	12	10	31	27	23	20	17	15	12	11	9	6	25	21	18	16	13	11	9	7	5	3	19	16	14	11	9	7	5	3	2	0
3.4	43	39	34	30	26	23	20	17	14	12	37	33	29	25	22	19	16	14	11	9	31	27	23	20	17	15	12	10	8	6	26	22	18	16	13	11	9	7	5	3	20	17	14	12	9	7	5	3	2	0
3.6	44	39	34	30	26	22	19	16	14	11	38	33	29	24	21	18	15	13	10	9	32	27	23	20	17	15	12	10	8	5	26	22	19	16	13	11	9	6	4	3	20	17	15	12	10	8	5	4	2	0
3.8	44	38	33	29	25	22	18	16	13	10	38	33	28	24	21	18	15	13	10	8	32	28	23	20	17	15	12	10	7	5	27	23	19	17	14	11	9	6	4	2	21	18	15	12	10	8	5	4	2	0
4.0	44	38	32	29	25	21	18	15	12	10	38	33	28	24	21	18	14	12	9	7	33	28	22	20	17	14	11	9	7	5	27	23	20	17	14	11	9	6	4	2	22	18	15	13	10	8	5	4	2	0
4.2	44	38	33	29	24	21	17	15	12	10	38	33	28	24	20	17	14	12	9	7	33	28	23	20	17	14	11	9	7	4	28	24	20	17	14	11	8	6	4	2	22	19	16	13	10	8	6	4	2	0
4.4	46	38	33	28	24	20	17	14	11	9	39	33	28	24	20	17	14	11	9	6	34	28	24	20	17	14	11	9	7	4	28	24	20	17	14	11	8	6	4	2	23	19	16	13	10	8	6	4	2	0
4.6	44	38	32	28	23	19	16	14	11	8	39	33	28	24	20	17	13	10	8	6	34	29	24	20	17	14	11	9	7	4	29	25	20	17	14	11	8	6	4	2	23	20	17	13	11	8	6	4	2	0
4.8	44	38	32	27	22	19	16	13	10	8	39	33	28	24	20	17	13	10	8	5	35	29	24	20	17	13	10	8	6	4	29	25	20	17	14	11	8	6	4	2	24	20	17	14	11	8	6	4	2	0
5.0	45	38	31	27	22	19	15	13	10	7	39	33	28	24	19	16	13	10	8	5	35	29	24	20	16	13	10	8	6	4	30	25	20	17	14	11	8	6	4	2	25	21	17	14	11	8	6	4	2	0
6.0	44	37	30	25	20	17	13	11	8	5	39	33	27	23	18	15	11	9	6	4	36	30	24	20	16	13	10	8	5	2	31	26	21	18	14	11	8	6	3	1	27	23	18	15	12	9	6	4	2	0
7.0	44	36	29	24	19	16	12	10	7	4	40	33	26	22	17	14	10	8	5	3	36	30	24	20	15	12	9	7	4	2	32	27	21	17	13	11	8	6	3	1	28	24	19	15	12	9	6	4	2	0
8.0	44	35	28	23	18	15	11	9	6	3	40	33	26	21	16	13	9	7	4	2	37	30	23	19	15	12	8	6	3	1	33	27	21	17	13	10	7	5	3	1	30	25	20	15	12	9	6	4	2	0
9.0	44	35	26	21	16	13	10	8	5	2	40	33	25	20	15	12	9	7	4	2	37	29	23	19	14	11	8	6	3	1	34	28	21	17	13	10	7	5	2	1	31	25	20	15	12	9	6	4	2	0
10.0	43	34	25	20	15	12	8	7	5	2	40	32	24	19	14	11	8	6	3	1	37	29	23	18	13	10	7	5	3	1	34	28	21	17	12	10	7	5	2	1	31	25	20	15	12	9	6	4	2	0

3.10 QUESTIONS

1. Briefly describe the operation of incandescent, fluorescent, mercury vapour and high pressure sodium vapour lamps.
2. Explain the two methods of lighting design.
3. What is meant by coefficient of utilisation, explain.
4. What is light loss and what are the parameters that contribute to light loss?
5. A point source has 3000 cd in the direction of interest. Determine the illumination at a distance of (a) 5 m and (b) 7.5 m form the source.
6. Determine the CCR, RCR and FCR for an office space 15m by 10m, with a ceiling height of 3.6 m. The work plane is 0.9 m above the floor. Assume that the lamps are mounted on the ceiling.
7. Determine the number of light fixtures required to provide an illumination level of 300 lux for an office space of 60 m^2. The coefficient of utilisation and light loss factor may be taken as 0.7 and 0.85 respectively. The light fixture contains two lamps each producing 2500 lumens. If the room is 10 m × 6 m in size, determine a suitable layout of the lamps. The maximum spacing to mounting height may be taken as 1.5.
8. An office floor 30 m × 15 m in size is to be provided with an illumination of 250 lux. The height of the ceiling is 3.9 m. The height of the work plane is 0.75 m from the floor level. The ceiling/wall/floor reflectences are 70%/50%/20%. Design a lighting scheme using general purpose 2 × 40 W fluorescent fixtures whose coefficient of utilisation chart is same as that given in Example 3.2. Assume that the light fixtures are suspended from the ceiling at 0.6 m below the ceiling. The light loss factor may be assumed to be 0.7. The initial lamp lumens = 4500 lumens.
9. An industrial work shop measuring 16 × 10 m is to be provided with an illumination of 200 lux. The height of the work plane is 1.0 m from the floor level. The ceiling height is 6 m. The ceiling/wall /floor reflectence are 70%/50%/10%. Design a lighting scheme using 2 × 40 W industrial purpose fluorescent light fixtures whose coefficient of utilisation chart is given below.

Coefficient of utilisation chart

Room reflectance			Room index									
Floor	**Ceiling**	**Wall**	**0.6**	**0.8**	**1.0**	**1.25**	**1.50**	**2.0**	**2.5**	**3.0**	**4.0**	**5.0**
10	70	50	0.36	0.46	0.51	0.57	0.61	0.66	0.70	0.72	0.75	0.77
		30	0.51	0.40	0.47	0.52	0.56	0.62	0.66	0.68	0.72	0.74
		10	0.28	0.36	0.42	0.48	0.52	0.58	0.62	0.65	0.70	0.72
10	50	50	0.35	0.45	0.50	0.56	0.60	0.64	0.68	0.70	0.73	0.75
		30	0.31	0.40	0.46	0.52	0.55	0.61	0.64	0.67	0.71	0.73
		10	0.28	0.36	0.42	0.48	0.52	0.58	0.62	0.64	0.68	0.71
10	30	30	0.31	0.40	0.45	0.51	0.54	0.59	0.64	0.66	0.69	0.71
		10	0.28	0.36	0.42	0.47	0.51	0.57	0.61	0.63	0.68	0.70

Assuming that the light fixtures are fixed at an height of 4.5 m from the floor, design a lighting scheme and show the light layout. The light loss factor may be assumed to be 0.65 and the initial lamp lumens = 4700. The spacing to mounting height may be limited to 1.5.

10. It is proposed to provide illumination to a room of size 24 × 9 m and 4.5 m high. The work plane may be assumed to be 0.75 m above the floor level. The ceiling/wall/floor reflectances

are 80%/50%/20%. Find the values of coefficient of utilisation if, 1) The lamps are mounted on the ceiling and 2). If the lamps are mounted 0.90 m below the ceiling. The CU table for the light fixture is given below.

Table of Coefficient of utilisation

Effective floor cavity reflectance = 20%																		
RC	**80**				**70**				**50**			**30**			**10**			**0**
RW	70	50	30	10	70	50	30	10	50	30	10	50	30	10	50	30	10	0
RCR																		
1	72	69	67	65	70	68	66	64	65	63	62	63	61	60	60	59	58	57
2	67	62	58	55	65	61	58	55	59	56	53	57	54	52	55	53	51	50
3	62	56	51	48	60	55	51	47	53	49	47	51	48	46	50	47	45	44
4	57	50	45	42	56	49	45	41	48	44	41	46	43	40	45	42	40	38
5	53	45	40	36	51	44	39	36	43	39	35	42	38	35	41	37	35	33
6	49	41	36	32	48	40	35	32	39	35	31	38	34	31	37	34	31	30
7	45	37	32	28	44	36	31	28	35	31	28	35	30	27	34	30	27	26
8	42	33	28	24	41	33	28	24	32	27	24	31	27	24	30	27	24	23
9	38	30	25	21	37	29	25	21	29	24	21	28	24	21	27	24	21	20
10	36	27	22	19	35	27	22	19	26	22	19	26	22	19	25	21	19	17

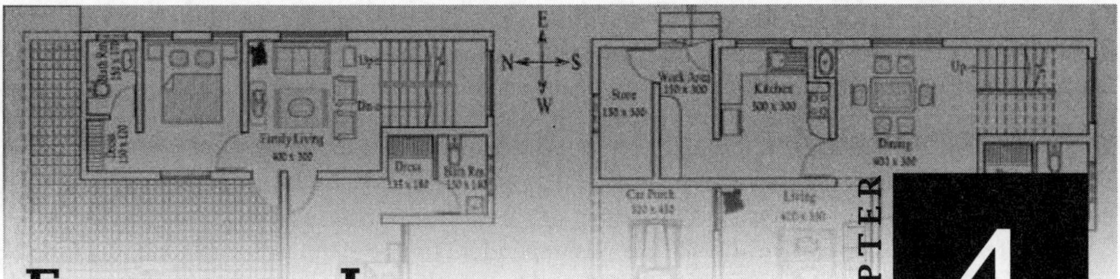

CHAPTER 4

Electrical Installations in Domestic Buildings

OBJECTIVES

At the end of this chapter you will

- Understand the requirements of light and socket points in a domestic building.
- Be able to decide the requirements of light and socket outlets for each area of domestic buildings.
- Be able to compute the connected load and hence the maximum demand for the building.
- Be able to decide the type of supply required.
- Understand how to select the distribution board for the installation.
- Know the different types of wiring methods.
- Understand the features of high rise residential buildings.
- Know about the pre-commissioning tests.

4.1 CLASSIFICATION

Domestic dwellings or Residential buildings are those buildings intended for normal residential purpose with sleeping, cooking and dining facilities. These buildings can be broadly classified into:

1. Stand alone buildings which is occupied by the members of a single family and accommodating not more than 20 persons.
2. Apartment buildings/flats where independent living spaces for three or more families are provided.

The primary objectives in planning the electrical system for a domestic building are safety and economy. Other factors like energy efficiency, reliability, convenience and provision for future expansion also merits consideration. The electrical installations of residential buildings are perhaps the simplest form of electrical system. But, this system needs careful planning because of the drastic life style changes that is sweeping across the world. Electricity has become part and parcel of normal human life and it is almost impossible to exist without electricity. It is in this context the design of electrical system for domestic buildings is to be approached.

The characteristic feature of the type of installation for domestic buildings is that the supply is derived at 240 V for single phase systems and at 415 V for three phase systems.

The safety level of an electrical installation is dependent on the standards in place at the time of construction, how carefully they were followed, and how well the installation has been maintained. Today, with increasing requirements for the electrical infrastructure, safety provisions should be given utmost attention. The safety and reliability of electrical installations requires the use of highly technical protective systems. Many such as circuit breakers have been used for decades and continuously developed to cope with modern requirements, others such as over-voltage protectors, have been used in industry but are new to domestic installations and some, such as arc detectors, are completely new.

A 'protective measure' is an electrical element that is intended to eliminate the risk of danger arising from the installation. In this way, it reduces the risk of fire, injury or material loss. These measures can be divided into the following categories:

Adequate earthing: This enables a fault current to flow safely and directly to the earth. Following the low resistance earthing path, a fault current will be sufficiently large to initiate the operation of over-current protective devices within the required time.

Adequate protection against earth leakage: Today, an increasing number of electrical appliances permit a small current to 'leak' into the protective conductor of the main wiring system of a home. This 'earth leakage' current is, for each individual device, quite small but, when there are many items of equipment, the sum of the leakage currents is large enough to be potentially dangerous. Due to deterioration of the insulation of the wiring also the leakage current can be present. To avoid the danger due to the leakage current, various design features (such as ELCBs RCCBs) should be installed in every domestic installation.

Over-current protection: Circuits are designed to carry the rated load and are fitted with a protective device, i.e., a fuse or a circuit breaker. When the demand for current exceeds the rating of the protective device the circuit is disconnected. Without this protection, the excessive current generates excessive heat in the wiring and can cause damage to the equipment or a fire. If protective devices operate frequently, it indicates that the circuit capacity is inadequate for the load and it should be re-wired with a cable of a larger conductor cross-sectional area. Increasing the rating of the protective device is sometimes permissible, but only after a complete appraisal of the conductor size and lengths installed. Low voltage circuit breakers are the mainstay of protection in domestic installations.

4.1.1 Low Voltage Circuit Breakers

Low voltage circuit breakers are common in domestic, commercial and industrial application, and are classified as:

1. Miniature circuit breakers (MCB)
2. Moulded case circuit breakers (MCCB) and
3. Air circuit breakers (ACB)

4.1.1.1 Miniature Circuit Breaker (MCB)

A miniature circuit breaker is a device designed to isolate a circuit during an over current event without the use of a fusible element. A breaker is a resettable protective device that protects against two types of over current situations; overload and short circuit. The circuit breaker plays an important role in providing over-current protection and a means of disconnection in electrical networks. Recent advancements in circuit breaker technology have increased breaker performance and protection.

Overload

Overload is a situation in which a slow and small over current causes the current carrying capacity and temperature of the circuit to gradually increase over time. This type of event is characterized by a slight increase in the load current (ampacity) on the circuit and is interrupted with a time delay by the thermal trip element of the breaker.

Short circuit

A short circuit is a rapid and intense over current situation that causes the current carrying capacity of the circuit to increase drastically. This type of event is characterized by a dramatic increase in the current (ampacity) on the circuit and is interrupted instantaneously by the magnetic trip element of the breaker. MCBs are available with rated current not more than 100A. Trip characteristics are normally not adjustable. MCBs are available in both thermal operation or thermal and magnetic operation. Figure 4.1 shows the constructional details of a typical MCB.

Figure 4.1 Cross-sectional view of a miniature circuit breaker (MCB) (courtesy M/s ABB)

1. Actuator lever – used to manually operate and reset the circuit breaker. Also indicates the status of the circuit breaker (On or Off/tripped). Most breakers are designed so they can still trip even if the lever is held or locked in the "ON" position. This is sometimes referred to as "free trip" or "positive trip" operation.

2. Actuator mechanism – Opens and closes the contacts.

3. Contacts – Permits the current to flow through with minimum resistance in the closed position.
4. Terminals – To connect the MCB with external circuits.
5. Bimetallic strip – Thermal trip element.
6. Calibration screw – allows the manufacturer to precisely adjust the trip current of the device after assembly.
7. Solenoid – magnetic trip element.
8. Arc divider/extinguisher – extinguishes the arc when the contact separates.

Low voltage power circuit breakers can be mounted in multi-tiers in distribution boards. They are modular in construction. Based on the tripping characteristics, MCBs are available in 'B', 'C', 'D', 'Z', 'K and 'S characteristics to suit different types of applications.

'B'- characteristic: These MCBs are mainly used for the protection of electrical circuits with equipment that do not have starting or inrush current. They are mainly used in lighting and distribution circuits in residential or similar applications. Short circuit release (magnetic tripping) is set to **(3–5)** I_n. Figure 4.2 shows the tripping characteristics of B characteristic MCB. The upper portion of the curve is the thermal protection region and the lower vertical portion, the magnetic protection region. I_1 is the non-tripping current in the thermal region and is equal to 1.13 I_n. I_2 is the tripping current in the thermal region and is equal to 1.45 I_n. Magnetic non-tripping region is below 3 I_n. I_3 and I_4 represent the magnetic non-tripping and magnetic tripping current.

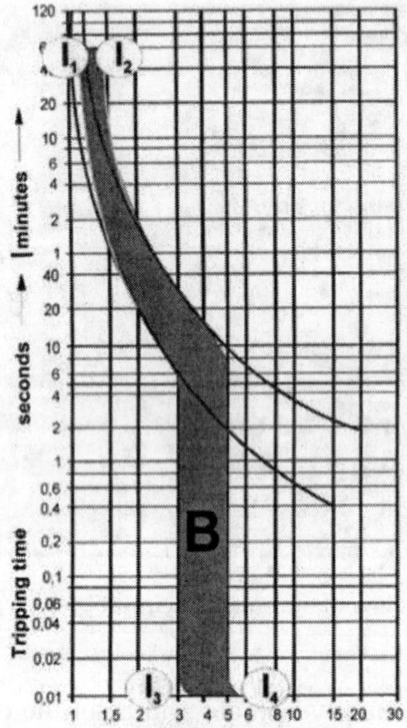

Figure 4.2 MCB tripping characteristics-"B"

'C' characteristic: Used for protection of electrical circuits with equipment that causes surge current (inductive loads and motor circuits with medium starting or inrush current). Short circuit release is set to **(5–10) I_n**. Figure 4.3 shows the tripping characteristics of C characteristic MCB. The upper portion of the curve is the thermal protection region and the lower vertical portion, the magnetic protection region. I_1 (refer figure 4.2) is the non-tripping current in the thermal region and is equal to 1.13 I_n. I_2 is the tripping current in the thermal region and is equal to 1.45 I_n. The tripping characteristics in the thermal region is exactly similar to that of "B" characteristic MCB. Magnetic non-tripping region is below 5 I_n. I_3 and I_4 represent the magnetic non-tripping and magnetic tripping current.

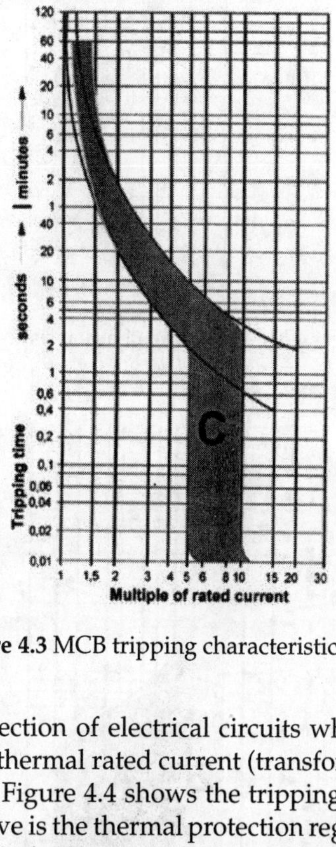

Figure 4.3 MCB tripping characteristics-"C"

'D' characteristics: Used for protection of electrical circuits which causes high inrush current at starting, typically 12–15 times the thermal rated current (transformers, X-ray machines, etc.). Short circuit release is set to **(10–20) I_n**. Figure 4.4 shows the tripping characteristics of D characteristic MCB. The upper portion of the curve is the thermal protection region and the lower vertical portion, the magnetic protection region. I_1 (refer figure 4.2) is the non-tripping current in the thermal region and is equal to 1.13 I_n. I_2 is the tripping current in the thermal region and is equal to 1.45 I_n. The tripping characteristics in the thermal region is exactly similar to that of "B" characteristic MCB. Magnetic non-tripping region is below 10 I_n.

'Z' characteristics: Used to protect highly sensitive equipment with semiconductor devices. Instantaneous trip current is **(23)** I_n. The short thermal delay and low magnetic trip point are ideal for applications where devices and components have low surge and short circuit tolerances. Figure 4.5 shows the tripping characteristics of Z characteristic MCB. The upper portion of the curve is the thermal protection region and the lower vertical portion, the magnetic protection region. I_1 (refer Figure 4.2) is the non-tripping current in the thermal region and is equal to 1.05 I_n. I_2 is the tripping current in the thermal region and is equal to 1.20 I_n. The tripping characteristics in the thermal region is not similar to that of "B" characteristic MCB. Magnetic non-tripping region is below 2 I_n.

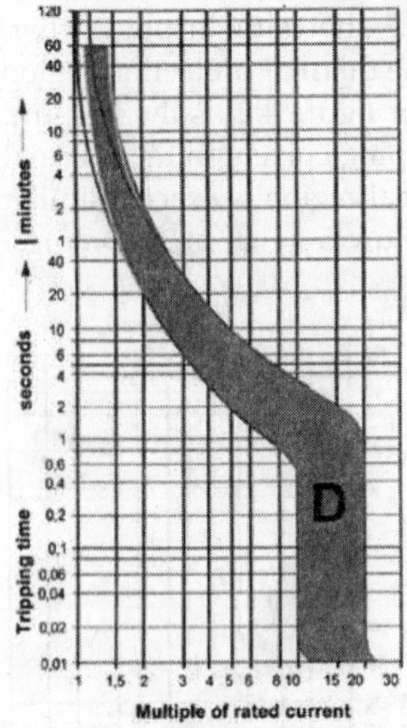

Figure 4.4 MCB tripping characteristics-"D"

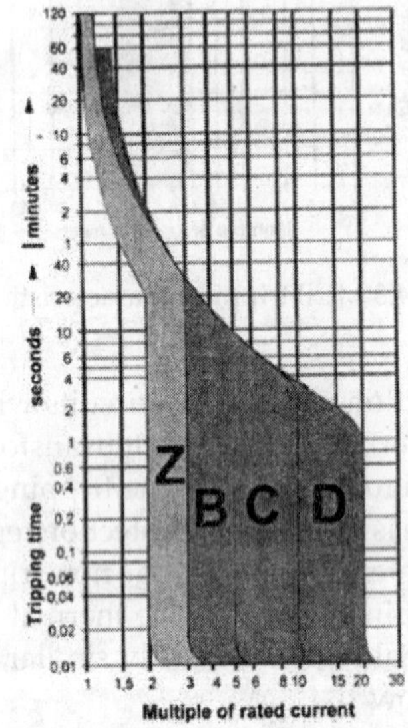

Figure 4.5 MCB tripping characteristics-"Z"

Other two MCB characteristics, even though not very popular are:

'K' characteristics: Suitable for applications where high levels of inrush current are expected. Instantaneous trip is **(812)** I_n. The high magnetic trip point is ideal for motors and transformers. The narrow range (compared with the type D curve) makes it ideal for applications where nuisance tripping is not an issue.

'S' characteristics: Suitable for applications with highly inductive loads, especially in control circuits with coils and light filaments. Instantaneous response between **(1317)** I_n.

Low voltage circuit breakers are available in the following configurations.

1 pole breakers in one module:

Compact 1P + N breakers in two modules:

Compact 2P breakers in two modules:

2P, 3P and 4P breakers:

Low voltage circuit breakers can be mounted in multi-tier form in factory assembled distribution boards. They are modular in construction and very easy to handle.

4.1.2 Residual Current Devices (RCD)/Residual Current Circuit Breakers (RCCB)

The overload currents and short circuit currents can be detected by circuit breakers like MCB's MCCB's & HRC Fuses, etc. But, circuit breakers cannot detect leakage currents, which are dangerous for humans and livestock and if not detected at the appropriate time, can lead to serious consequences like electric shock, fire hazards, etc. Residual current devices are the solution that detects such leakages currents and disconnects the circuits from the power supply. Here comes the use of RCCB (Residual Current Circuit Breaker) also known as ELCB (Earth Leakage Circuit Breaker) which provides protection against direct and indirect contact of personnel or livestock and against probable fires. Figure 4.6 shows a typical SPN residual current circuit breaker.

Figure 4.6 Typical SPN residual current circuit breaker (RCCB)

RCCB/ELCB works on the principle that in an electrical circuit the incoming current is the same as outgoing current. RCCB incorporates a core balance transformer having primary and secondary windings and a sensitive relay for instantaneous detection of fault signal. The primary winding lies in series with the supply mains and load. Secondary winding is connected to a very sensitive relay. In a faultless situation, the magnetizing effects of the current carrying conductors cancel each other. There is no residual magnetic field that could induce a voltage in the secondary. During flow of leakage current in the circuit an imbalance is created in the circuit which gives rise to leakage flux in the core. This leakage flux generates an electrical signal that is sensed by the relay and it trips the mechanism thereby disconnecting the supply. When pressing the TEST button 'T', a fault is simulated via the test resistance and the RCCB trips, indicating healthy condition of the device.

As per the Rule 61A of the Indian Electricity Rules, 1956, the supply of energy to the following installations shall be controlled by the earth leakage protective devices so as to disconnect the supply instantly on the occurrence of earth fault or leakage current.

1. Installations having load above 5 kW.
2. Luminous Tube Installations.
3. XRay Machines.

These devices are available in two-pole as well as four-pole versions. They are available in the range of 16–63 as well as 80–125 Ampere.

The sensitivity of an RCD is expressed as the rated residual operating current, noted as $\mathbf{I}_{\Delta n}$. Preferred values have been defined by the International Electrotechnical Commission (IEC), thus making it possible to divide RCDs into three groups according to their $I_{\Delta n}$ value.

High sensitivity (**HS**): 6 – 10 – 30 mA (for direct-contact / life injury protection)

Medium sensitivity (**MS**): 100 – 300 – 500 – 1000 mA (for fire protection)

Low sensitivity (**LS**): 3 – 10 – 30 A (typically for protection of machines)

The nominal value of residual current indicated above is not an absolute value.

4.1.3 Residual Current Breakers with Over Current Protection (RCBOs)

RCBOs are tripped by both current leakage to earth and overloads as well as short-circuits and they are self-protecting up to a maximum short-circuit current value indicated on the label. They are available in the range of 16–40A with a current sensitivity of 30100 and 300 mA. They are manufactured in SPN and TPN versions. Figure 4.7 illustrates the photo of a two pole RCBO.

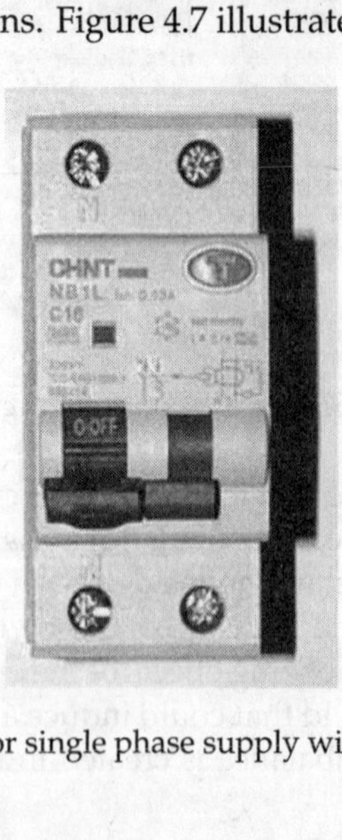

Figure 4.7 A RCBO–suitable for single phase supply with current sensitivity of 30 mA

4.1.4 MCCBs (Molded Case Circuit-Breakers)

MCCBs are available up to a rated current of 2500 A. They are ideally suited for incoming and outgoing feeder circuits of distribution boards. They are available both in thermal or thermal-magnetic operation. Trip current may be adjustable in larger ratings. The MCCBs provide overload and short circuit protection for all circuit elements. They are designed for use in switchboards, control panels, combination starters, in separate enclosures and meet the requirements of lighting, distribution and power circuits. The DPX model of MCCBs manufactured by M/s Legrand is available in thermal-magnetic and electronic versions; DPX MCCBs protect and control circuits up to 1600 A. They can incorporate control and signaling auxiliaries and can be used with earth leakage modules or residual current relays. Figure 4.8 shows a 250A, four pole MCCB.

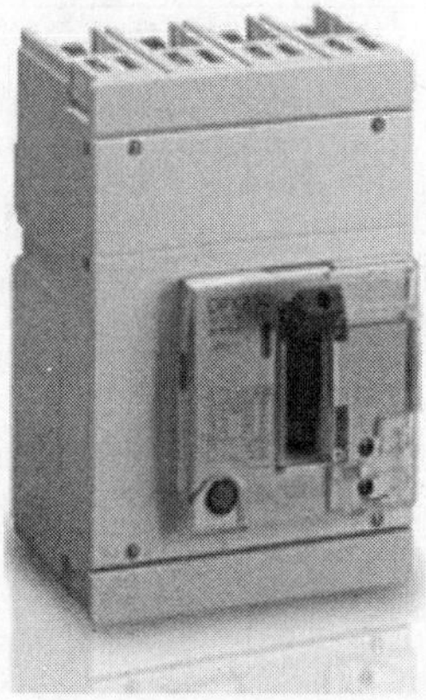

Figure 4.8 Three-phase, 250 A, four-pole MCCB (Courtesy M/s Legrand)

4.2 ESTIMATION OF LOAD REQUIREMENTS

Electrical installations in domestic buildings basically cater to the lighting, ventilation, domestic appliances and gadgets. Lighting and ventilation aspects are discussed in detail in chapter 3. If the actual wattages of the appliances and light points are not known, the load requirements can be

Table 4.1 Recommended ratings of light and socket points (NEC-2011)

Specification	Rating (Watts)
Incandescent lamps	60
Fans (ceiling / table)	60
6A three-pin sockets	100
16A three-pin sockets	1000
Fluorescent lamp 600 mm	25
Fluorescent lamp 1200 mm	50
Exhaust fan 300 mm	45

calculated based on the recommended ratings as listed in Table 4.1. The National Electrical Code also recommends the minimum number of points to be provided in various domestic buildings. Table 4.2 lists the number of points to be provided on the basis of plinth area of the building. National Electrical Code also provides the details regarding the minimum number of socket outlets required for various sizes of residential buildings. Table 4.3 lists these details.

Table 4.2 Recommended number of points in dwelling units (NEC-2011)

Description	Area of the residential building m^2				
	35	**45**	**55**	**85**	**140**
Light points	7	8	10	12	17
Ceiling fans*	2–2	3–2	4–3	5–4	7–5
6A sockets	2	3	4	5	7
16A sockets	–	1	2	3	4
Call bell	–	–	1	1	1

*indicates the number of fan points wired and the actual number of fans provided.

Table 4.3 Details of socket outlets required for domestic buildings (NEC-2011)

Description	Socket outlets	
	6A	**16A**
Bedrooms	2–3	1
Living rooms	2–3	2
Kitchen	1	2
Dining room	2	1
Garage	1	1
For refrigerator		1
For air-conditioner		1
Verandah	1 for every 10 m^2	
Bathrooms	1	1

The details given are the minimum requirements and the owner of the building is at liberty to decide to opt for higher number of points for their buildings in consultation with the design engineer so as to meet the present as well as the future anticipated loads. Once all the light points, fan points and the socket points are decided, the total connected load of the building can be calculated.

The total installed apparent power (kVA) is assumed to be the sum of the kilovolt ampere rating of the individual loads. The maximum demand kVA however need not be equal to the installed kVA. This is because of the fact that all individual loads need not necessarily operate at full rated power or at the same time. Therefore, a suitable correction factor is to be applied to the installed capacity to determine the maximum kVA demand of the building. The two factors that are useful are:

- Factor of maximum utilisation k_u
- Factor of simultaneity k_s.

4.2.1 Factor of maximum utilisation – k_u

In normal situation, power consumed by a load is less than the normal rating of the load. For example, a motor may not be operating at its rated capacity. Hence, it is justifiable to apply a utilisation factor k_u to get a reliable estimate of the load. This factor must be applied to individual loads which are very rarely operated at full load. Standard values of k_u are given below.

For fluorescent lighting	$k_u = 1.00$
For water heaters	$k_u = 1.00$
For industrial loads (motor)	$k_u = 0.75$
For socket outlets	k_u – depends entirely on the type of apparatus being supplied.

4.2.2 Factor of simultaneity – k_s

It is a well known fact that the simultaneous operation of all the connected loads of a given installation never occurs in practice. In other words, there is always some degree of diversity and this factor is taken into account for the purpose of estimating. The simultaneity factor k_s is applied to a group of loads (e.g. loads being supplied from a distribution board or switchboard). The determination of this factor is the sole responsibility of the system designer, since it requires the detailed knowledge of the working of the installation. For this reason it is not defined precisely. The term diversity factor as defined in IEC and simultaneity factor k_s are one and the same. However, in India, diversity factor is the reciprocal of simultaneity factor and its value is always greater than one. The use of k_u and k_s will be more relevant to industrial installations and will be discussed in a later chapter.

For domestic installations, the diversity factor can be assumed to be 2.00 to 2.50 on the assumption that only 40 to 50 percent of the connected load will be used simultaneously.

EXAMPLE 4.1

The floor plan of a two storied residential building is shown in Figure 4.9. Calculate the number of light, fan, and socket and exhaust fan points required for the building as per NEC norms. Also find out the type of supply required and the size of the distribution board.

Solution:

Determination of the connected load and maximum demand
A very detailed lighting calculation is not required for domestic buildings and the lighting requirements can be worked out based on the guidelines in the NEC. The number of light, fan and socket points required for each area is tabulated and is indicated in Table 4.4.

Table 4.4 Details of light, fan and socket points for example 4.1

Room space	1 x 40 W	WB	CF	6A	16/6A	EF	Others
Bedrooms + dressing (3)	1 x 3	(2 + 1)3	1 x 3	3 x 3	1 x 3		
Living room	1 x 2	2 x 2	1 x 2	3 x 2	1 x 2		
Dining room	1	1	1	2	1		
Kitchen	1	1	1	2	3	1	
Verandah / sit out		1					Call bell-1
Toilets		1 x 3		1 x 3	1	1 x 3	
Store		1					
Work area		1			1		
Peripheral lights							Bulkhead-3
Car porch		1		1			
Total points	7	22	7	21	11	4	3(1)

(1) Call bell not included in connected load calculation.

1 × 40 W	40 W fluorescent fixtures
WB	Wall brackets with 11/15 W CF lamps
CF	Ceiling fans (1200 mm or 1500 mm)
6A	6A light socket points
16/6A	16/6A power socket points
EF	220/300 mm fresh air fans.

Once the points have been finalised, the total connected load of the building can be calculated using the NEC guidelines (Table 4.1).

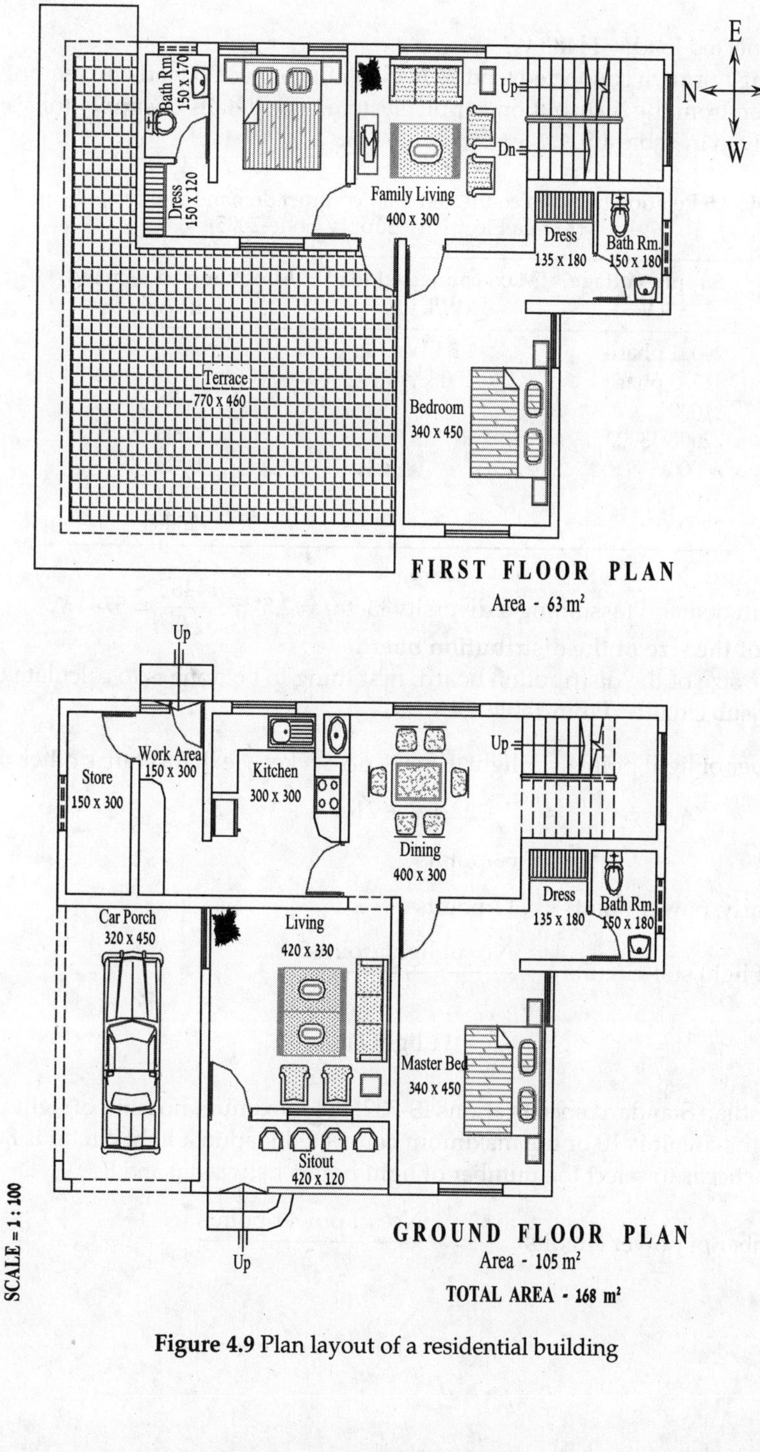

Figure 4.9 Plan layout of a residential building

Room space	1 x 40 W	WB	CF	6A	16/6A	EF	Others
Connected load-W	7 x 50	22 x 15	7 x 60	21 x 100	11 x 1000	4 x 60	3 x 15

The total connected load = 14485 W.

As per the existing norms, a connected load of 14485 W of connected load will require a three-phase supply to be taken from the distribution supply system. The details of supply for various types of consumers are given in Table 4.5.

Table 4.5 Relation between connected load/contract demand and type of supply (Kerala Electricity Supply Code-2005)

Supply voltage V	Max. connected load kW/kVA	Max. contract demand kVA
240, 1 phase	5 kW	
415, 3 phase	100 kVA	
11000		3000
22000/33000		6000
66000		8000
110000		20000
220000		>20000

The maximum demand (assuming a diversity factor = 2.5) = $\frac{14485}{2.5}$ = 5794 W

Determination of the size of the distribution board

To determine the size of the distribution board, first thing to be done is to calculate the number of light and power sub circuits. From Table 4.4,

The number of light points = light + fan + 6A socket + exhaust fan + other points

= 7 + 22 + 7 + 21 + 4 + 3

= 64 points

Similarly, power points = 11 points

$$\text{The number of light sub-circuits} = \frac{\text{No. of light points}}{6}.$$

$$= \frac{64}{6} \approx 11 \text{ light circuits}$$

(According to Indian Standard specifications IS-732, the maximum number of light points that can be wired in a light circuit is 10 or the maximum connected load in a light circuit is 800 W). A good engineering practice is to select the number of light points between 6 and 8.

$$\text{The number of power circuits} = \frac{\text{No. of power points}}{2}.$$

$$= \frac{11}{2} = 6 \text{ power circuits.}$$

The total number of sub-circuits required = No of light circuits + Power circuits.
= 11 + 6 = 17 sub-circuits.

A three-phase 8-way MCB distribution board will provide 24 sub-circuits of which 17 can be used now and the remaining 7 can be used as spare capacity for future expansion.

Specification of the distribution board

Three-phase distribution board conforming to IS-8623, universal mounting type with 100A copper bus bar, neutral bar, earth bar and cable ties for cable management with top and bottom removable gland plates with knockouts and provision for four pole RCD/MCB/ISOLATOR as incoming and single pole MCBs as outgoing. The protection class of the distribution board may be either IP20 (single door) or IP42 (double door) with acrylic or metal door.

Incomer : 25 A, 4 pole, RCBO, 415 volt, with 30 mA sensitivity
Outgoing : 6A single pole MCB with B-curve 11 Nos
and 16A single pole MCB with C-curve 6 Nos.

The B-curve MCBs are to be used for lighting loads and C-curve MCBs for highly inductive loads. Most of the power devices other than the water heaters or cooking ranges will be inductive in nature. RCBO devices provide triple protection against earth leakage, over current and short circuits. If ELCB + MCBs are used, protection is guaranteed for earth leakage and short circuits only. MCBs provide protection against short circuit only. No protection is guaranteed if only MCB isolators are used. It will serve only as an isolating device.

Figure 4.10 shows the plan layout of the example 4.1 with all electrical points marked on the drawing. The locations of the light fixtures, ceiling fans, socket points etc are clearly marked on the drawing which can serve as a document to be used at the construction site. Such drawing will also be required for getting official sanctions for the work from the Electricity Supply Departments.

EXAMPLE 4.2

The floor plan of a five-bed residential building is shown in Figure 4.11. Locate the light and other electrical points required for electrification of the building as per NEC requirement. Find also the connected load of the building and the type of supply required.

Solution:

Calculation of connected load and maximum demand

Just as discussed in example 4.1, a very detailed lighting calculation is not required for domestic buildings and the lighting requirements can be worked out based on the guidelines in the NEC. The number of light, fan and socket points required for each area is tabulated and is shown in Table 4.6. These numbers may vary according to the requirements of the clients and shall always be decided in consultation with the owner of the buildings. Even the location of the light, fan and socket points may be fixed in consultation with the owners, because some of them may have specific ideas of using certain areas inside the house.

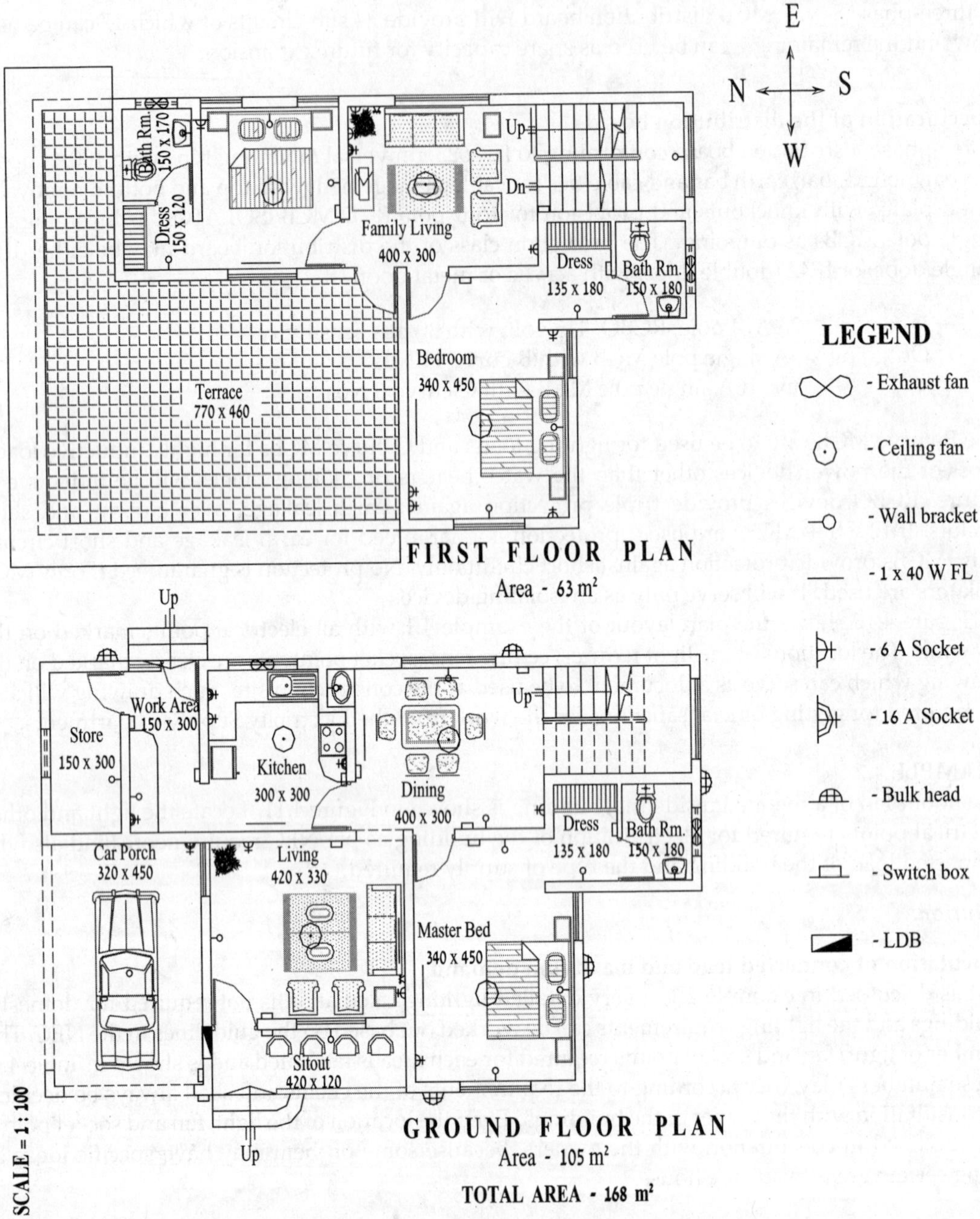

Figure 4.10 Plan layout of a residential building

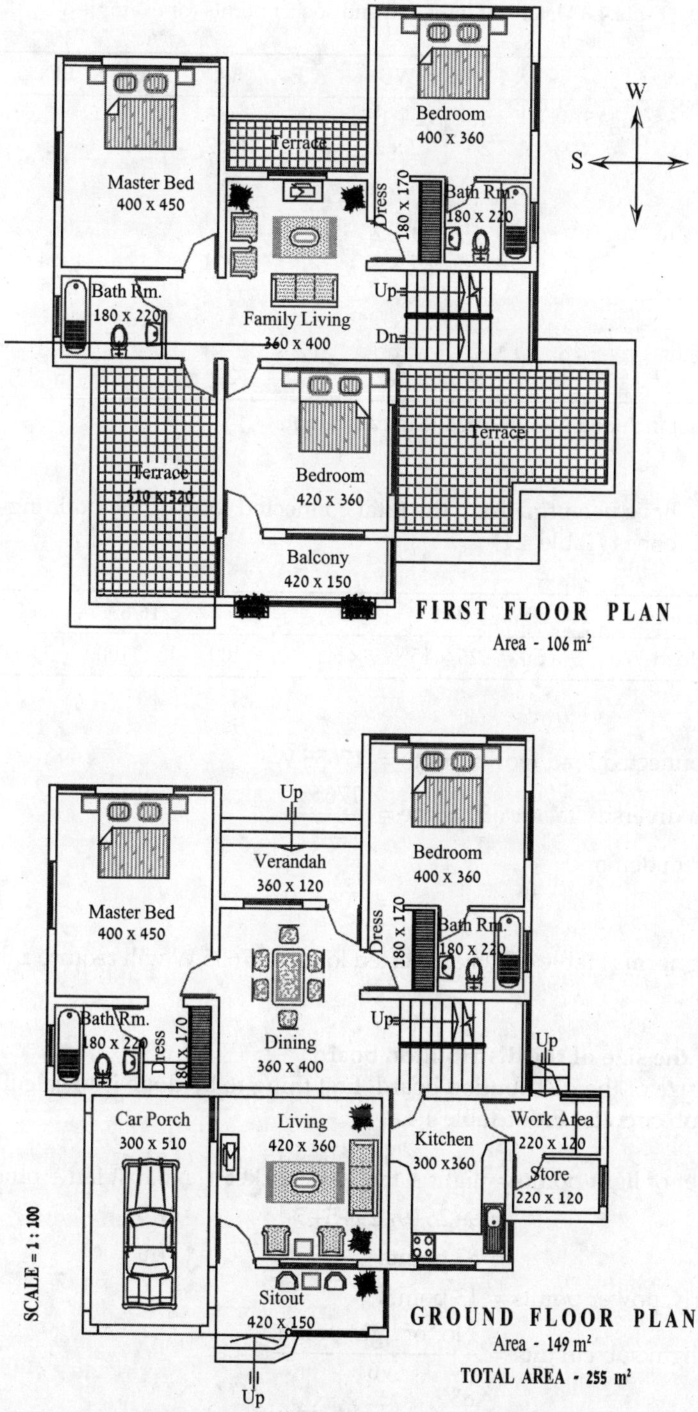

Figure 4.11 Plan layout of a residential building

Table 4.6 Details of light, fan and socket points for example 4.2

Room space	1 × 40W	WB	CF	6A	16/6A	EF	Others
Bedrooms + dressing (5)	1 × 5	(2 + 1)5	1 × 5	3 × 5	1 × 5		
Living room (2)	1 × 2	2 × 2	1 × 2	3 × 2	1 × 2		
Dining room	1	1	1	2	1		
Kitchen	1	1	1	2	3	1	
Verandah/sit out (2)		1 × 2					Call bell-1
Toilets		1 × 4		1 × 4	1	1 × 4	
Store		1					
Work area		1			1		
Peripheral lights							Bulkhead-5
Total points	9	26	9	29	13	5	5[1]

[1] Call bell not included in connected load calculation.

Once the points have been finalised, the total connected load of the building can be calculated using the NEC guidelines (Table 4.1)

Room space	1 × 40W	WB	CF	6A	16/6A	EF	Others
Connected load-W	9 × 50	26 × 15	9 × 60	29 × 100	13 × 1000	5 × 60	5 × 15

The total connected load works out to = 17655 W

Assuming a diversity factor of 2.5, the $= \dfrac{17655}{2.5}$

the maximum demand

$= 7062$ W.

As per the existing norms (Table 4.5), a connected load of 17655 W will require a three phase supply to be taken.

Determination of the size of the distribution board

To determine the size of the distribution board, first thing to be done is to calculate the number of light and power sub circuits. From Table 4.4,

$$\text{The number of light points} = \text{light + fan + 6A socket + exhaust fan + other points}$$
$$= 9+26+9+29+5+5$$
$$= 83 \text{ points}$$

$$\text{Similarly, power points} = 13 \text{ points}$$

$$\text{The number of light sub circuits} = \frac{\text{No. of light points}}{6}.$$
$$= \frac{83}{6} \approx 14 \text{ light circuits}$$

(According to Indian Standard specifications IS-732, the maximum number of light points that can be wired in a light circuit is 10 or the maximum connected load in a light circuit is 800 W). A good engineering practice is to select the number of light points between 6 and 8.

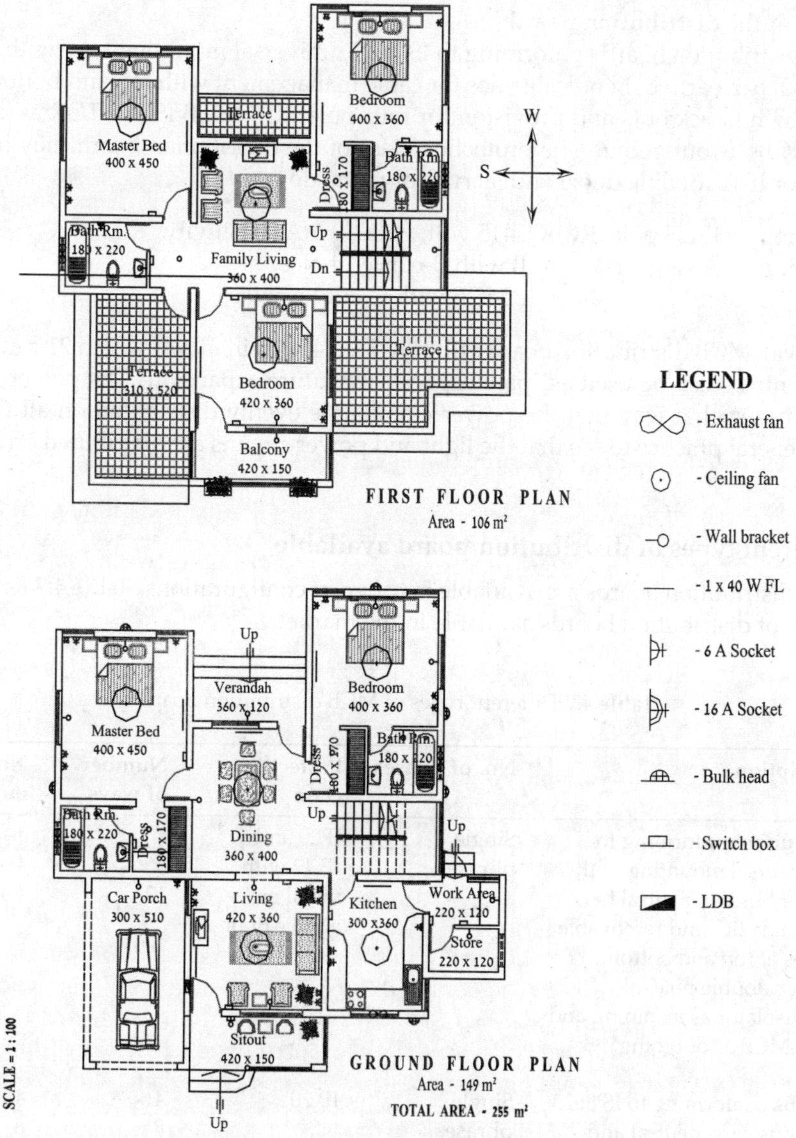

Figure 4.12 Plan layout showing the location of electrical points.

$$\text{The number of power circuits} = \frac{\text{No. of power points}}{2}$$

$$= \frac{13}{2}$$

$$= 7 \text{ power circuits.}$$

The total number of sub-circuits required = No. of light circuits + Power circuits.

= 14+7 = 21 sub-circuits.

A 3-phase 8-way distribution board can be selected to meet the requirement.

Specification of the distribution board

Three-phase distribution board conforming to IS-8623, universal mounting type with 100A copper bus bar, neutral bar, earth bar and cable ties for cable management with top and bottom removable gland plates with knock outs and provision for four pole RCD/MCB/ISOLATOR as incoming and single pole MCBs as out going. The protection class of the distribution board may be either IP20 (single door) or IP42 (double door) with acrylic or metal door.

Incoming : 40 A, 4 pole, RCBO, 415 volt, with 30 mA sensitivity
Outgoing : 6 A single pole MCB with B-curve 14 Nos.
and 16 A single pole MCB with C-curve 7 Nos.

A 3-phase 8-way MCB distribution board will provide 24 sub circuits of which 21 can be used now and the remaining 3 can be used as spare capacity for future expansion. The power socket points may be wired in such a way that the power circuits are evenly distributed on all the phases. It should be a general practice to see that the light and power circuits are distributed on all the phases in a balanced manner.

4.2.3 Different types of distribution board available

Factory built distribution boards are available in different configurations. Table 4.7 lists some of the popular types of distribution boards available in the market.

Table 4.7 Different types of MCB distribution board

Type description	No. of phases	Protection class	Number of ways	No. of MCB modules
Consumer units conforming to IS-8623, universal mounting with 100A copper bus bar, neutral bar, earth bus, cable ties and removable gland plates at top and bottom. Provision for double pole RCD/MCB/Isolator as incoming and single pole MCB as outgoing.	Single phrase	IP20 and IP 42 with metal or acrylic door	4 8 12	4 + 4 4 + 8 4 + 12
Regular DBs conforming to IS 8623 with 100A bus bars, neutral and earth bus bars.	Single phrase	IP 20	4 8 12	4 8 12
Flush mounting DBs conforming to IS-8623 and with reversible pan assembly for top and bottom cable entry, masking sheet for protection against entry of concrete, and with provision for double pole RCCB/MCB/Isolator as incomer and MCBs as outgoing.	Single phrase	IP 20	4 8 12	2 + 4 2 + 8 2 + 12

Continued...

Type description	No. of phases	Protection class	Number of ways	No. of MCB modules
Consumer units conforming to IS-8623, universal mounting with 100A copper bus bar, neutral bar, earth bus, cable ties and removable gland plates at top and bottom. Provision for four pole RCCB as incoming and three single pole MCBs as outgoing.	Three Phrase	IP 20 and IP 42 with metal or acrylic door	3 × 3	6 + 9
ETPN DBs conforming to IS-8623, universal mounting, with 100A copper bus bars, neutral and earth bars, cable ties, top and bottom removable gland plates with knock outs and suitable for fixing four pole RCD/MCB/Isolator as incoming and single pole MCB as outgoing.	Three Phrase	IP 20 and IP 42 with metal or acrylic doors	4 6 8 12	8 + 12 8 + 18 8 + 24 8 + 36
Per phase isolation DBs conforming to IS-8623, universal mounting, with 100A copper bus bars, neutral and earth bars, cable ties, top and bottom removable gland plates with knockouts and suitable for fixing four pole RCD/MCB/isolator as incoming, 3 double pole RCD/MCB/isolator as sub incomer and single pole MCB as outgoing with separate neutral bars. (The advantage of per phase isolation DBs are that only the phase where there is an earth leakage fault is isolated and the other two phases are in operation, avoiding total darkness.)	Three Phrase	IP 20 and IP 42 with metal or acrylic doors	4 6 8 12	8 + 12 8 + 18 8 + 24 8 + 36

Continued. . .

Type description	No. of phases	Protection class	Number of ways	No. of MCB modules
Phase selector DBs as per IS 8623, universal mounting with 100A copper bus bar, neutral and earth bar, top and bottom removable gland plates with provision for 3 Nos. 63A selector switches, 3 Nos. piano type indicating lamp switch, four-pole RCD/MCB/Isolator as incoming and SP MCBs as outgoing.	Three Phrase	IP 20 and IP 42 with metal doors	4 6 8 12	6 + 12 6 + 18 6 + 24 6 + 36
Seven segment DBs as per IS 8623, with 100A copper bus bar, neutral and earth bar, top and bottom removable gland plates with provision for pole RCD/MCB/ Isolator as incoming, DP RCDs/ MCBs/isolator as sub-incomer and SP MCBs as outgoing. (These DBs provide phase segregation and separation between incomer and outgoing)	Three Phrase	IP 20 and IP 42 with metal doors	4 6 8 12	4 + 12 + 12 4 + 12 + 18 4 + 12 + 24 4 + 12 + 36
VTPN DBs as per IS 8623, universal mounting with 200 A copper bus bars, earth and neutral bars, cable ties, top and bottom removable gland plates, and with provision for four-pole RCD or RCBO/MCB/ isolator as incoming and SP/TP MCBs as outgoing. (The advantage of this DB is that both single and three phase outgoings can be taken from this DB)	Three Phrase	IP 20 and IP 42 with metal or acrylic door	4 8 12	8 + 12 8 + 24 8 + 36

4.3 SELECTION OF TYPE OF WIRING

Different types of wiring as specified in the BIS specification IS-732-1989 are available for wiring of domestic installations. However, one of the following methods is more suitable for residential buildings.

- PVC insulated and PVC sheathed wires on teak wood batten
- PVC insulated wires in steel/PVC surface conduits and
- PVC insulated wires in concealed conduits.

If economy is the only criterion, the first type of wiring is the best option. It is economical as well as functional and is the best choice for low cost housing. On the other hand if aesthetics is the most important consideration, the third type of wiring is the best option. The second type of wiring is functional as well as easy to maintain. Hence this type of wiring is more suitable for locations where functionality and maintenance are more important. The first and second type of wiring has the advantage that these wiring can be taken up after completing the building structure. Whereas, concealed wiring is required to be carried out along with the building construction. And this requires proper co-ordination between the agencies involved in the building construction. Concealed conduit wiring with rigid PVC conduits is the most accepted form of wiring for modern residential buildings.

4.3.1 General rules applicable for conduit wiring using rigid PVC conduits

- Only rigid PVC conduits shall be used for the wiring.
- The conduits may be either threaded type or pasted type.
- Conductors of different supply systems shall not be bunched in the same conduit.
- Phase segregation in separate conduits is recommended for lighting and small power applications.
- The number of insulated wires inside a conduit shall be as per Table 4.8.
- Conduits shall be interconnected by means of threaded or pasted couplers depending on whether the conduits used are threaded type or pasted type.
- For long straight runs, inspection type couplers should be used at regular intervals.
- Conduits shall be supported by saddles with a minimum spacing of 600 mm.
- Necessary conduit specials like bend, inspection bend, elbow and T's shall be used wherever appropriate.
- In case of concealed conduits, the chase in the walls shall be made neatly and with sufficient size to permit the conduits to be laid properly. If the building is under construction, the chase shall be made before plastering of the walls.
- Conduits in ceiling shall be laid while concreting the ceiling slabs.
- Suitable inspection boxes shall be provided to facilitate periodical inspection and removal of wires if necessary. The minimum size of the inspection box shall be 75 × 75 mm.
- All outlets like switch, socket, ceiling rose, etc., may be either of surface type or flush type.
- All flush mounting outlets shall be made of mild steel or aluminium boxes with a cover plate of insulating material and with provision for earthing.

The details of number of wires that can be drawn through different sizes of conduits are listed in Table 4.8.

4.3.2 Selection of conductor size (Reference-IS-732)

All conductors used for domestic installations shall be PVC insulated copper or aluminium. The conductors used in the final sub-circuits for light and fans shall not be less than 1.00 mm^2 copper or 1.5 mm^2 aluminium. The cross-sectional area of power sub-circuits shall not be less than 2.5 mm^2 copper or 4 mm^2 aluminium. The minimum cross section of the flexible cord shall not be less than 0.5 mm^2 copper. The standardisation in the size of conductors is done with a view to limiting the voltage drop within the specified limits.

Table 4.8 Maximum permissible number of 1.1 kV grade cables in metallic conduits (NEC-2011)

Size of cable mm^2	Size of conduits - mm				
	20	**25**	**32**	**40**	**50**
1	7	13	20		
1.5	6	10	14		
2.5	5	6	10	14	
4	3	6	10	11	
6	2	5	9	11	
10		4	7	9	
16		2	4	5	12
25			2	2	6

4.3.3 Selection of switches and sockets

Switches are the most frequently used item in the domestic wiring. The reliability and useful life are the main features in selecting the switches for domestic applications. 6A switches are normally used in light, fan, exhaust fans and light socket points. 6A switches are used for power socket points. Switches with higher ratings may be necessary for controlling room air conditioners, water heaters and cooking ranges with capacity above 2000 Watts.

Different types of switch in different deco'r are available in the market. Their prices range from Rs.15 to Rs.120 depending upon the type and brand. Switches are generally classified as tumbler, flush, plate and modular switches. Tumbler type switches are now obsolete and were used in CTS wiring systems with exposed switch boards. Flush type switches are cheap and reliable and have very good service life. Flush type switches and sockets are independent and in case of damage to any of these, only the damaged component needs to be replaced. They are mounted on switch boxes with hylam or acrylic sheet covers. Plate switches gives aesthetic looks and are comparatively costlier. Plate type sockets come as an integral unit and in case of failure of switch or socket, the entire unit will have to be replaced. Hence, plate type switches and sockets are costlier in terms of initial cost and maintenance cost, when compared to flush type switches and sockets. Modular switches, sockets and other gadgets are very popular now because of their flexibility in arrangement in modular form. The sizes of switches, sockets and other gadgets are standardised in multiples of modules and different combinations are possible to meet the individual requirements. They are very aesthetic and reliable. Their maintenance is very easy. However, the initial cost of modular switches and sockets are comparatively costlier than the other two types.

4.3.4 Selection of light fittings and fans

Selection of the right type of light fitting and the right size of fans does not require an elaborate calculation. The market is flooded with different kinds of light fixture and various sizes of fans. It is preferable to buy only ISI branded products (especially with energy star rated products if available) to make sure that the products are able to provide guaranteed service and energy efficiency.

4.3.5 Earthing of domestic installations

The earthing of domestic installations is to be carried out as per the requirements specified in IS-3043. One pipe earth station for single phase connection and two pipe earth stations for three phase connections are the mandatory requirements. A detailed earthing design is not required for domestic buildings.

A typical schedule of work relating to the electrical installations for residential buildings is shown in Table 4.9.

Table 4.9 Typical schedules of quantities for the electrical installations of a residential building

S.No.	Specifications	Unit	Qty	Rate	Amount
1	**Light distribution boards:** Installation, testing and commissioning of 500V double insulated IP 42 modular flush mounting DBs conforming to IS-8623 and with reversible pan assembly for top and bottom cable entry, masking sheet for protection against entry of concrete, and with provision for double pole RCCB/MCB/isolator as incomer and MCBs as outgoing etc. complete with 25A, 4 pole RCBO as incomer and 6 A B-curve SP MCBs- XX Nos. 16A C-curve MCBs - YY Nos. as outgoing	E			
2	**Point wiring:** Point wiring for lights, fans, exhaust fans, etc. consists of 1. All wiring in ISI marked rigid PVC conduits of approved brand with junction boxes, pull through boxes etc. on walls, ceiling, columns, etc., as required. 2. 2 × 1.0 sq.mm. PVC insulated flexible copper conductors of approved brand. 3. 14 G earth continuity conductor or 1 sq.mm PVC insulated (green stripes) flexible copper cable for earth continuity conductor 4. Terminal MS/AL boxes to house modular switches, sockets, fan regulators of approved brand. 5. Modular switches, sockets and regulators of approved brand. 6. All points terminated with end connectors/ceiling roses. 7. And all interconnections.				
2.1	One light point controlled with one switch	E			
2.2	Two light points controlled with one switch	E			
2.3	Three light points controlled with one switch	E			
2.4	Call bell pointss	E			
2.5	Staircase light point	E			
2.6	Ceiling fan points with regulators	E			
2.7	Exhaust fan points without regulator	E			

Continued...

S.No.	Specifications	Unit	Qty	Rate	Amount
3	**Point wiring for sockets**				
3.1	Wiring for 16/6A dual socket points generally as specified in point wiring section but with 2 × 2.5 sq.mm PVC flexible copper conductor of approved brand and provided with 16/6 A switch and sockets of approved brand.	E			
3.2	6A. socket points generally as specified in 2.0 and provided with 6A switch and socket of approved brand away from the switch board	E			
3.3	6A socket points in the existing switchboard	E			
3.4	Wiring for 20A metal clad socket points with 2 × 4 sq.mm PVC flexible copper conductor of approved brand in concealed rigid conduits and terminated with factory built socket unit controlled by 20A DP MCB and with 14g earth continuity conductor etc. complete. (for air conditioners)	E			
3.5	Wiring for telephone socket points to the respective telephone sockets with modular socket outlets of approved brand using 0.5 sq.mm PVC insulated two pair telephone cable in recessed conduits.	E			
3.6	Wiring for TV sockets with modular TV socket outlets and wired with 75 ? TV cable for cable TV connection	E			
4	**Light fixtures and fans** Testing and commissioning of light fittings, fans and exhaust fans complete with a) Lamps and fittings b) Earthing of fittings and c) fixing accessories.				
4.1	**Ceiling mounted fittings**				
4.1.1	1 × 36 W fluorescent fixture	E			
4.1.2	2 × 36 W	E			
4.2	**Wall mounted fittings**				
4.2.1	1 × 36 W fluorescent fitting	E			
4.2.2	23/15 W CFL fitting	E			
4.3	**Decorative false ceiling mounted fittings**				
4.3.1	1 × 36 W	E			
4.3.2	2 × 36 W	E			
4.4	**Decorative wall mounted fittings**				
4.4.1	1 × 36 W	E			
4.4.2	2 × 36 W	E			
4.5	**Fans and exhaust fans**				
4.5.1	1200 mm sweep ceiling fans with regulator	E			
4.5.2	225 mm fresh air fan with gravity louvres	E			
4.5.3	300 mm exhaust fans-light duty	E			

Continued...

S.No.	Specifications	Unit	Qty	Rate	Amount
4.5.4	External security light fittings (bulk head)	E			
4.5.5	Electronic (musical) call bell E				
5.0	**Earthing and miscellaneous items**				
5.1	Providing pipe earthing stations as per IS 3043–1987 for equipment earthing complete with excavation in hard soil, pipe electrode of approved size, brick masonry with hinged cover for watering, charcoal and salt as required, and back filling pipe earthing.	E			
5.2	Laying of copper earth wires/strips for interconnections of earth stations, distribution board, switchboards etc. complete with fixing accessories.				
5.2.1	No. 10 swg	m			
5.2.2	No. 12 swg	m			
6	**Meter board and accessories**				
6.1	Fabrication, supply & installation of dust and vermin proof, wall mounting, thermoplastic meter box of following size suitable for fixing SEB meter, fuse units, CT, etc, hinged door construction for the front side with compartments if necessary, providing vision panel for the meter compartment, knock out for cable entry and exit etc, complete. 600 × 300 × 214 mm	E			
6.2	Supply and fixing the following accessories in the existing meter board and giving connections etc.				
6.2.1	S & F switch disconnector fuse unit complete with handle assembly, door interlock and padlocking arrangement of the following ratings: 63A, AC 23 A, TPN SDF for DIN fuse	E			
6.2.2	63 A, 415 V Backelite fuse fitting Set				
6.2.3	63 A, 415 V, 2 way neutral link mounted on DMC/SMC base	E			
7	**Total amount**				

4.4 SPECIAL FEATURES APPLICABLE FOR HIGH-RISE APARTMENT BUILDINGS

High-rise residential apartments are those buildings having a height above ground level of more than 15 m. Lift machine rooms or the water tanks will not be taken into account in fixing the height of the building. These buildings come under the section 54 of the Electricity Act, 2003 and rule 50(A) of the Indian Electricity Rules 1956. The electric power supply arrangement in the high-rise apartment buildings are normally provided through a transformer substation which receives power at 11 kV on the high voltage side.

4.4.1 Substations

Normally, outdoor type transformer substations are not permitted at the premises of high-rise apartments. However, outdoor transformers with cable end boxes on both HT and MV sides can be permitted with special emphasis on safety measures. Only dry type transformers will be permitted to be installed in the basement floor or above the ground floors. If the total connected load of the apartment is 50 kVA and above, separate transformer shall be installed by the promoter/owner at his cost. If the common and essential loads exceed 100 kVA, it is advisable to install a separate transformer for this purpose by the promoter/owner. Fire pump capacity is not to be included in assessing the total connected load. For calculating the connected load, a minimum of 1.2 kW for single phase connection and 5 kW for three phase connection shall be used if exact information is not available. Otherwise, a uniform rate of 50 W/m^2 shall be used for working out the total connected load. A diversity factor not exceeding 2 is included in deciding the maximum demand for the apartment. Emergency supply disconnection facility using push button shall be provided at a conspicuous place in the ground floor, preferably at a height of 2.75 m from the floor level.

4.4.2 Primary and secondary protection

If there is only one transformer, an HT load break switch fuse unit or a vacuum circuit breaker may be used as the primary controlling unit. HT load break switch is adequate for capacities below 1000 kVA. For higher ratings, vacuum circuit breakers are to be used. If there are more than one transformer in the same premise, the HT switchboard shall have an HT load break isolator of adequate capacity on the incomer side as a group controller and HT SFU (switch fuse unit) or VCB (vacuum circuit-breaker) as outgoing for controlling individual transformers. The HT metering arrangement shall be arranged on the primary control unit. Only tamper proof digital TOD (Time of Day) meters shall be provided for metering the energy consumption. The metering current transformer shall be of single ratio; single core type and no extra tapping points are permitted on this CT. Tappings from the metering Potential Transformer (PT) can be permitted through separate fuses for connection to the indicating lamps, volt meter, power factor meter and watt hour meter.

Acircuit breaker of adequate capacity shall be provided on the secondary side of the transformer. Oil filled circuit-breakers or oil filled switch fuse units shall not be used in high rise buildings. Low set earth fault protection using neutral CT need not be insisted for transformers feeding high-rise buildings where, low tension metering is provided for individual consumers. However ELCB (earth leakage circuit breakers) shall be provided on all consumer distribution boards and other motor loads excepting the fire pump motors.

4.4.3 Metering panel

For high-rise apartment buildings, the metering panel shall be located at the ground floor level. The maximum number of outlets from a metering panel will be limited to eight in the case of three phase supply and twelve in the case of single-phase system. The energy meters for the individual consumers are to be provided by the promoter of the building and these meters have to be certified by the competent authority before being installed in the metering panel. Fuse/MCB protection shall be provided for every energy meter. Locking and sealing facility shall also be provided on all metering panels, main switch boards, HT switchboard, etc., to prevent theft of energy or to prevent pilferage. Unmetered spare outlets are not permitted in MSB, SSBs and in the metering panels.

4.4.4 Cabling

Cables from the metering panel to various floors shall be routed through cable ducts only. There shall be access to these cable ducts from all floors. Fire barriers shall be provided in the cable ducts at all floor levels inside the ducts. This is to prevent the spread of fire to upper floors in the event of a fire breaking out in a lower floor. The grid power cables and the generator cables shall be segregated inside the duct. Other service cables like telephone network, CCTV cable etc shall not be provided in the cable ducts for power. Aluminium armoured cables are preferred up to the distribution boards of the individual consumers. Beyond the distribution board copper wires are preferred for reasons of convenience and reliability.

4.4.5 Distribution boards

Factory assembled distribution boards with IP42 protection (Double door) is the best choice for apartments. The incomer device for these distribution boards shall be an RCBO or ELCB + MCB (30 mA sensitivity) of appropriate rating depending on the connected load. The outgoing MCBs shall be of B-characteristics, for light and resistive loads and of C-characteristics, for highly inductive loads. The maximum number of ways permitted for three phase distribution board is only 8.

4.4.6 Emergency power supply

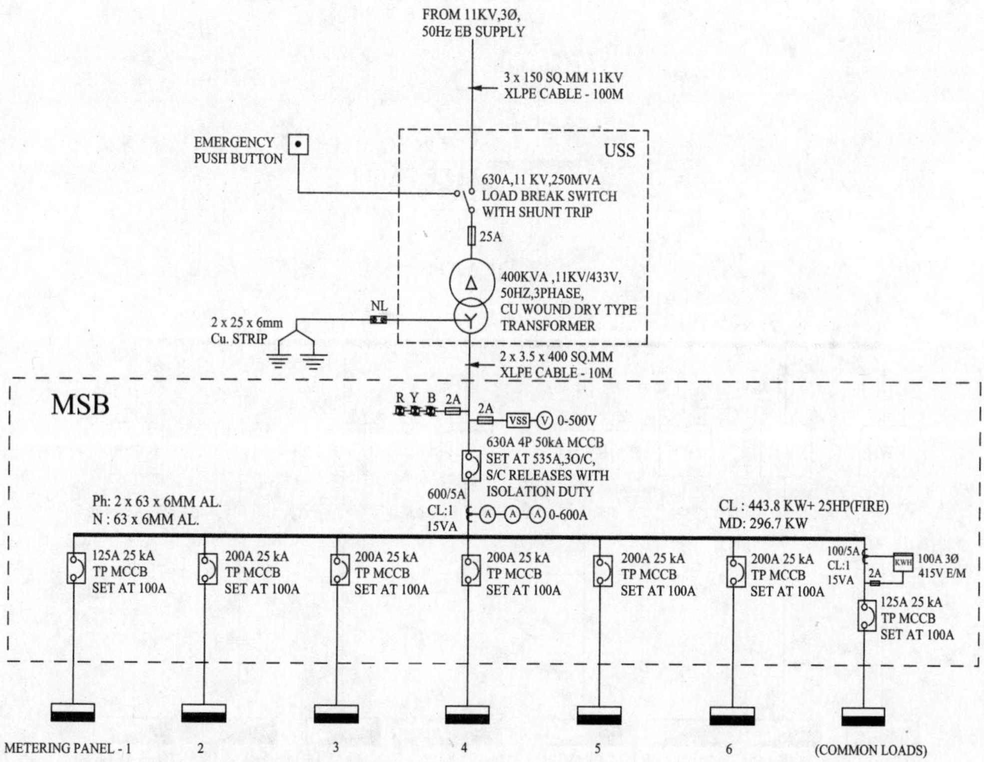

Figure 4.13 Typical single line layout transformer sub station and MSB

A standby generator of adequate capacity is to be provided for every apartment building to meet the emergency needs and essential loads in case of a power failure. The capacity of the generator shall be a minimum of 20% of the connected load of the apartment. When generator supply is extended to individual consumers, it shall be carried out through a separate emergency distribution board.

The generators shall conform to the requirements of the Pollution Control Board and shall be housed in an acoustic enclosure to limit the sound level below 70 dB. The exhaust pipe of the diesel generator set shall also be extended to the top of the building. If the generator is to be installed on any floor above the ground level, structural fitness of the building will have to be ascertained to check whether the building will be able to carry the static and vibratory loading of the generator. A separate room shall be provided with proper oil drainage facility. An emergency engine stopping arrangement also is to be provided outside the generator room.

All electrical rooms accommodating the transformer, switchboards, metering panels, generator, etc., shall be located with easy access and proper ventilation. Adequate clearances are also to be guaranteed for easy maintenance work.

Figures 4.13 to 4.19 shows, the details of electrical installations of a high-rise apartment building with twelve floors (G + 11) which includes forty residential apartments and three penthouses. The ground floor is set apart for common facilities and each upper floor has four flats up to tenth floor and the eleventh floor has three penthouses. Figure 4.13 shows the typical transformer substation layout with the main switchboard (MSB). The details of metering panels are shown in Figures 4.14 and 4.15. Since each metering panel can accommodate only up to eight outgoing feeders, there has to be five such metering panels to meet the requirements of the forty flats.

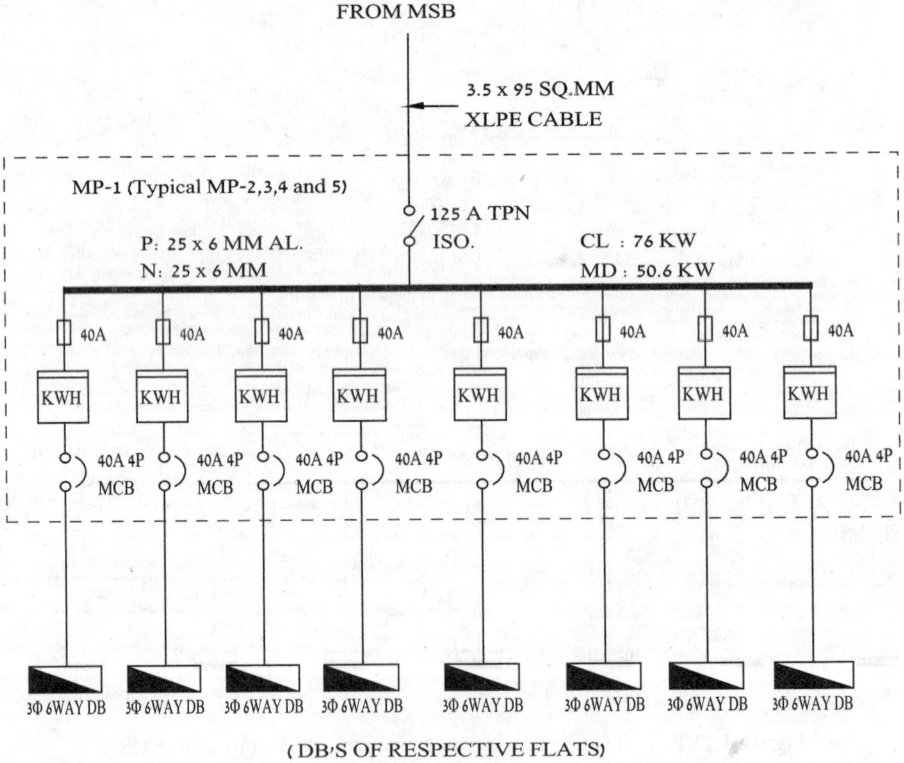

Figure 4.14 Typical metering panel

The three penthouses are supplied from the metering panel No.6 (as shown in Figure 4.15). The outgoing cables from the metering panels are connected to the respective distribution boards of the apartments using four core cables.

The details of the generator control panel (GCP) and generator switchboard (GSB) is shown in Figure 4.16. The capacity of the generator is selected in such a way as to meet the common essential loads and emergency lighting requirements in each flat. The generator control panel (GCP) consists of visual indication for both, generator ON and load ON conditions. The metering arrangement includes volt meter with selector switch, frequency meter, three numbers of ammeter, and energy meter. The generator is protected by a suitable moulded case circuit breaker (MCCB).

The details of the common service panel are shown in Figure 4.17. This panel caters to the lighting requirements of common areas like entrance lobby, stairways, car parks, equipment rooms, external lights, etc. Supply to the lifts, domestic pumps, fire pumps and other building services are also provided from this panel.

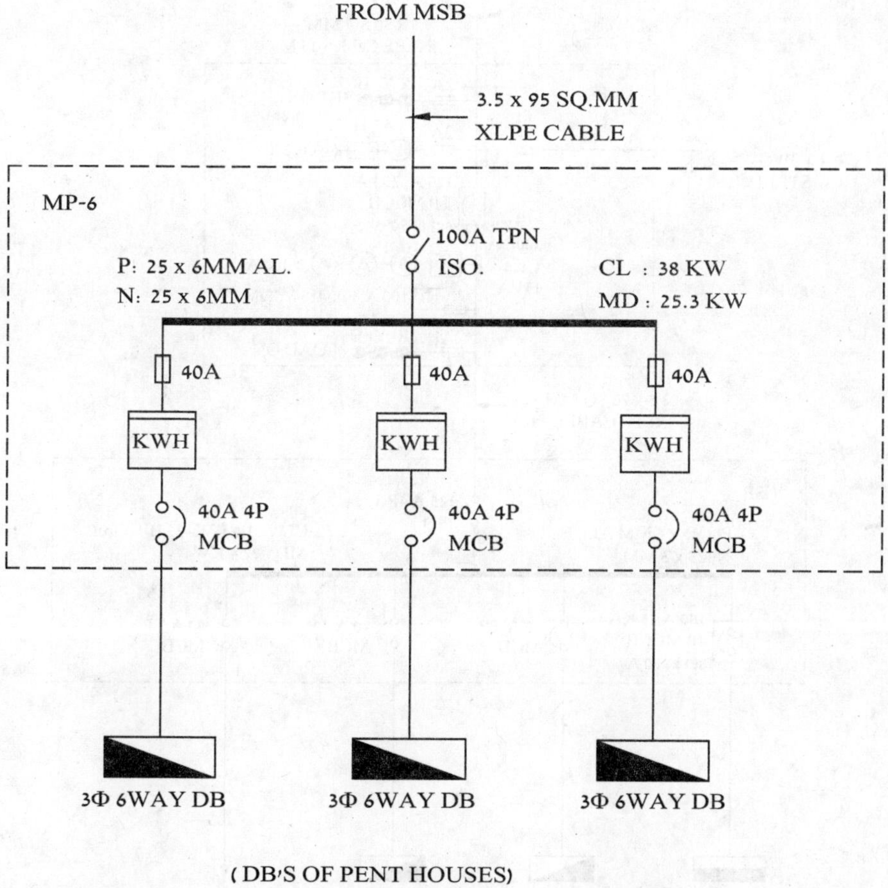

Figure 4.15 Metering panel for pent houses

The total energy consumption from this panel is separately metered and the total cost is shared by the owners of the residential flats as part of the monthly maintenance charges.

Figure 4.18 shows the details of the lift machine panel. As per statutory regulations, a minimum of two lifts are mandatory in high rise buildings. The lighting requirement of the lift car as well as the lift well and lift machine room is provided through the lift light distribution boards (LLDB).

The wiring details for the emergency light points in the individual flats are shown in Figure 4.19. A power circuit of the distribution board in a typical flat and one single phase circuit from the emergency DB on the ground floor are connected to a changeover switch (COS). All the emergency light points are fed from this change over switch. When the normal supply fails, the power is changed over to the generator supply manually by the occupants of the flat. More complex and automatic changeover facilities are available with additional cost. Other options are also available depending on the percentage of backup required in case of power failure.

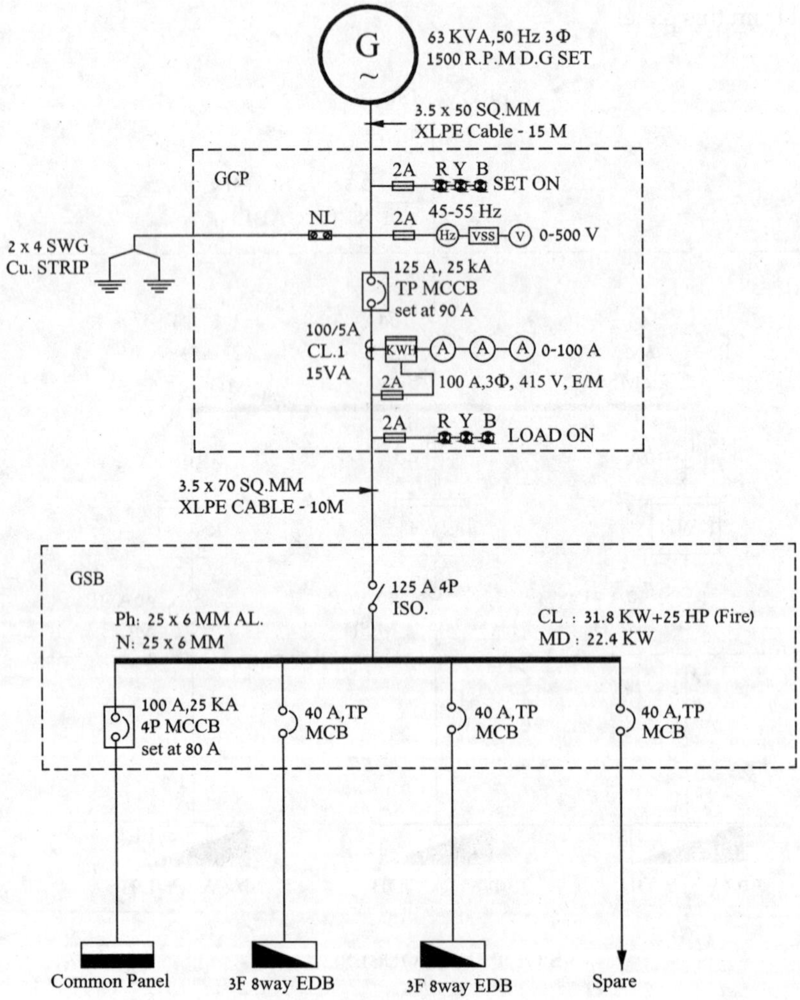

Figure 4.16 Typical layout of generator and GSB

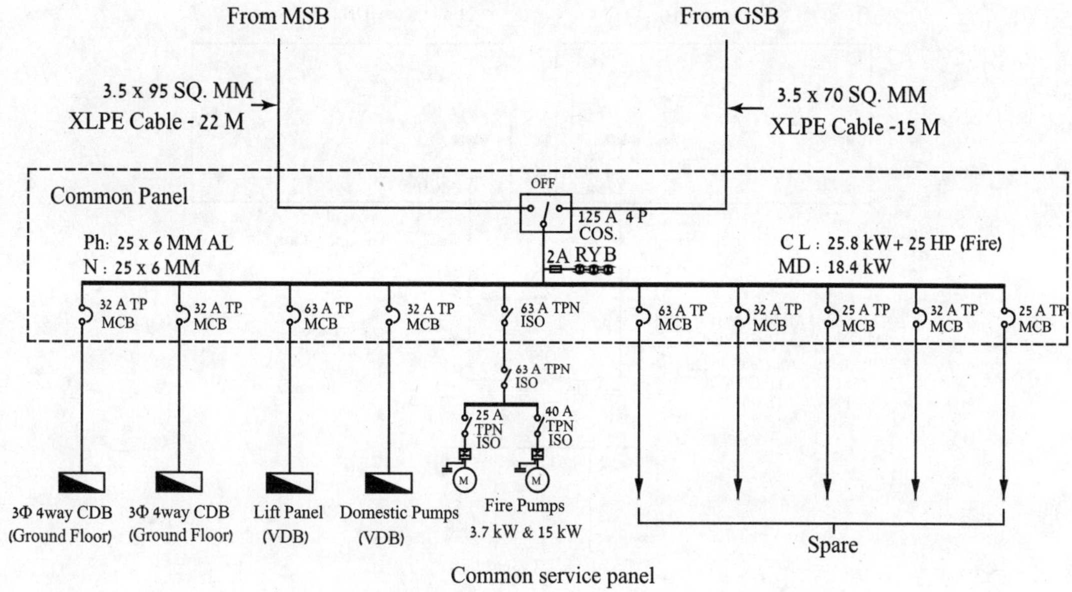

Figure 4.17 Details of common service panel

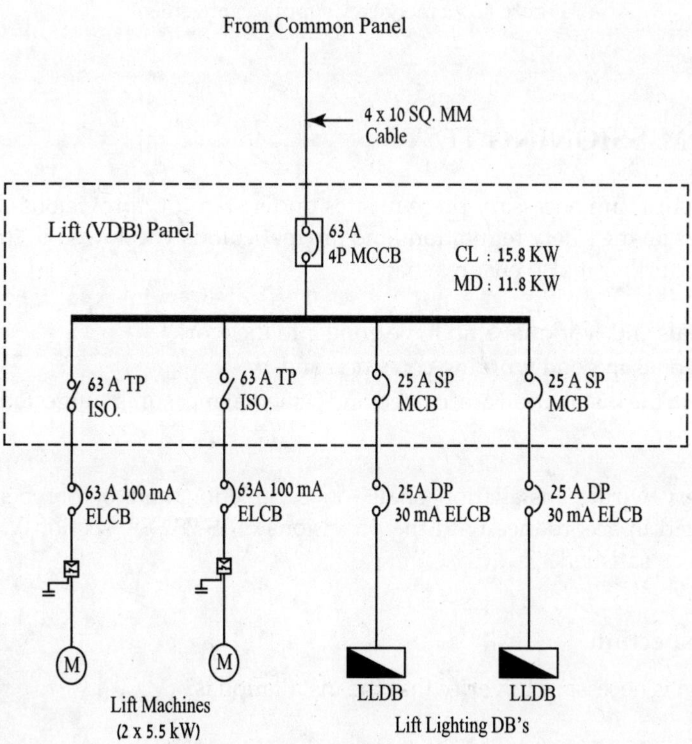

Figure 4.18 Lift machine panel

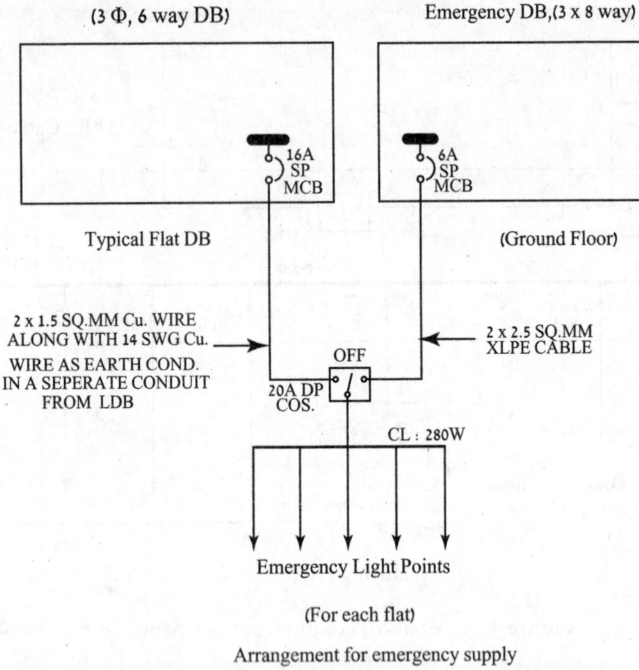

Figure 4.19 Emergency supply arrangement

4.5 PRECOMMISSIONING TESTS

Precommissioning tests are necessary prerequisites under statutory provisions. It ensures that the installation satisfies all the safety regulations and are meticulously followed in the implementation of the installation. Further it also ensures that:

- Good materials and workmanship have gone into the work
- The installation is in good working condition and
- Faults if any in the installation are located and remedial measures suggested before energising the installation.

Therefore, every wiring installation shall, on completion and before being energised, be inspected and tested in accordance with the provisions of IS-732-89, section V, to verify that the provisions have been satisfied.

4.5.1 Visual inspection

A visual inspection is necessary to verify that the installation is:

- In compliance with the applicable Indian standard specifications
- Correctly erected in accordance with the relevant code and
- No visual damage is seen outside.

4.5.2 Testing

Testing of the installation is to be carried out in the following sequence.

1. Continuity of the final circuit conductors
2. Continuity of protective conductors, including the main and supplementary equipotential bonding
3. Earth electrode resistance
4. Insulation resistance
5. Polarity
6. Earth fault loop impedance and
7. Operation of residual current devices (RCDs)

Continuity test is to be carried out in all final circuits to make sure that there is no discontinuity in the circuit. Similarly, every protective conductor also is to be tested separately to make sure that they are electrically sound and are correctly connected. This will include all protective conductors and extraneous conductive parts used for providing equipotential bonding. The resistance of the earth electrode is to be measured as per the methods specified in the IS-3043-Code of practice for earthing.

Insulation resistance measurements of larger installations are to be carried out by dividing the installation into smaller units, each containing not less than 50 outlets. The test voltage to be applied for the insulation measurement is selected from the following list

For voltages up to 250 V	500 volts
250 to 500 V	1000 volts
For circuits above 500 V	2500 volts

Insulation measurements are to be recorded between:

1. The installation (live conductors) and the earth
2. Live conductors.

4.5.2.1 Insulation resistance between live conductor and earth

The following steps are involved in the measurement of insulation resistance between conductor and earth.

a. The main switch is positioned in the OFF position
b. All fuses, MCBs beyond the main switch are in place and ON
c. All lamps and other accessories are in place
d. The phase conductors and the neutral conductors are connected together near the main switch (on the consumer side)
e. The insulation resistance between the conductors (phase and neutral conductors connected together) and the earth is measured using a megohm meter of appropriate voltage class.

The measured value of the insulation resistance shall not be less than 1 MΩ.

4.5.2.2 Insulation resistance between conductors

The steps involved in the measurement of insulation resistance between conductors are:

1. The main switch is placed in the OFF position
2. The phase and neutral conductors are not connected together
3. All switches are placed in the ON position
4. All lamps and accessories/equipment are removed
5. All fuses/MCBs beyond the main switch on the consumer side are ON.

The insulation resistances between any two phase conductors are recorded using a megohm meter of appropriate voltage class. The insulation resistance recorded between any two phase conductors shall not be less than 1MΩ.

When the equipment are disconnected from the installation for the above test, and if the equipment have exposed conductive parts to be connected to the protective conductor, the insulation resistance between the exposed conductive parts and all live conductors of the equipment shall be measured separately and the insulation resistance value shall not be less than 0.5MΩ.

The steps involved in conducting the polarity tests are:

1. All equipment/apparatuses are disconnected
2. All switch covers are open
3. Use a test lamp with long leads
4. Touch one end of the lead to the neutral point
5. Touch the other lead to the input terminal of the switch. If the lamp burns, the polarity is correct. In case there is no power supply available for this test. A battery source can also be used.

When protective measures are incorporated in the installation which operates on the principle of leakage currents, the earth fault loop impedance plays an important role in the satisfactory operation of the protective device. Therefore, the earth fault loop impedance shall be measured with the help of a standard method of measuring the loop impedance.

In installations where residual current devices are mandatory for providing protection against indirect contact, the effectiveness of the residual current devices are to be tested by simulating an artificial fault condition on the installation. This testing is independent of any test facility provided on the device by the manufacturers.

4.6 QUESTIONS

1. What are the basic requirements of electrification of domestic buildings?
2. Explain how will you calculate the connected load of a residential building?
3. What are the NEC guidelines for calculating the connected load of a residential building?
4. Prepare a building plan of your own house, identify the electrical points for the building and determine the connected load of the building.
5. Check whether the electrical installation in your house is carried out as per the guidelines of the NEC. If not suggest the modifications required.
6. What are the important features of electrification of a high rise apartment building?

7. In what ways the electrical installations of a high-rise apartment building defers from that of an ordinary residential building.
8. Explain briefly the procedure of selection of a transformer substation for a high rise apartment building.
9. Explain briefly the pre-commissioning tests for a domestic wiring installation.
10. The plan layout of a two bedroom domestic building is shown in Figure 4.20.

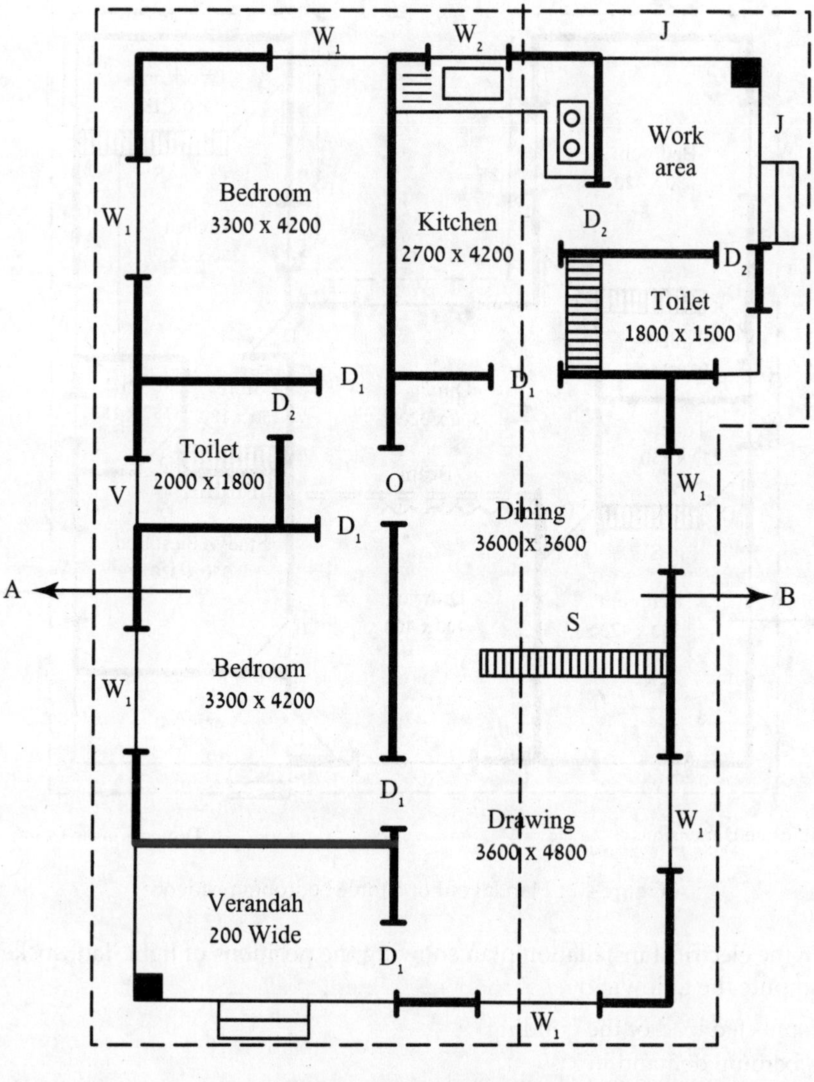

SKETCH PLAN

Figure 4.20 Plan layout of a two bedroom residence

Prepare the electrical installation plan showing the positions of light, fan, socket points, etc., and compute the following:

(a) Connected load of the building
(b) Maximum demand in kW
(c) Type of supply required
(d) Number of lights and power circuits and
(e) The details of the distribution board.

11. The plan layout of a three bedroom domestic building is shown in Figure 4.21.

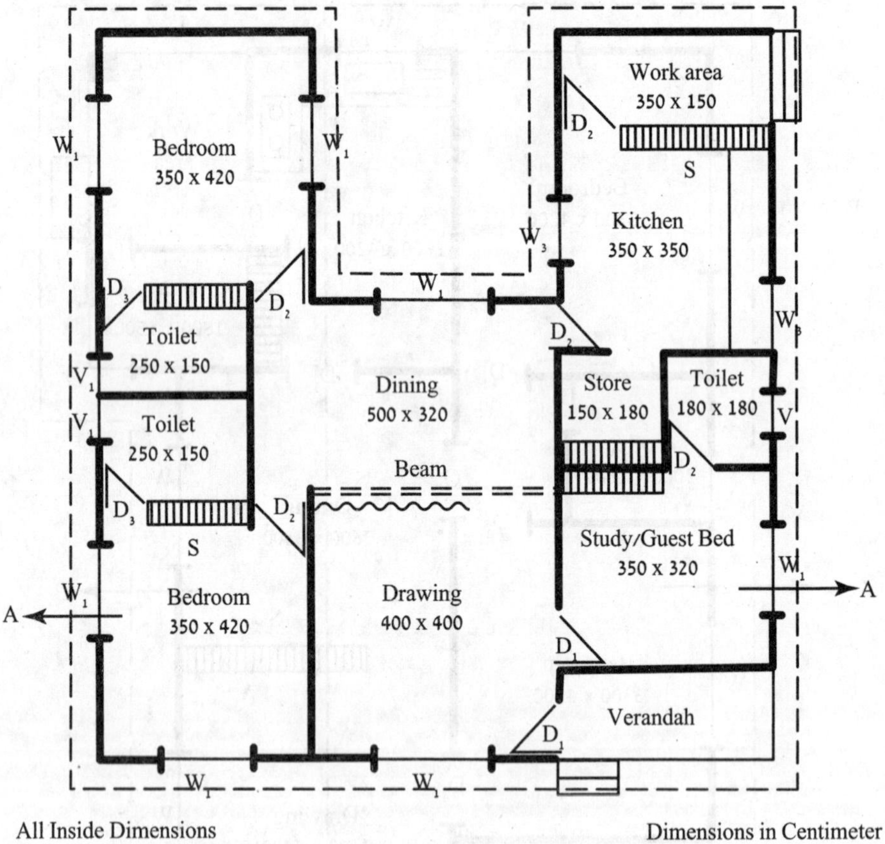

Figure 4.21 Plan layout of a three bedroom residence

Prepare the electrical installation plan showing the positions of light, fan, socket points, etc., and compute the following:

(a) Connected load of the building
(b) Maximum demand in kW
(c) Type of supply required
(d) Number of lights and power circuits and
(e) The details of the distribution board.

12. The plan layout of a two bedroom domestic building is shown Figure 4.22.

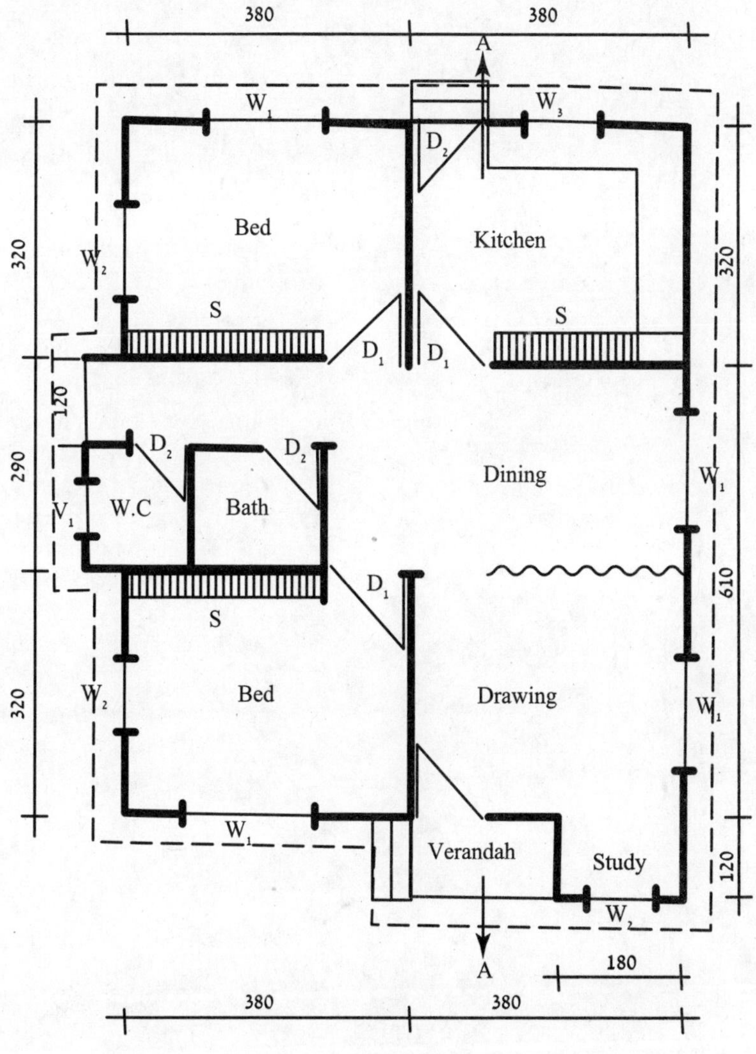

Figure 4.22 Plan layout of a two bedroom residence

Prepare the electrical installation plan showing the positions of light, fan, socket points, etc., and compute the following:

(a) Connected load of the building
(b) Maximum demand in kW
(c) Type of supply required
(d) Number of lights and power circuits and
(e) The details of the distribution board.

Prepare also the schedule of the work showing the detailed specification of the work (all dimensions are in centimeters)

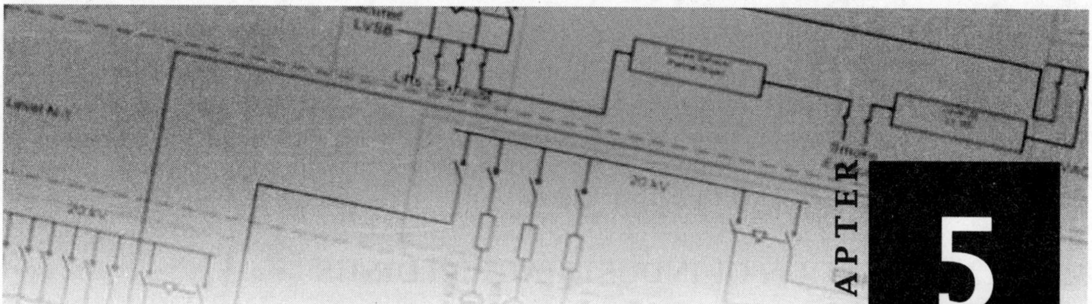

CHAPTER 5

INDUSTRIAL INSTALLATIONS

OBJECTIVES

At the end of this chapter you will

- Understand the classification of industries based on NEC 2011.
- Know about the general characteristics of the industrial installations.
- Be able to select the architecture and supply arrangements.
- Be able to select the type of transformer substations.
- Be able to understand the principles of selection of drives.
- Be able to conduct the short circuit studies on the power distribution systems.
- Understand the significance of X/R ratio of the power system.
- Be able to determine the peak asymmetrical fault current.
- Be able to determine the equivalent system impedances.
- Be able to solve problems in finding the short circuit currents at different points in a power system for three-phase faults.
- Understand the design of earthing system for transformer substations.
- Understand the operation of different types of switchgear used in electrical systems.
- Know how to coordinate the operation of switchgear components.

5.1 INTRODUCTION

The design of electrical installations for industrial premises is more sophisticated when compared to non-industrial buildings. Industrial installations will have to take care of the diverse load requirements in an economic manner, while preserving all the safety and reliability associated with electrical installations. Another important aspect of industrial installation is the easiness of maintenance and fault location features. Economic considerations and energy conservation also form an important feature of industrial installations.

5.2 CLASSIFICATION OF INDUSTRIAL BUILDINGS

Industrial installations are of various types with unique features of their own. In a particular industrial site, the requirements of electrical load may be of varying nature and hence a unified approach is not possible in the design of industrial installations. Each installation has to be designed by taking into account the characteristics and the specific need of that industry. For the purpose of evaluating the electrical requirements as per the National Electrical Code, industries are classified taking into consideration three basic criteria.

- Fire hazards
- Power consumption and
- Pollution hazards.

5.2.1 Classification based on fire hazard

The National Building Code (NBC-2005) classifies the industrial buildings in group G from the fire safety point of view. The sub divisions are:

Group G1 – Buildings used for low fire hazard industries
These buildings include those industries in which the contents are of such low combustibility and the process/operations conducted in those buildings are of such a nature that there are no possibilities of self-propagation of fire to occur and the only danger in the case of a fire and the only danger to life and property may arise out of panic, fumes or smoke or fire from an external source, e.g., cement factories, glass factory, rice mills, soap and detergent factory, sugar industry, etc.

Group G2 – Buildings of moderate fire hazard
Buildings under this category include those industries in which, the industrial process or operations are liable to give rise to a fire that will burn with moderate rapidity and may give rise to considerable volume of smoke. These fires will generate only non-toxic fumes and will not result in an explosion, e.g., ship repairing docks, coir industry, chemical manufacturing, flour mills, fertiliser plants, etc.

Group G3 – Buildings with high fire hazard
This category includes all industrial buildings in which, the contents or the process/operations are liable to give rise to a fire which will burn with extreme rapidity or will generate poisonous fumes or may lead to explosions, e.g., alcohol distillery, explosive manufacturing, fire work industry, oil mills, petrochemicals, paint factory, etc.

5.2.2 Classification based on power consumption

The electrical systems in an industry will largely depend on the quantum of electric power required for its functioning. The NEC-2011 classifies the industrial buildings according to the power requirements as:

- Light industries (small)
- Average industries (medium) and
- Heavy industries (large).

Table 5.1 shows the classification of industrial buildings based on their power consumption.

Table 5.1 Classification of industries based on power consumption

Description	Average power requirement (1 kVA)	Examples
Light industry	Up to 50 kVA	Hosiery, garment making, ornament making, etc.
Average industry	Above 50 up to 2000 kVA	Furniture, pottery, glass, tobacco, electrical gadgets, textiles, engine fitting, light pressings, etc.
Heavy industry	Above 2000 kVA	Heavy electrical equipment, steel mills, foundries, ship building, chemical plants, fertilizer plants, metal extraction, petrochemicals, paper mills, etc.

5.2.3 Classification based on pollution level

A significant aspect of certain industries is the level of pollution created by the industry. This pollution is detrimental to human life and the nature. The classification of industrial buildings based on their level of pollution given in NEC 2011 is as follows.

For the purpose of evaluating creepage distances and clearances, the following four degrees of pollution in the micro-environment are established:

a) *Pollution degree 1* – No pollution or only dry, non-conductive pollution occurs. The pollution has no influence.

b) *Pollution degree 2* – Only non-conductive pollution occurs, except that occasionally a temporary conductivity caused by condensation is to be expected.

c) *Pollution degree 3* – Conductive pollution occurs or dry non-conductive pollution occurs, which becomes conductive due to condensation – which is to be expected.

d) *Pollution degree 4* – Continuous conductivity occurs due to conductive dust, rain or other wet conditions.

5.3 GENERAL CHARACTERISTICS

These are factors that define the fundamentals and details of the power distribution architecture of the industrial installations.

5.3.1 Activity

Activity refers to the type of activity carried out in the building. An indicative list of industrial buildings include:

- Manufacturing buildings
- Food and beverage processing
- Logistics, etc.

5.3.2 Site topology

The site topology refers to the architectural characteristics of the buildings taking into consideration the number of buildings, number of floors in each building and the floor area of each building. Examples are:

- Single-storey buildings
- Multistorey buildings and
- High-rise buildings.

5.3.3 Layout latitude

Layout latitude is the characteristic of the building taking into account the constraints in terms of the layout of electrical equipment in the building. Some of the parameters taken into consideration for identifying the layout latitude are:

- Aesthetics
- Presence of dedicated location
- Use of technical corridor (horizontal service ducts) per floor
- Use of technical service ducts (vertical).

There are three layout categories of industrial buildings. They are:

- Low
- Medium and
- High.

The layout latitude is low when the positions of the electrical equipment are virtually imposed on the building and cannot be changed to achieve some of the parameters. Whereas the layout latitude is high if there are no constraints in positioning the electrical equipment in the building to satisfy the parameters mentioned above. Layout latitude will be medium if there is some flexibility in positioning the electrical equipment.

5.3.4 Service reliability

Service reliability is defined as the ability of the power supply system to meet the supply function under stated conditions for a specified period of time. There are three categories of service reliability. They are:

- Minimum service reliability
- Standard and
- Enhanced service reliability.

The service reliability will be minimum if the installation is subjected to interruptions related to geographical, technical or economic factors. Geographical factors refer to the positioning of the buildings with reference to the power production centres. If the industrial building is located far from the power production centres or if the supply is delivered from an isolated power system, the service reliability will be minimum.

Technical factors take into account whether the supply is through overhead line or from a purely radial network. The drawback of such a supply is that they are subjected to all sorts of natural fury and any fault on the system will reduce the availability of the supply.

The economic factors refer to the maintenance aspects (lack of proper maintenance) and the capability of the system to meet short-term overloading.

The service reliability can be enhanced by taking measures to reduce the power interruptions (by using underground service connections or providing ring main supply system), and providing overloading capability at design stage itself and providing timely maintenance.

5.3.5 Maintainability

This refers to the design considerations taken during the design stages to limit the impact of maintenance on the operation of the whole or part of the installation. Maintainability will be minimum if the installation must be stopped to carry out the maintenance operations. Maintainability will be standard if maintenance can be carried out during the operations, but with reduced performance levels. In this case maintenance activities must preferably be carried out during low level of activity. Enhanced maintainability means the maintenance actions can be carried out without disturbing the operations of the production. This is normally achieved by providing double ended supply configuration.

5.3.6 Installation flexibility

This is the possibility of easily moving the power delivery points within the installation or the facility to increase the power availability at certain points. Different categories of installation flexibility are:

- Non-flexible
- Design flexibility
- Implementation flexibility and
- Operation flexibility.

If the position of the equipment is rigidly fixed throughout its life cycle, may be because of their weight or constraints relating to the building, the installation is said to be, non-flexible. Design flexibility means that if the number of power delivery points, their power rating and locations are not known at the design stage and if the design is capable of providing satisfactory performance, such a design assures design flexibility. Implementation flexibility in a power distribution system means that the power delivery installation is completed before the installation of the loads and a satisfactory performance is guaranteed. In the case of operation flexibility, the position of the loads can be changed according to the process reorganisation or extension of the existing facility. This aspect is very important in the case of industrial installations as there can be an expansion of the existing facility or a change in the production process.

5.3.7 Power interruption sensitivity

This is the capability of the process or system to accept power interruptions. The different categories are:

Sheddable loads: In this case it is possible to shut down a portion of the loads at any time for an indefinite duration.

Long interruption acceptable loads: When the system is capable of accepting interruptions for more than three minutes, such a system is classified as long interruption acceptable system.

Short interruption acceptable: If the acceptable interruption time is less than three minutes, the system is classified as short interruption acceptable system.

No interruption acceptable: When a power interruption cannot be accepted by the industry, the system is referred to as, no interruption acceptable.

The importance of power interruption on the operation of the industry is that it causes several consequences like loss of production, deterioration of production quality or even mortal danger to the persons. The severity of these consequences are categorised as:

- No notable consequences
- Loss of production
- Deterioration of production facility, and
- Mortal danger.

All these consequences can be expressed in terms of the criticality of the supplying the loads.

Non-critical loads: These are the loads which can be shed at any time. For example, water heating loads for domestic purpose.

Low criticality loads: A power interruption to these loads will cause temporary discomfort without much financial implications. Prolonging the interruption beyond the critical time may cause loss of productivity, e.g., HVAC in the industrial offices.

High criticality loads: Any power interruption to these loads will be mortal or may cause unacceptable financial losses, e.g., underground mining operations, operation theatre in a hospital, etc.

5.3.8 Disturbance sensitivity

Disturbance sensitivity is the ability of an industry to work properly in the presence of power disturbances. Power disturbances can be any one of the following:

- Brown outs
- Over voltages
- Voltage distortion due to harmonics
- Voltage fluctuations, or
- Voltage unbalance.

Disturbance sensitivity is classified as:

- Low sensitivity
- Medium sensitivity, and
- High sensitivity.

When the disturbances in the supply system have very little effect on the operation of the process or on the production, the system is said to be low sensitive to disturbances, e.g., heating devices.

If the voltage disturbances cause a notable deterioration in the operation of the industry, it is a case of medium sensitivity, e.g., induction motors, lighting loads, etc.

If the disturbances in the supply system cause the stoppage of operation or even damage to the equipment, it is classified as a high sensitive load, e.g., IT equipment, sensitive instruments, etc.

5.4 SELECTION OF DISTRIBUTION ARCHITECTURE

The electric power distribution architecture in an industry involves the spatial configuration, the choice of power sources, the definition of different distribution levels, the single line diagram and the choice of equipment.

The best architecture will always be a choice expressed in terms of seeking a compromise between the various performance criteria that interest the consumer who uses the installation at different phases of its life cycle. The success of arriving at an optimal solution is strongly associated with a committed interaction between the various players involved in the designing of the various section of the industry, viz:

- The architects who define the space in terms of buildings to house the equipment.
- The designers of technical sections like lighting, heating, air-conditioning, etc.
- The professionals who design the process.

The designing of electrical distribution architecture can be represented by a flow chart consisting of three major sections. Figure 5.1 shows the flow chart of the design process. The process described in the flow chart is not the only solution for the problem. There are other methods and the designers are at liberty to choose the process of their choice.

5.4.1 Stage 1: Choice of the power distribution architecture fundamentals

This is the stage at which the general feature of the electrical installation is defined. The macroscopic characteristic of the industry and its usage is taken into account for this purpose. Because these characteristics have a formidable impact on the type of electrical connection to the upstream network, number of transformer substations, etc. At the end of this stage there will be several distribution schematic diagram solutions. These schematic diagrams are used as a starting point for finalising the single-line diagram. The single line diagram can be broken down into several key parts; such as

- Connection to the power utility
- Configuration of the HV circuits
- Number of distribution transformers and substations.

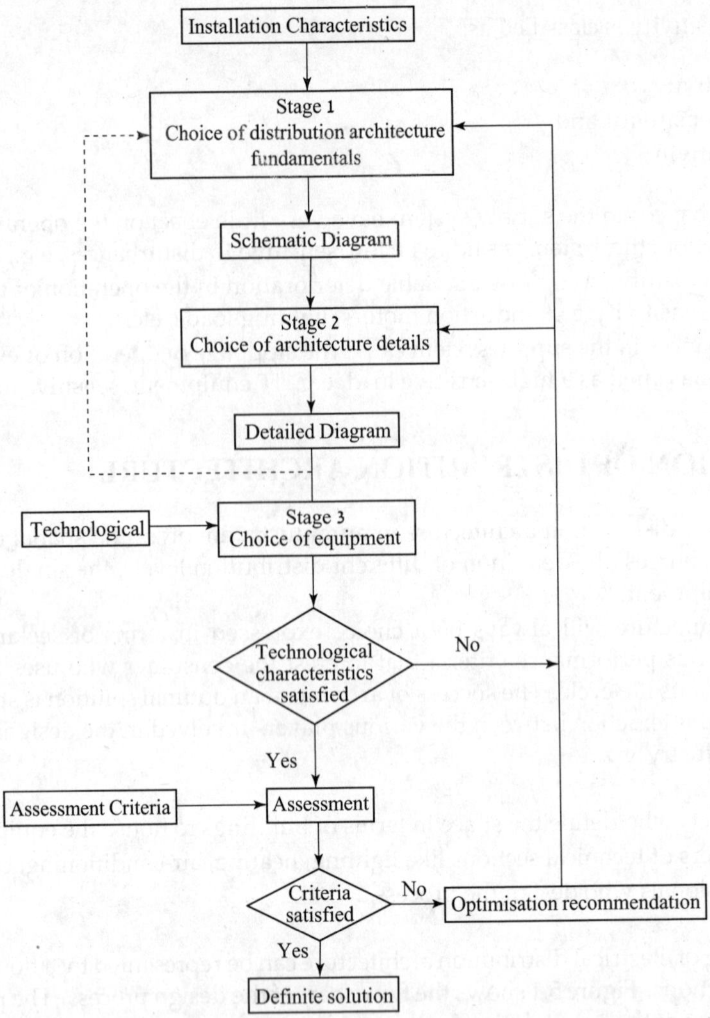

Figure 5.1 Flow chart of the design process (Courtesy: Schneider Electric)

5.4.1.1 Connection to the power utility network

The different types of service connection are as follows:

- Medium voltage service connection
- High voltage single line service connection
- High voltage ring main service connection
- High voltage duplicate supply connection
- High voltage duplicate supply with double bus bar.

Figure 5.2 shows the different possible stages of upstream power utility connection.

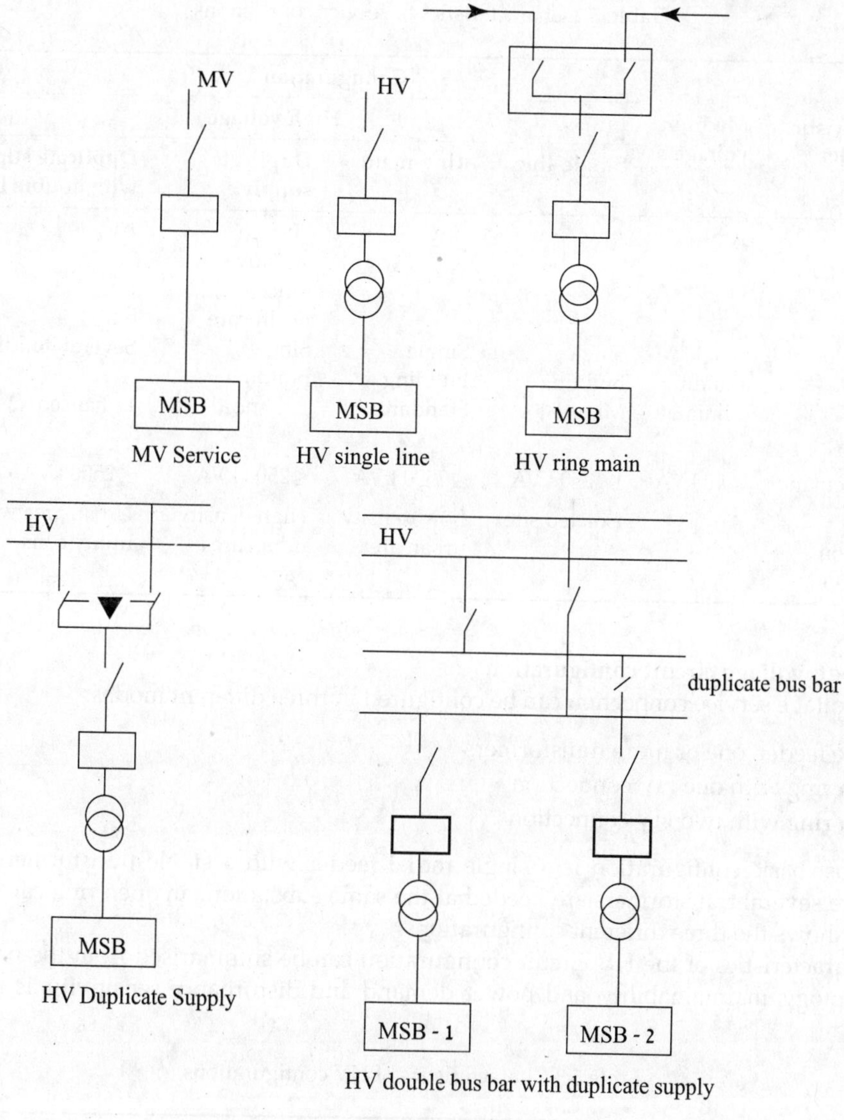

(MSB - Main MV Switch Board)

Figure 5.2 Different types of connection to utility network

The basic characteristics considered are the site topology, service reliability, power demand and any other geographical constraints. For the above mentioned service connections the most probable and reliable set of characteristics is given in Table 5.2.

Table 5.2 Characteristics of service connections

Characteristics to consider	Configuration				
	Medium voltage	High voltage			
		Single-line	Ring-main	Duplicate supply	Duplicate supply with double bus bars
Activity	Any	Any	Any	Hi-tech, sensitive office, healthcare	Any
Site topology	Single building	Single building	Single building	Single building	Several buildings
Service reliability	Minimal	Minimal	Standard	Enhanced	Enhanced
Power demand	< 50 kVA	≤ 1250 kVA	≤ 2500 kVA	≤ 2500 kVA	≤ 2500 kVA
Other connection constraints	Any	Isolated site	Low density urban area	High density urban area	Urban area with utility constraint

5.4.1.2 High voltage circuit configuration

The high voltage service connection can be configured in three different modes:

- Single feeder, one or more transformers
- Open ring with one HV connection
- Open ring with two HV connections.

The most basic configuration is a single radial feeder with a single transformer. In a situation where several transformers are needed at the same substation, an open ring can be realised. Figure 5.3 shows the three different configurations.

The characteristics of the HV circuit configuration can be summarised from the point of view of site topology, maintainability, and power demand and disturbance sensitivity as indicated in Table 5.3.

Table 5.3 Characteristics of HV configurations

Characteristics considered	Single feeder	Open ring one HV substation	Open ring two HV substations
Site topology	Any less than 25000 m^2	Building with one level or several buildings ≤ 25,000 m^2	Several buildings ≥ 25,000 m^2
Maintainability	Minimal or standard	Enhanced	Enhanced
Power demand	Any	>1250 kVA	>2500 kVA
Disturbance sensitivity	Long interruption acceptable	Short interruption acceptable	Short interruption acceptable

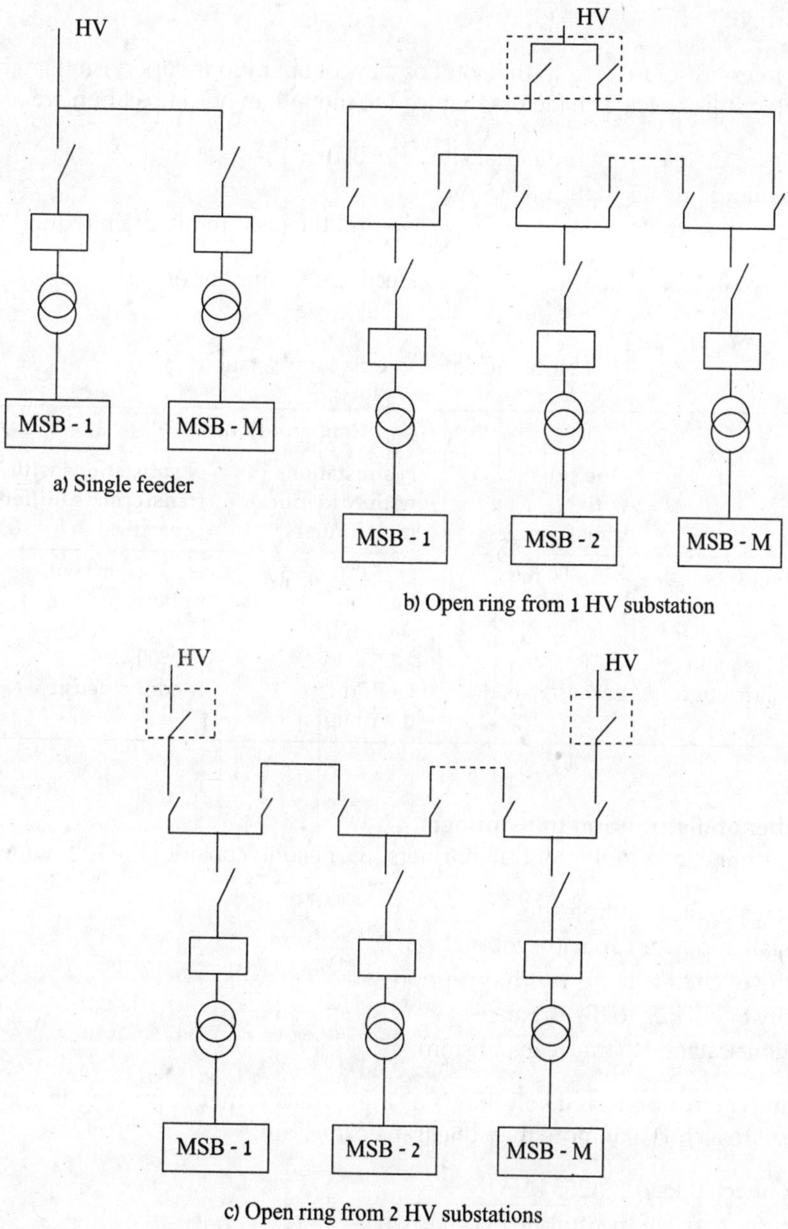

Figure 5.3 HV circuit configurations

5.4.1.3 Number and distribution of transformer substations

The main characteristics to be considered for deciding the number of substations are:

- Surface area of the industrial building or site
- Total power demand in comparison with standard transformer capacities

- Load distribution.

The most preferred solution from the point of view of operations supervision, maintenance and control is a single substation. The factors that lead to more than one substation are:

- A large surface area of the industrial site (> 25,000 m^2)
- Power demand greater than 2500 kVA
- Sensitivity to interruption—due to hazards—and the need to maintain redundancy.

Table 5.4 shows the basic characteristics of selecting the number of substations and their distribution.

Table 5.4 Selection criteria for substations

Characteristics	Configurations		
	One substation with N transformers	**N substations with N identical transformers**	**N substations with N transformers (different capacities)**
Building configuration	<25, 000 m^2	≥ 25,000 m^2 one building with several floors	≥ 25,000 m^2 with several buildings
Power demand	<2500 kVA	≥ 2500 kVA	≥ 2500 kVA
Load distribution	Localised loads	Uniform distribution	Medium density

5.4.1.4 Number of distribution transformers

In order to determine the number of transformers, we need to consider the following:

- Surface area of the building site
- Total installed power capacity
- Sensitivity of circuits to power interruption
- Sensitivity of circuits to disturbances
- Installation scalability (future expansion).

Here again the most preferred solution is a single transformer supplying the entire load. The factors that lead to a choice of more than one transformer are:

- Large connected load (>1250 kVA)
- Large surface area of the buildings (>5000 m^2)
- Need for partial redundancy
- Separation of disturbing loads.

Even though distribution transformers are available up to 2000 kVA capacity, it is preferable to have multiple units of smaller capacity transformers for ease of replacement and less space requirement. A large surface area of the industrial building also demands more than one transformer located close to load centres, from the point of view of reduced cable costs on the MV network.

Increasing the number of transformers also brings in redundancy in the system, either partial or total redundancy in the case of a transformer failure. If the load mix includes large volume of sensitive loads (IT related loads) along with conventional disturbing loads (large motors started frequently), it is preferable to have more than one transformer.

5.4.2 Stage-2 choice of architecture details

It is in this stage the details of the power distribution architecture are defined. The process involves the use of the result of stage 1 and satisfying the criteria relating to the implementation and operation of the industrial installation. The process can loop back to stage 1 if the criteria are not satisfied. It also involves an iterative process allowing several assessment criteria combinations to be analysed. At the end of this stage a detailed single line diagram of the installation will be ready.

A decision on the following choices is made at this stage:

- Layout
- Whether centralised or decentralised distribution
- Presence of backup generation
- Presence of uninterrupted power supply
- Configuration of LV circuits
- Architecture combination.

5.4.2.1 Layout

For most of the industries the layout of MV and HV equipment are more or less decided by the type of manufacturing process and very little changes are possible on this layout. However, there are other industries where some amount of flexibility exists on the location of equipment. A basic guideline for positioning the MV switchboard (power source) is to locate the power source as close as possible to the Bary centre (centre of gravity) of the power consuming equipment. The next consideration is to locate the power sources free of atmospheric constraints like high temperature, presence of dust, vibration, etc. The next consideration is to position all heavy equipment like transformers, etc., at a location with ease of access and maintenance.

5.4.2.2 Centralised or decentralised layout

In this centralised layout all power consuming devices are connected directly to this power source in a star configuration. Cables are used with point-to-point connection between MSB and the power consuming load or sub-switchboards (SSB). Figure 5.4 shows centralised layout of power distribution.

In the decentralised layout, power consuming equipment are connected to the power source (MSB) via a bus bar banking system. This method is most suitable for supplying many loads that are spread out, making it easy to change, move and add more connections.

Centralised layout is most suited for localised loads (with high unit power loads); but have very little installation flexibility, whereas decentralised layout is more suited for uniformly distributed low power loads and have good implementation flexibility. Power supply by cables yields greater independence of circuits (like light, power, HVAC, motors, and auxiliaries) reducing the consequences of a fault from the point of view of power availability. The use of bus bar trunking allows power circuits to be combined and saves on conductors/cable costs, by taking advantage of the clustering coefficient. Each layout has its own advantages and in many places, they are used in combination to take advantage of both systems. Figure 5.5 shows the schematic diagram of a decentralised distribution layout.

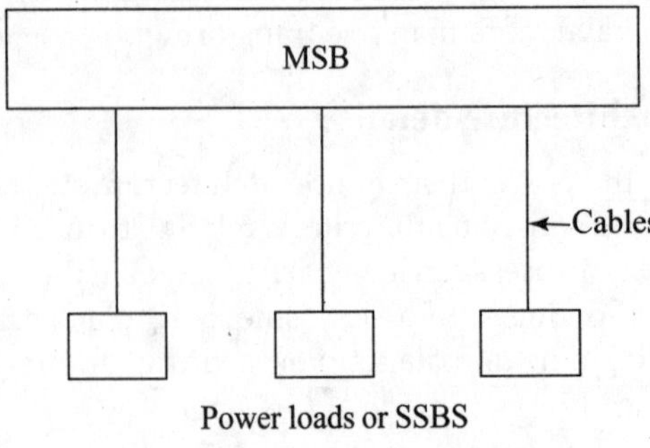

Figure 5.4 Centralised layout

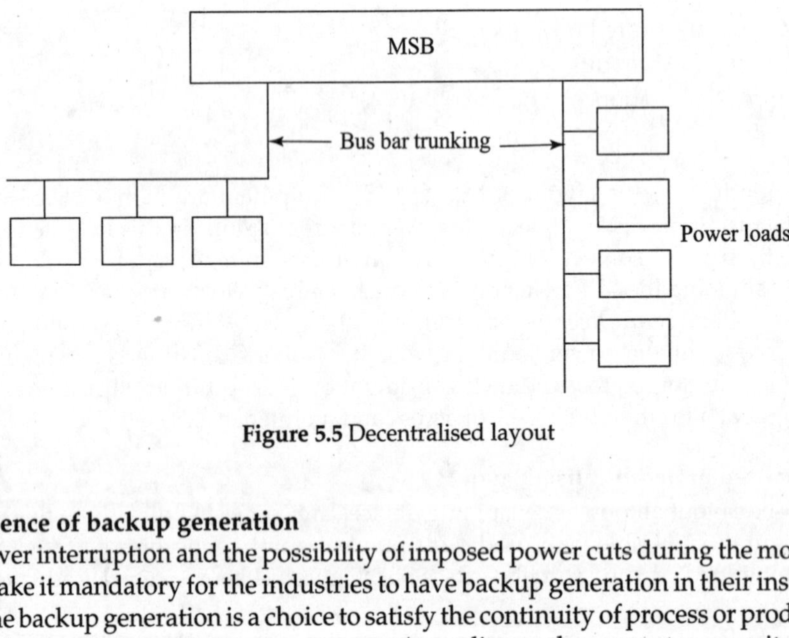

Figure 5.5 Decentralised layout

5.4.2.3 Presence of backup generation

Frequent power interruption and the possibility of imposed power cuts during the months of power shortages; make it mandatory for the industries to have backup generation in their installations. The capacity of the backup generation is a choice to satisfy the continuity of process or production during the periods of power interruption or power cuts. According to the generator capacity to meet all or part of the loads in the installation, total or partial redundancy is built into the system. The backup generator normally operates as disconnected from the utility network. An interlock switching is required to selectively switch either the utility power or generate power.

The backup generators can function permanently or intermediately, depending on whether the generator is rated for its prime power rating or not. A typical connection of a backup generator is shown in Figure 5.6.

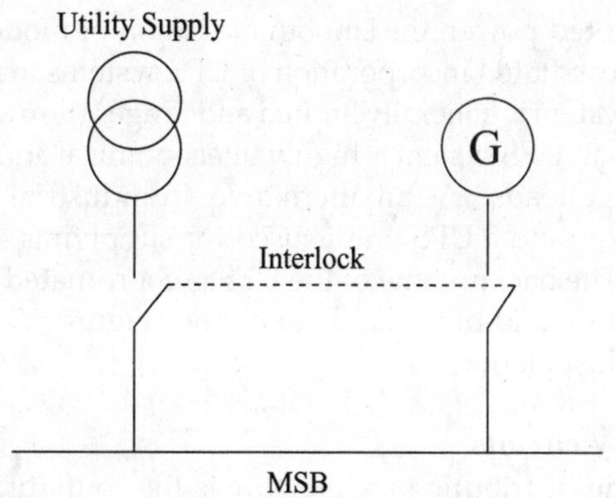

Figure 5.6 Connection of backup generators

The main considerations for implementing the backup generators are:

- Sensitivity of loads to power interruption
- Availability from the public network
- Compulsion from the service point of view (e.g.: hospitals, high-rise buildings, etc.)

5.4.2.4 Presence of uninterrupted power supply (UPS)

Uninterrupted power supply systems have became integral part of electrical installations almost in every field of application.

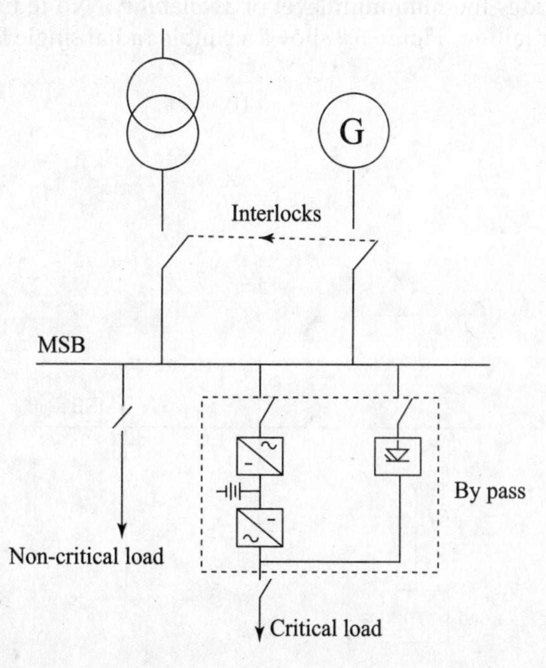

Figure 5.7 Connection of critical loads connected through UPS

The need for non-interrupted power, the smooth operation of modern electronic gadgets (that require clean voltage) has necessitated incorporation of UPS systems in industrial installations also. The backup time of the UPS system is generally limited and ranges from a few minutes to a few hours. The continuous operation on an UPS system is highly uneconomical and will be recommended only in cases of extremely sensitive loads. As an alternative, in industrial installations, simultaneous pressure of a backup generator and a UPS unit is used for supplying sensitive loads for which no power failure is acceptable. The backup time of the UPS is coordinated with the maximum startup time of its generator for starting and bringing them online. Figure 5.7 shows a typical layout of a UPS system in an industrial installation.

5.4.2.5 Configuration of MV circuits

The last stage of finalising the distribution architecture is the configuration of MV circuits. There are different possibilities of configuring MV Circuits. They are:

- Radial single feeder
- Two pole configuration
- Two pole with two 1/2 MSBs
- Sheddable switchboard
- Interconnected switch boards
- Double ended power supply.

Radial Single Feeder Configuration

This is the most basic configuration and the simplest configuration. The load is connected to only a single source and provides the minimum level of availability. No redundancy is built into the system in case of a power failure. Figure 5.8 shows a simple radial single feeder configuration.

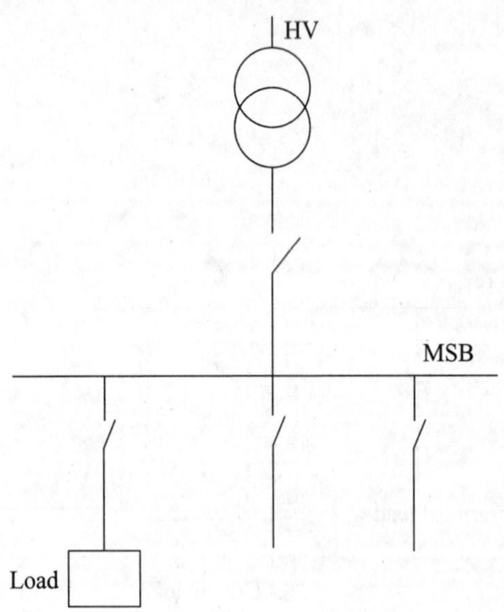

Figure 5.8 Radial single feeder configuration

Two-Pole Configurations

In this case power is made available from two transformers and connected to the same MSB. When the transformers are located close to each other, they are operated in parallel. Figure 5.9 shows a typical double-pole configuration.

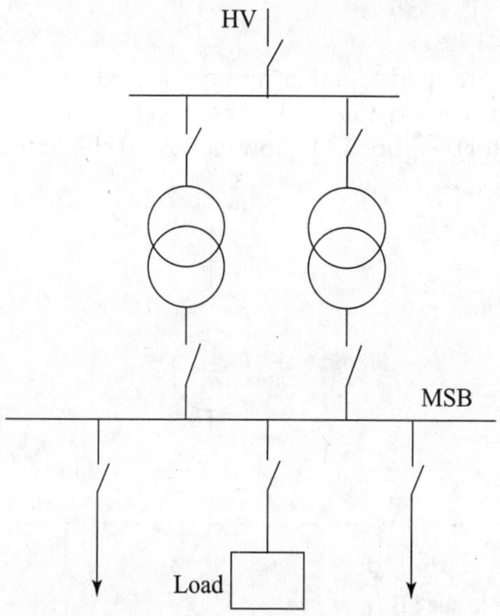

Figure 5.9 Two-pole configuration

Two-Pole with Two 1/2 MSB Configuration

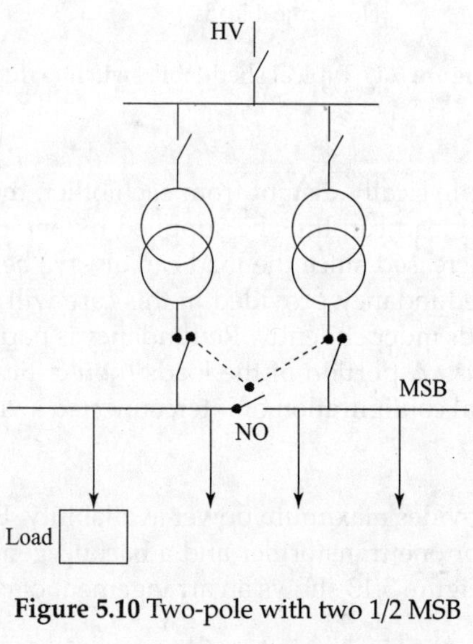

Figure 5.10 Two-pole with two 1/2 MSB

This is a variant of the two-pole configuration which increases the power availability in case of bus bar failure or allows maintenance on one of the transformer.

The MSB bus bar is split into two sections with an open link (Bus Coupler-NO). This configuration requires proper interlocking (two key three lock) arrangement on the MV side or an automatic transfer switching (ATS). Figure 5.10 shows a typical configuration.

Sheddable Switchboard

A set of sheddable loads are grouped together and connected to a dedicated switchboard. Thus, power to this switchboard can be interrupted when needed (i.e., in case of an overload on the system or when operating on generator). Figure 5.11 shows a typical sheddable switchboard arrangement.

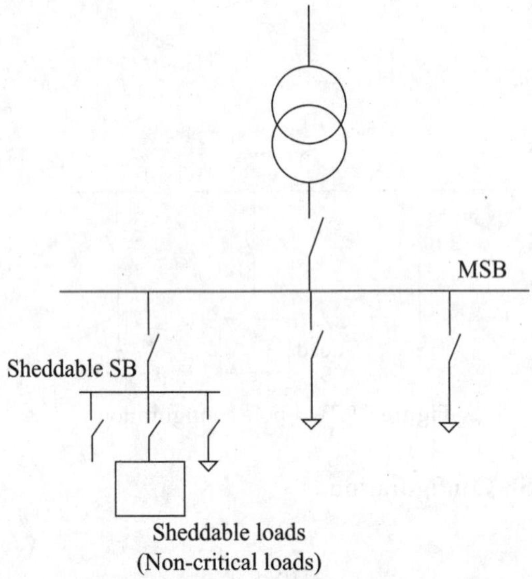

Figure 5.11 Typical sheddable switchboard

Interconnected switchboards

If transformer substations are physically distant from each other, they can be connected by a bus bar trunking system. A critical load can thus be supplied by one or the other transformer. The availability of power is thus increased since the load can always be supplied in case of failure of one of the transformers. The redundancy provided in this case will be total if each transformer is capable of supplying all the loads independently. Redundancy is partial if each transformer is rated to supply only part of the loads. A portion of the loads (non-critical loads) must be shed in this case. Figure 5.12 shows a typical configuration of interconnected switchboards.

Double-ended power supply

This is the arrangement that provides maximum power availability. Either two independent power sources by different HV lines or one transformer and a backup generator or a transformer and a UPS is employed in this case. Figure 5.13 shows an arrangement of double-ended power supply.

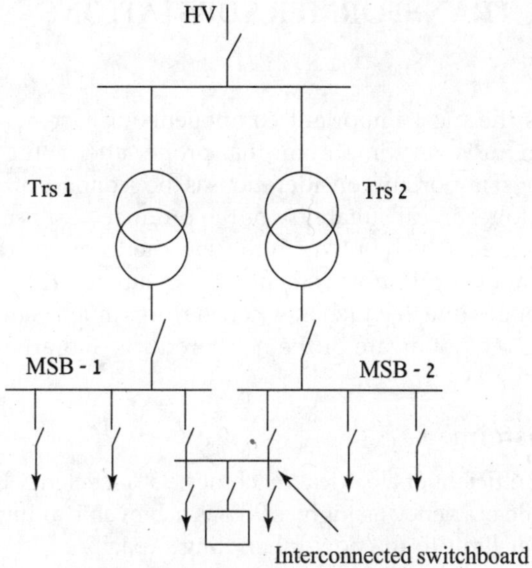

Figure 5.12 Interconnected switchboard

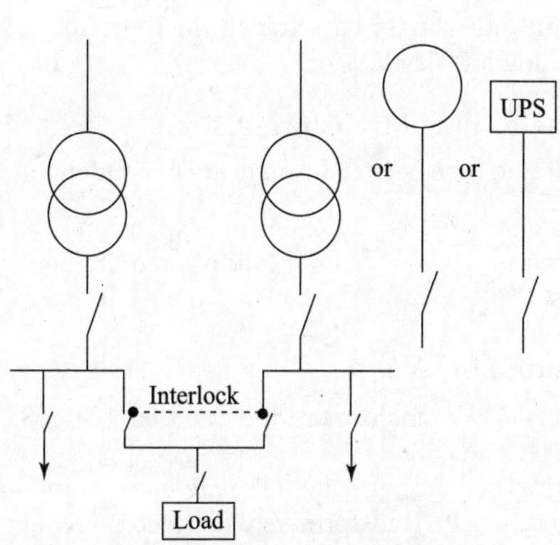

Figure 5.13 Double-ended configurations

Proper interlocking arrangement or automatic transfer switch (ATS) will be employed to avoid the sources being connected in parallel. This configuration also allows preventive and curative maintenance to be carried out on all electrical systems upstream without interrupting the supply.

A combination of all or some of the above-mentioned configurations can be employed in finalising the detailed single line diagram for an installation. The next step in the process of designing the industrial installation is the selection of equipment.

5.5 SELECTION OF TRANSFORMER SUBSTATIONS

5.5.1 Introduction

Transformers are perhaps the most important components of a power transmission and distribution network. Therefore, it is very important that proper attention is made in the selection of transformers. One of the most important considerations is the economy of operation. While selecting transformers, in addition to technical suitability of the product, cost is also a major factor. Short-term gains and long-term gains which outweigh the short-term ones, are to be carefully analysed before selection. With technological improvements, the manufacturing processes of transformers has undergone several changes due to which they can perform much better than their predecessors. However, many engineers are not aware of the new products and are reluctant to change to the latest products and practices.

5.5.2 Selection of transformers

Transformers are the most efficient of electrical machines, 95% efficiency being very common in the lower capacity ranges while efficiency as high as 98% is achievable at higher capacity ranges. Theoretically, there is no upper limit to the power handling capacity of transformers, but in practice the capacity is limited by transportation constraints, handling facilities, and availability of suitable switchgear.

Transformer industry in India has been able to meet the growing transmission voltage and power handling capacity. In late 1960's 220 kV class transformers were produced and 400 kV class in late 70's. The highest unit rating in 3-phase transformer, produced in the country is 315 MVA and highest single-phase rating is 200 MVA.

5.5.3 Factors deciding selection of transformers

Selection of the capacity and the type of transformers are dependent on the following factors.

1. Maximum demand
2. Future expansion
3. Spare capacity
4. Statutory requirements, and
5. Site conditions.

5.5.4 Standard rated output in kVA

The standard rating for 11/0.433 kV transformers are 50, 63, 80, 100, 160, 200, 250, 315, 400, 500, 630, 800, 1000, 1250, 1600 and 2000 kVA.

Transformers with rated impedance voltage = 4% are used mainly in distribution networks in order to keep the voltage drop small. Transformers with rated impedance voltage = 6% are used in industrial networks and major substations in order to limit the short circuit current.

5.5.5 Cooling

The method of cooling employed for the cooling of the transformer is indicated on the name plate of the transformer by the manufacturer in the form of four capital letters, the first two letters denote the type of coolant and the manner of circulation of the coolant for the windings, and the last two letters indicate the coolant and manner of circulation of the coolant for cooling the outside of the transformer. Tables 5.5 and 5.6 show the details of coolants and the manner in which they are circulated.

Table 5.5 Details of cooling medium

S.No.	Coolant name	Letter symbol
1	Mineral oil	O
2	Clophen (Askerel)	L
3	Gas	G
4	Water	W
5	Air	A

Table 5.6 Manner of circulating the cooling medium

S.No.	Coolant circulation	Letter symbol
1	Natural	N
2	Forced circulation	F

Example of cooling designation

ONAN: Oil immersed self cooled (Oil Natural Air Natural)
ONAF: Oil immersed forced air cooled (Oil Natural Air Forced).

Transformers are very versatile in handling overloading. The maximum permissible overloading which the transformers may be subjected to, is shown in Tables 5.7 and 5.8.

Table 5.7 Overloading capacity of oil immersed transformers

Previous continuous loading as a % of the rating	Overloading duration				
	10%	20%	30%	40%	50%
50	3 hrs	1.5 hrs	60 min	30 min	15 min
75	2 hrs	1.0 hr	30 min	15 min	8 min
90	1 hr	30 min	15 min	8 min	4 min

Table 5.8 Overloading capacity of dry type transformers

Previous continuous loading as a % of the rating	Overloading duration				
	10%	20%	30%	40%	50%
50	60 min	30 min	20 min	15 min	12 min
75	55 min	23 min	15 min	11 min	9 min
90	45 min	16 min	10 min	7 min	5 min

5.5.6 Parallel operation of transformers

Transformers can be operated in parallel for meeting higher demands. A pair of transformers is said to be in parallel if they are connected in parallel on at least two sides. The following conditions must be satisfied in order to avoid dangerous transient currents while operating transformers in parallel.

1. Transformers are of the same vector group and the terminals of the same designation must be connected together on the high voltage and low voltage sides.
2. The turns ratios should be as similar as possible and the windings shall be of the same voltage rating on the high voltage and low voltage sides.
3. The transformers shall have approximately the same impedance voltage. The difference shall not be more than 10%. In the event of larger difference in impedance voltages, an inductance can be connected in series to the transformer with the lower impedance voltage.
4. The ratios of the rated capacities of the transformers shall not be more than 3:1.

EXAMPLE 5.1

Three transformers of rated capacities 100 kVA, 250 kVA and 500 kVA are to be connected in parallel. The detailed specifications are given below. Calculate: 1) the load shared by individual transformers when the total load is 850 kVA and 2) if any of the transformer is not to be overloaded, what will be the maximum loading possible on the parallel combination.

Transformer-1, kVA rating-S_{n1} = 100 kVA and % impedance voltage-U_{k1} = 4%
Transformer-2, kVA rating-S_{n2} = 250 kVA and % impedance voltage-U_{k2} = 6%
Transformer-3, kVA rating-S_{n3} = 500 kVA and % impedance voltage-U_{k3} = 4.5%

Total capacity of the three transformers together S_n = 850 kVA
Find the load shared by each transformer if the total load is 850 kVA

$$\frac{S_n}{U_k} = \frac{S_{n1}}{U_{k1}} + \frac{S_{n2}}{U_{k2}} + \frac{S_{n3}}{U_{k3}}$$

$$\text{Resultant impedance voltage is: } U_k = \frac{850}{\frac{100}{4} + \frac{250}{6} + \frac{500}{4.5}}$$

$$= 4.78\ \%$$

The load shared by the individual transformers is

$$S_1 = \frac{S_{n1} \times U_k}{U_{k1}} = \frac{100 \times 4.78}{4} = 120\ \text{kVA}$$

$$S_2 = \frac{S_{n2} \times U_k}{U_{k2}} = \frac{250 \times 4.78}{6} = 199\ \text{kVA}$$

$$S_3 = \frac{S_{n3} \times U_k}{U_{k3}} = \frac{500 \times 4.78}{4.5} = 531\ \text{kVA}$$

To avoid overloading, transformers may be partially loaded such that the impedance voltage of each is 4% as in the case on the transformer-1.

Therefore, individual loading on each transformer will be,

$$S_1 = \frac{100 \times 4}{4} = 100 \text{ kVA}$$

$$S_2 = \frac{250 \times 4}{6} = 167 \text{ kVA}$$

$$S_3 = \frac{500 \times 4}{4.5} = 444 \text{ kVA}$$

Total loading capacity of the three transformers together = 711 kVA.

Hence, we see that, unless proper care is made in the choice of impedance voltage of the transformers operated in parallel, the total output is to be limited to 711 kVA, leading to underutilisation of the capacity.

Effect of dissimilar transformation ratio on the performance of transformers connected in parallel

Dangerous transient currents can occur if transformers with different voltages between taps are operated in parallel. Disregarding any dissimilarity in impedance, phase angle, the voltage difference Δ_u drives through both sides as circulating current of

$$I_s = \frac{\Delta_u}{\left[\frac{U_{k1}}{I_{n1}} + \frac{U_{k2}}{I_{n2}}\right]}$$

For example;

If, $U_{k1} = U_{k2} = 6\%$, $I_{n1} = 1000$ A, $I_{n2} = 533$ A and $\Delta_u = 4\%$

$$I_s = \frac{4}{\left[\frac{6}{1000} + \frac{6}{533}\right]} = 232 \text{ Amps.}$$

This transient current is superimposed on the transformer rated currents I_{n1}, I_{n2} which are fed into the network.

5.5.7 Installation of transformers

Indoor and outdoor oil filled transformers require no special protection against climatic conditions. Dry type transformers however, must always be installed in dry rooms, if necessary, with additional protection against moisture.

Statutory requirements for the installation of transformers are given in section 5.5.12.

5.5.8 Developments in Transformers

M/s Crompton Greaves Limited manufactures "rectangular core-cast aluminium foil wound transformers" in collaboration with Westinghouse Electric Corporation of USA. Foil wound transformers conform to IS - 2026/1977 and are compatible to run in parallel with transformers of conventional design. Rectangular core coil construction LV foil type winding results in improved space factor, reduced dimension, increased efficiency, reduced operational cost and uniform quality.

Aluminium sheet or foil is used as L.V. conductor. This results in reduced overall dimension since layer and turn insulations are one and the same. Aluminium foil has a larger surface area for a given cross section which increases heat dissipation. Advantages of DC foil wound transformers are:

1) Greater short circuit current withstanding capacity.
2) Greater overload capacity
3) Greater impulse withstand capacity
4) Higher life expectancy
5) Reduced floor area requirement
5) Minimum maintenance.

5.5.9 Standard rating of transformers

500 kVA - 5000 kVA - 33 kV class,
63 kVA - 2000 kVA - 11 kV class

5.5.10 Dry type transformers

Dry type transformers are most suited in locations where conventional oil filled transformers present a safety hazard. Some of these locations are:

1) Oil refineries
2) Chemical plants
3) Marine applications
4) Metro railways
5) High-rise buildings.

Advantages of Dry type Transformers

1) No fire risk
2) Lighter in weight
3) Less floor area
4) No toxicity unlike synthetic liquid filled transformers
5) Efficient noiseless operation
6) Reduced installation cost
7) Maintenance free.

Types of dry type transformer

1) Cast resin type
2) Resin impregnated type

Cast resin type transformers are class F glass fibre reinforced epoxy resin. It can withstand 155 °C continuously. The disadvantages of cast resin type transformers is that in the event of failure, the complete unit including HV and LV winding is to be replaced, whereas, with resin impregnated transformer, only the affected winding is required to be replaced. The details of the absolute temperature withstand capacities of various insulation class are shown in Table 5.9.

Winding hot spot temperature = (temp. rise + ambient temp + 30) °C.

Table 5.9 Absolute temperature withstand capacity of different insulation classes

Insulation class	Winding temperature rise in °C as per IS 2026-part-2	Absolute temperature withstand capacity in °C as per IS 1271
A	50	105
E	65	120
B	70	130
F	90	155
H	115	>180

5.5.11 Cable connections on power transformers

While selecting cables on HV side, in addition to current carrying capacity, short circuit withstand capacity may also be verified. The minimum size of the cable for short circuit withstand capability is given by

$A = 14.14\, I_{sc}\, \sqrt{t}$ for paper insulated cables
$A = 13.2\, I_{sc}\, \sqrt{t}$ for PVC cables
$A = 11.1\, I_{sc}\, \sqrt{t}$ for XLPE cables.

Where, A = area of cross section in mm.
I_{sc} = prospective short circuit current.
t = duration of fault = 1 sec.

Size of cables on the secondary side of the transformer is given in Table 5.10. When the secondary current exceeds 1000 amperes, it is preferable to use truncated bus bars or bus ducts for connecting the transformers to the switch gears.

5.5.12 Installation of transformer substations: Statutory requirements

There are three different types of transformer substations for industrial purposes. They are:

- Outdoor substations
- Indoor substations, and
- Unitised substations.

The statutory requirements of these types of transformer substations are listed below.

Outdoor installations

1. Minimum clearance between suppliers and consumers 11 kV structure shall be 3 m. If it is 5 m or less, they shall be braced together at the top and middle of the poles.
2. Minimum size of RSJ poles shall be 175 × 85 mm and channel size shall be 75 × 50 mm.
3. Phase-to-phase and phase-to-earth clearances at AB switch shall be minimum 915 mm and 610 mm respectively.
4. Minimum size of 11 kV jumpers shall be number '0' SWG copper.
5. Maximum length of unsupported 11 kV jumpers shall be 2.44 m for solid conductors and 1.5 m for stranded conductors.

6. Minimum safety working clearance of 2.6 m shall be maintained for bare conductors and live parts.
7. Lightning arrestors shall be provided between AB switch and D.O/H.G fuse.
8. Group control AB switch shall be provided for multiple transformer installations.
9. The minimum clearance from ground to the 11 kV bushings of the transformer shall be 2.6 m.
10. AB switch operating handle shall be fixed at 1 m to 1.25 m from ground level.
11. 11 kV AB switches shall be three insulator types.
12. Fencing with a height of 1.8 m shall be provided, keeping a minimum clearance of 1.5 m from the transformer body.
13. Minimum clearance between 11 kV up and down jumpers shall be 75 cm.
14. Outdoor transformer is not permitted in corporation areas and at the premises of high-rise apartment buildings; however outdoor transformers with cable end boxes at both sides and indoor breaker/SFU at primary side may be permitted judiciously.
15. Indoor transformer is recommended for rating above 630 kVA.
16. If more than one transformer, having more than 2000 litres of oil, is installed side by side, baffle walls of minimum 4 hr fire rating shall be provided in between transformers.

Indoor Installations

1. Only dry type transformers shall be installed inside residential/commercial buildings (Rule 64(2) (e) (iv)). Oil filled transformers are permitted outside residential and commercial buildings subject to the following conditions.
   (a) Only indoor type transformers are permitted in corporation areas and at the premises of residential buildings.
   (b) There shall be free access for fire engines to the transformer stations.
   (c) The building in which transformer is installed shall have minimum clear distance of 1.5 m from the residential/commercial buildings.
   (d) The substation building shall be of weatherproof and fireproof. Walls shall be of 355 mm thick brick or 230 mm thick RCC. If there is a window/door opening towards the residential/commercial building, there shall be a fire resistant wall between the substation building and residential/commercial building keeping a minimum of 1.5 m from both the buildings.
   (e) Transformers may be installed outdoors, keeping a minimum distance of 6 m from commercial buildings in areas other than municipal corporations.
   (f) If the transformer is installed outdoors, within 6 m from commercial buildings, there shall be a fire resistant wall in between the transformer and the building, keeping a minimum distance of 1.5 m both from the building and the transformer.
2. Only dry type transformers shall be installed in cellars and above ground floor. IE Rule 64(2) (f) (iii).
3. Air break switch fuse unit/vacuum circuit-breaker shall be provided on the primary side of the transformers. For hospital premises, breakers shall be recommended.
4. Minimum size of the cable on the 11 kV side shall be 150 sq.mm XLPE. But the size may be 70.00 sq.mm XLPE if backup fuse is provided.
5. Fire separation walls shall be provided in between oil filled transformers. Thickness of the wall shall be minimum 30 cm (brick) or 20 cm (RCC).

6. Oil filled transformers shall be segregated from HT and MV switchgears using fire separation wall.
7. Oil drainage facility/soak pits shall be insisted for oil filled transformers if quantity of oil is more than 2000 litres.
8. All-round clearance shall be provided for indoor transformers. It shall be minimum 1.25 m if walls are provided at four sides.
9. Transformers shall be installed at the periphery of the building for getting easy access and adequate ventilation.
10. Doors and windows of transformer room shall be of non-inflammable material.
11. Minimum 50 cm headroom clearance shall be provided for the transformer.
12. (a) If dry type transformers are installed above ground level, 11 kV AB, SFU/breaker shall be provided at ground floor for completely isolating the supply as envisaged in rule 50 A with lockable push button near the transformers and a mechanical interlock between AB SFU/breaker and transformer enclosure.
    (b) Structural fitness certificate from a competent structural engineer shall also be furnished.
13. Minimum phase to phase and phase to earth clearance of bus bars in 11 kV panel shall be 130 mm and 80 mm respectively.
14. Minimum 1.0 m clear space shall be available in front of H.T panel in breaker drawn out position.

Unitised Substation

1. Transformer shall be of dry type only.
2. Switchgears attached to the unit shall not be oil filled type.
3. Transformer and control gears in the unitised substations shall be clearly segregated using metallic/FRP barriers.
4. Earthing switch if any attached to the 11 kV ABSF unit shall be at the outgoing side. This earthing switch and the 11 kV ABSF unit shall be properly interlocked to prevent accidental closing of earthing switch when ABSF unit is in the ON position.
5. There shall be proper interlocking to prevent opening of the enclosure of the unit when the 11 kV control gear is in the ON position.
6. There shall be a metallic/FRP barrier between supply incoming live parts and HRC fuse units in the ABSF unit.
7. The incoming supply shall preferably be connected to the fixed contact of 11 kV ABSF unit.
8. The interconnection between transformer and control gear shall be with rigid strips of adequate normal current rating and short time current rating.
9. Phase-to-phase clearance of 11 kV interconnecting strips shall be minimum 13 cm and phase-to-earth clearance shall be minimum 8 cm.
10. 11 kV control gear and tap changer of the transformer shall be properly interlocked to prevent changing of tap position when 11 kV control gear is in the ON position.
11. If potential transformer is provided in the unit, separate chamber shall be provided for housing the same.
12. Seal-off bushings shall be provided when 11 kV buses pass from one chamber to another.

Table 5.10 Cables on secondary side of transformers (11 kV/433 V)

Capacity of transformer (kVA)	Primary current (A)	Secondary current (A)	Minimum size of neutral earthing conductor (mm)	Minimum size of AYFY cable (mm^2)
63	3.3	84	25 × 3	50
100	5.25	133.3	25 × 3	95 (2 × 50)
160	8.40	213.3	25 × 3	185 (2 × 95)
200	10.49	266.6	25 × 3	300 (2 × 120)
250	13.12	333.0	25 × 3	(2 × 185)
315	16.53	420.0	31 × 3 or 25 × 4	(2 × 300) or (3 × 185)
400	21.80	533.0	38 × 3	(3 × 300) or (2 × 400)
500	26.20	666.5	25 × 6	(3 × 400) or (4 × 240)
630	33	840.0	31 × 6	(4 × 400)
750	39.36	1000.0	50 × 4	Bus bar trunking
1000	52.50	1333.0	210 mm^2	Bus bar trunking
1250	65.60	1667.0	290 mm^2	Bus bar trunking
1600	83.98	2133.0	380 mm^2	Bus bar trunking
2000	105.00	2666.0	450 mm^2	Bus bar trunking

General

1. If there are more than one transformer and the aggregate transformer capacity exceeds 100 kVA, a circuit breaker of adequate rating shall be provided after point of commencement of supply (IE Rule50 (1) (a) (iii)).
2. Minimum 2 m clearance shall be provided between two dry type transformers installed in the same room. Statutory clearance between dry type transformer and DG set of acoustic enclosure shall be 2 m.
3. Transformer installation shall conform to I.S 10028 (Part II) - 1981.
4. Earthing shall conform to I.S 3043/1987.
5. Minimum values of insulation resistance of transformer shall be as per standards.
6. Minimum protection required for transformers of different ratings are given in Table 5.11.
7. Minimum size of cables laid underground at secondary side of transformers is given in Table 5.10.
8. Transformer oil shall conform to Indian standard specifications.
9. Maximum rating of 11 kV /433 volts transformers is limited to 2000 kVA.
10. MV side installation shall conform to I.S. 732/1989.

Above sizes are recommended assuming the following derating factors. Enclosure effect = 0.75, Temperature effect = 0.88, Impurity effect = 0.94. Bus bar trunking shall be designed for minimum fault current of 25 kA or 50 kA depending on the actual fault level. 50 kA Short circuit current rating is recommended for all short circuit currents above 25 kA. Maximum distance between bus supports shall be 50 cm with central stiffeners and bolting for phase bus bars.

Table 5.11 Protective measures on transformers (11 kV/433 V)

Rating	Method of Installation	Primary control and protection	Secondary control and protection
Up to and including 400 kVA	Outdoor	A.B. Switch, lightning arrestor, D.O. Fuse	MCCB having isolation duty and E/F release and standby Low set E/F relay using CT at neutral earthing conductor or ACB with proper current setting facility with 3 O/C and 1E/F release.
Up to and including 400 kVA	Indoor	11 kV, ABSF unit or VCB/GCB with 2 O/C and 1E/F relays preferably with DC trip	MCCB having isolation duty and E/F release and standby Low set E/F relay using CT at neutral earthing conductor or ACB with proper current setting facility with 3 O/C and 1E/F release.
Above 400 kVA and up to and including 630 kVA	Outdoor	A.B. Switch, lightning arrestor, D.O. Fuse	ACB draw out type with 3 O/C and 1 E/F release and standby low set E/F relay using CT at neutral earthing conductor
Above 400 kVA and up to and including 630 kVA	Indoor	11 kV, ABSF unit or VCB/GCB with 2 O/C and1E/F relays with DC trip	ACB draw out type with 3 O/C and 1 E/F release and standby low set E/F relay using CT at neutral earthing conductor
Above 630 kVA and below 1000 kVA	Indoor	11 kV, ABSF unit or VCB/GCB with 2 O/C and 1E/F relays preferably with DC trip	ACB draw out type with 3 O/C and 1 E/F release and standby low set E/F relay using CT at neutral earthing conductor
Above 1000 kVA and up to and including 2000 kVA	Indoor	VCB/GCB with 2 O/C relays with high set elements, one instantaneous E/F relay, Buchholz relay, oil and winding temperature high alarm and trip. For dry type transformers, Buchholz relay, oil temperature relay and trip are irrelevant.	ACB draw out type with 3 O/C and 1 E/F release, 2 O/C and 1 E/F relay, standby low set E/F protection using CT at neutral earthing conductor, REF protection relay with primary trip. If breaker is having microprocessor based programmable releases having different time settings, separate O/C and E/F relay need not be provided. But low setting E/F relay and REF relay are to be provided.

5.5.13 Transformer inrush current

When transformers are energised, a transient inrush current up to 10–50 times the rated transformer current can flow and may last for several cycles. The worst condition exists when the transformer is switched on at or around the zero crossing of the voltage wave form. This inrush current can cause malfunctioning or unwanted tripping of the primary protective device. Hence, it is very essential that a careful analysis of the characteristics is made in selecting the primary protective device for transformers. Tables 5.12 and 5.13 give the relationship between the capacity of the transformers with the time constant of the inrush current and the ratio of inrush current to the normal current for oil filled and cast resin type transformers respectively.

The relation between the primary current threshold setting and the time delay setting of the protective device is shown in Figure 5.14. Where the y-axis represents the ratio of time delay setting to the inrush time constant and the x-axis represents the ratio of setting threshold of primary current to the inrush current.

Table 5.12 Inrush time constants for oil filled transformers

Transformer rating kVA S_{tr}	$k_i = \frac{i_{P_{inrush}}}{I_{1_{tr}}}$	τ_{inrush}
50	15	0.10
100	14	0.15
160	12	0.2
250	12	0.22
400–500	12	0.25
630	11	0.30
1000	10	0.35
1600	9	0.40
2000	8	0.45

Table 5.13 Inrush time constants for cast resin transformers

Transformer rating kVA S_{tr}	$k_i = \frac{i_{P_{inrush}}}{I_{1_{tr}}}$	τ_{inrush}
200	10.5	0.15
250	10.5	0.18
315	10	0.20
400–500	10	0.25
630	10	0.26
800–1000	10	0.30
1250	10	0.35
1600	10	0.40
2000	9.5	0.40

(Source: ABB Ltd)

where,

S_{tr} = rated capacity of the transformer - kVA
$i_{p_{inrush}}$ = inrush current in primary winding - A
$I_{1_{tr}}$ = primary rated current - A

Consider, for example, a 500 kVA oil filled transformer with a voltage ratio of 11 kV/433 V and suitable for 50 Hz three phase operation.

From Table 5.12,

The ratio of inrush to rated primary current k_i = 12
Time constant of the inrush current = 0.25 sec
Full load rated current on the primary = 26.2 A (from Table 5.10)
The inrush current = 12 × 26.2
= 314.4 A

Assuming a setting threshold for primary current I_t' of 35 A

$$\frac{It'}{i_{P_{inrush}}} = \frac{35}{314.4} = 0.111$$

From Figure 5.14, corresponding to $\frac{It'}{i_{P_{inrush}}} = 0.111$, the ratio $\frac{t_r}{\text{inrush}} = 1.8$

Therefore, time delay setting $t_r = 1.8 \times 0.25 = 0.45$ seconds.

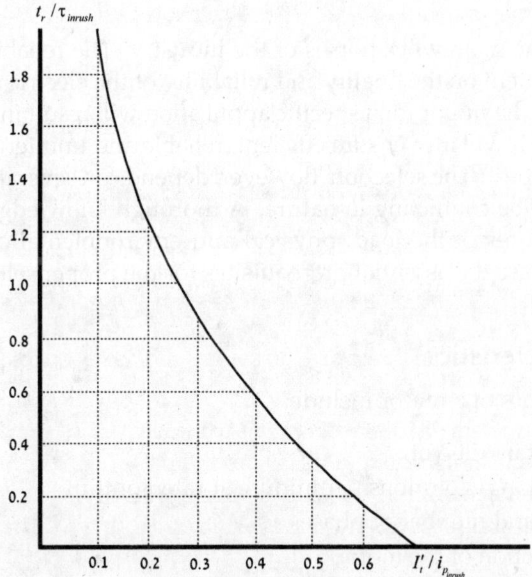

Figure 5.14 Relation between threshold setting and time delay setting

Table 5.14 Fuse ratings on the primary side of transformers (11 kV/433 V)

Capacity of transformer (kVA)	Primary current (A)	Secondary current (A)	HT fuse rating (A)	
			Minimum	Maximum
63	3.3	84	10	16
100	5.25	133.3	16	25
160	8.40	213.3	16	40
200	10.49	266.6	25	40
250	13.12	333	32	40
315	16.53	420	40	63
400	21.80	533	40	63
500	26.20	666.5	50	100
630	33	840	63	100
750	39.36	1000	75	160
1000	52.50	1333	100	160
1250	65.60	1667	100	200
1600	83.98	2133	160	250
2000	105.00	2666	160	250

This is the minimum time delay required by the primary protective device to avoid unwanted tripping or malfunctioning of the protective device on the primary side.

In the case of transformers with fuse protection on the primary side, the fuse manufacturers provide information regarding the minimum and maximum size of the high voltage fuse ratings. Table 5.14 provides a typical data of the fuse ratings on the primary side of transformers.

5.6 SELECTION OF DRIVES

The electric motors are the main work horses of the industry. The reliability of production in any industry is mainly dependent on the quality and reliability of the electric motors used in the industry. Judicious selection of the motor for a specific application will result in lower capital investment and lower running costs. It will also ensure efficient, reliable and uninterrupted operation over the useful life period and beyond. The selection, however, depends on several factors, the requirements of which sometimes may be conflicting in nature. A thorough knowledge of the electrical characteristics, mechanical features of the load, physical and environmental conditions, efficiency and economic considerations are the essential prerequisites for the proper selection of the motors.

5.6.1 Electrical characteristics

The electrical characteristics of a motor include:

- Rated motor output in kilowatts
- Approximate speed in revolutions per minute at rated output
- Voltage, frequency and number of phases
- The type of construction of the motor
- Method of starting employed, and
- Duty type of the motor.

5.6.1.1 Rated motor output

The output of a motor required for driving a machine can be determined by knowing the load details of the driven machine. The motor rating must be slightly in excess of the workload plus the friction. Otherwise the motor will get overloaded and thereby overheated. The life of a motor greatly depends on the temperature rise the motor is subjected to and the life gets reduced to a great extent if the permissible temperature limits are exceeded. On the other hand, increasing the capacity of the motor far above the actual requirement would unnecessarily increase the initial cost as well as the running cost.

It is not just sufficient that the motor meets the output requirement at rated full load conditions. But, matching the torque-speed characteristics of the motor and that of the driven machine is very important till the motor reaches the full load speed. In other words, the torque developed by the motor at any speed must be in excess of the resisting torque due to the load and inertia of the revolving parts as well as friction. In fact the difference between these two will decide how soon the motor accelerates to the full speed, when started. This is otherwise known as the acceleration period. A wrong selection of the size of the motor or the starting method can lead to total failure of the motor.

5.6.1.2 Approximate speed of the motor

The speed at which the driven machine is to be run will be available from the specification of the driven machine. The machine manufacturers are aware of the limitations of the induction motors with regard to their speed limitations and most of the machine designers, design the machines to be driven at the standard speeds of the induction motors, i.e., 2880, 1440, 720 rpm etc., which correspond to a synchronous speed of 3000, 1500, 750 rpm. As far as DC motors are concerned, the motors can be run at a wider range of speeds from their design speed. With the availability of modern solid state variable speed drives, induction motors now can be run at any speed to meet the optimum running conditions.

5.6.1.3 Voltage, frequency and the number of phases

Except in special cases, all industrial drives utilise three-phase 415 volt motors for equipment of moderate ratings. Three-phase 415 volt motors can be selected for capacities up to 200 kW. It is preferable to use higher voltage ratings for drives requiring higher capacities. Some of the standard voltages for high voltage motors are 3.3 kV, 6.6 kV and 11 kV.

Variation in supply voltage, especially on the lower side of the rated voltage, affects the performance of the induction motors adversely. The adverse effects are reduced starting torque, increased full load current and increased temperature rise. Therefore, the supply voltage variation must be kept as minimum as possible to achieve optimum performance. The permissible voltage variation allowable within the distribution system is ±3% as per Indian Standard Specification IS 325 for medium voltage systems. If the supply voltage variations are more than the specified limits, it is preferable to choose motors with extended voltage range of 400–440 volts.

The induction motors are also sensitive to voltage unbalance between phases. The phase unbalance must be less than 1% for proper operation. If the unbalance is more than 2%, the motor is likely to become overheated when operating near full load capacity. A phase unbalance of 3.5% will result in a temperature rise of 25% and a current increase of 6–10 times the voltage unbalanced. These effects are due to the flow of negative sequence currents in the rotor. The voltage unbalance can be calculated using the following formula

$$V_{un} = \frac{\text{Maximum deviation from average voltage}}{V_{av}} \times 100 \tag{5.1}$$

EXAMPLE 5.2

The line voltage measurements on the terminals of an induction motor over a period are, 410, 418, 420, 405 and 402. Calculate the percentage voltage unbalance.

Solution:

$$\text{Average voltage across the induction motor} = \frac{410 + 418 + 420 + 405 + 402}{5} = 411 \text{ V}$$

$$\text{Substituting in equation (5.1)} \qquad V_{un} = \frac{420 - 411}{411} = 2.19\%$$

If a motor is constrained to operate with voltage unbalance greater than 1%, then the motor shall be derated in accordance with Table 5.15.

Table 5.15 Derating factors of induction motors due to voltage unbalance

% Voltage unbalance	1.5	2.0	2.5	3.0	3.5	4.0	4.5	5.0
Derating	0.99	0.97	0.93	0.9	0.87	0.835	0.80	0.755

As far as frequency of supply is considered, 50 Hz is accepted frequency in our country. The permissible variation in frequency is again ±3%. However, a variation up to 5% is permissible. Frequency variation normally will not be a problem, if the motor is supplied from the utility supply.

Polyphase motors are generally preferred to single-phase motors because of their starting and performance superiority. Motors rated below 2.2 kW are generally single-phase motors.

5.6.1.4 Types of motor

Most of the industrial applications which do not require starting very frequently or with high inertial loads may be satisfactorily driven by squirrel cage induction motors (SCR motors) with low slip. Loads which are characterised by high inertia and frequent starting may require the use of slip ring motors or squirrel cage motors with high slip. A typical selection guideline is given in Tables 5.16, 5.17 and 5.18.

5.6.1.5 Method of starting

The appropriate selection of starting method is a very important aspect of selection of drives. The basic requirements of a starting method are:

- The method should be capable of accelerating the motor along with the coupled load to the normal speed within a reasonable period of time.
- The method should not be producing unnecessary mechanical stresses on the machine components like shaft, pulleys, gears, etc.
- There should not be undue voltage drop on the supply system which is likely to affect the other consumers connected to the system.

Table 5.16 Guidelines for selection for general applications

Application	Type of motor	Precautions	Remarks
Pumps: centrifugal and reciprocating	Standard squirrel cage or slip ring	To be started with closed valve	If star-delta starting with open valve is required, starting torque requirement should be taken into account
Compressors: Rotary	Standard squirrel cage or slip ring	Unloaded starting may be started with partial load	When using squirrel cage motor check the starting torque requirement
Reciprocating	Standard slip ring		
Centrifugal fans/blowers	Standard squirrel cage or slip ring	To be started with closed dampers or closed valves	If squirrel cage motor is used, the acceleration time should not exceed 15 seconds on DOL starting
Agitators and mixers for liquids	Standard squirrel cage or slip ring	Light starting duty	

Table 5.17 Guidelines for selection for heavy starting duty and high peak load applications

Application	Type of load	Characteristics	Type of motor	Remarks
Power mixers, pulp grinders, pulverizers, loaded conveyors	Heavy starting	Started on load, Normal starting torque about 200% FL torque	High starting torque squirrel cage or slip ring	
Flour, paper and rubber mills, Rolling mills Crushers and hammer mill (without flywheel)	High peak load	Pull out torque about 200–250% FL torque	High starting torque squirrel cage or slip ring.	Additional information required: Load characteristics, starting and pullout torques, Load cycle, method of braking if any

Two basic categories of starters normally used for induction motors are:

1. Full voltage starters otherwise called direct on line starters (DOL) and
2. Reduced voltage starters.

Full voltage starters

In this method the motor is started with the rated voltage applied to the motor at starting. This method guarantees very favourable starting torque conditions and accelerates the motor from standstill condition to the rated speed in a very short time. The main disadvantage is that the starting current is of the order of 5–6 times the rated full load current, while the starting torque developed is only of the order of 1.5–2 times the rated full load torque. If the capacity of the motor is greater than 3.7 kW, it will create objectionable voltage drop in the supply lines during starting and thereby cause undesirable effects on the equipment connected to the system.

Table 5.18 Guidelines for selection for high torque and high slip application

Application	Type of load	Characteristics	Type of motor	Remarks
Crushers, Power Hammers, Rolling mills, Riveting machine, Pellet presses etc. (with flywheel)	High peak load	Heavy starting, High full load torque	High slip, high torque squirrel cage or slip ring motor	Additional information required: Inertia of load and fly wheel, number of working strokes per hour, energy absorbed per stroke, duration of blow, starting and pull out torque, continuous or intermittent service.

Figure below shows an induction motor connected to a public distribution system. If the motor is started using a direct on line starter, it will draw a very high starting current and will cause a large voltage drop, considerably higher than the normally permitted level, along the line.

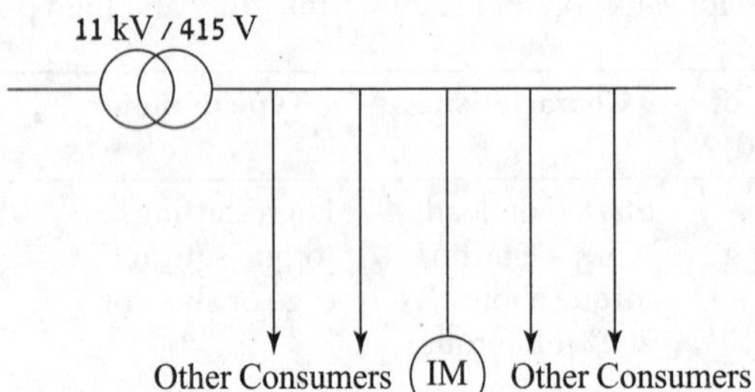

An induction motor connected to a public distribution system.

This undue voltage drop will adversely affect almost all the consumers connected to the substation. It is for this reason that the supply authorities insist on using a reduced voltage starting method for all motors rated above 3.7 kW and above and especially if the motors are connected to a public medium voltage distribution system. This restriction is not applicable to industries having their own transformer substations.

Reduced voltage starting

The starting current in an induction motor is proportional to the input voltage per phase to the motor. If V_{LL} is the line to line voltage applied to the motor,

$$I_{ST} \propto E_{LL} \tag{5.2}$$

where,

I_{ST} = starting current and
E_{LL} = induced voltage (line to line).

The voltage drop in the stator impedance is small compared to the applied voltage and can be neglected.

$$|V_{LL}| = |E_{LL}| \tag{5.3}$$

Therefore, the starting current in an induction motor is proportional to the supply voltage. If a reduced voltage is applied to the motor at starting, the starting current also will be reduced accordingly. While the starting current is reduced, causing lesser voltage drop during starting, a lower starting current means a lower starting torque.

The most popular and effective means of reduced voltage starting is achieved by means of star delta starters. In this method the motor windings are connected in star configuration during starting and then switched over to delta once the machine has picked up sufficient speed. By connecting the motor in star configuration, the reduction in voltage corresponds to $\frac{1}{\sqrt{3}} = 0.577$. The reduction in starting current when compared to DOL starting is 33.3%. But the starting torque also gets reduced by 33.3% and this is a disadvantage. If this method is to be used, we must make sure that the load torque and loss torque taken together is less than the starting torque provided by this method. Being a very simple and cost effective method, star delta starting is the most favoured starting method for induction motors.

Two variants of the star delta starters are the semiautomatic star delta starter and the automatic star delta starter. In both semiautomatic and automatic star delta starters, a set of contactors are used to switch from star to delta connections. The ON/OFF operation is through push buttons as in the case of DOL starters. In the semiautomatic star delta starter, when the green button is pushed the motor is started with the windings connected in star. It remains in star connection until the push button is released and then it is switched to delta connection. The transition from star to delta is controlled by the operator and a correct judgement is required to time the transition from star to delta as the transition takes place at the instant of release of the green button. In the case of automatic star delta starter, starting is initiated by pushing the green push button and the motor starts in star connection. When the green button is pushed, a built in timer takes control and the transition to delta is timed by the built in timer which can be set according to the requirement. The operator has no control in the transition from star to delta in this case. The operations of both semiautomatic and automatic star delta starters are similar to that of DOL starters and are very convenient to operate.

Autotransformer starter is another form of reduced voltage starting of induction motors. A three-phase autotransformer is used to reduce the voltage applied to the motor at starting. The ratio of the autotransformer is selected such that a fraction of the supply voltage is available as output of the autotransformer. The ratio of out put to input voltage of the autotransformer is, x, the value of which lies between 0.0 and 1.0 ($0 < x < 1$). The starting current and the starting torque will be x^2 times the values of starting current and starting torque with DOL starting. If the autotransformer ratio $x = 0.707$, both starting current and starting torque are $x^2 = (0.707)^2 = 0.5$ times the values of starting current and torque with DOL starting. Star delta starting can be considered equivalent to an autotransformer starter with $x = 0.577$ (57.7%) so that starting current and torque will be $x^2 = (0.577)^2 = 0.333$ (33.3%) of the starting current and torque with DOL. The advantage of the autotransformer starter is that when the star delta starter is not capable of providing the required starting torque, by suitably selecting the ratio x, and using an autotransformer starter the problem can be solved. The main disadvantage is the high cost of the autotransformer which is in operation only when the motor is started.

Slip ring induction motors are started with external resistance included in the rotor winding at starting. Rated voltage is applied to the stator windings and the external resistance is gradually cut out as the speed picks up. The required level of starting torque can be obtained by suitably selecting the value of the external resistance. The permissible value of the external resistance R_{ext} lies between the limits $[r_2 < (r_2 + R_{ext}) < x^2]$ where, r_2 is the rotor resistance and x_2 is the rotor inductive reactance per phase. The range of external resistance is between 0.0 and $(x_2 - r_2)$ and the maximum starting torque is obtained when $R_{ext} = (x_2 - r_2)$.

5.6.1.6 Duty type of the motor

Electric motors are called upon to discharge different types of duties ranging from operation for a few seconds (as in the case of a municipal siren motor) to several days or months continuously (as in the case of blowers or fans). IS-325 classifies the operation of motors into eight types. Table 5.19 lists the duty types as defined in IS-325 and shows the corresponding types of application.

Table 5.19 Duty types of motors (Reference: IS-325)

Duty type symbol	Duty type	Applications
S1	Continuous duty	Pumps, blowers, compressors, fans etc.
S2	Short time duty	Siren, capstan, flood relief gates etc.
S3	Intermittent periodic duty	Valve actuators, wire drawing machine
S4	Intermittent periodic with starting	Hoists, cranes, lifts, escalators
S5	Intermittent periodic with starting/braking	Hoists, cranes (with electric braking) rolling mills
S6	Continuous duty with intermittent periodic loading	Machine tools, conveyors
S7	Continuous duty with starting and braking	Machine tools
S8	Continuous duty with periodic load changes	Pole changing applications

Continuous duty type S1

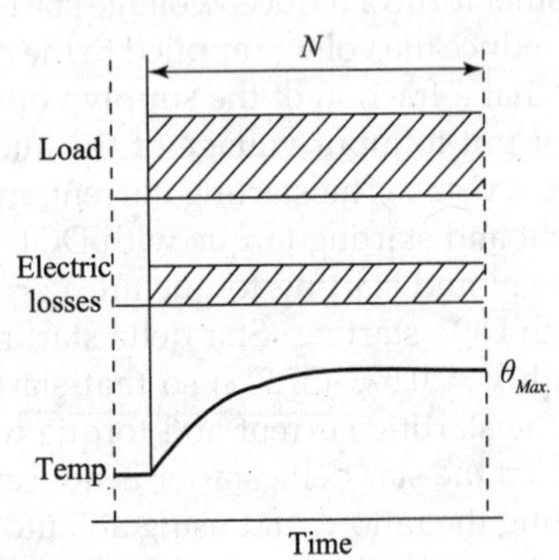

Figure 5.15 Characteristics of S1 duty type motors

These motors are designed for operation at constant load for sufficient duration of time for thermal equilibrium to be reached. Figure 5.15 shows the time Vs load, time Vs losses and time Vs temperature rise characteristics of S1 type motors.

N = Time of operation at constant load
θ_{max} = Maximum temperature attained during the operation.

Short time duty type S2
Motors designed for operation at constant load for a time much less than the time required for reaching thermal equilibrium are classified as type S2 duty. The operation time is so short and the rest time is sufficiently large to establish equality of temperature with the surrounding medium. The recommended values of short time duty are 10, 30, 60 and 90 seconds, E.g. S2-10, S2-60, etc. Figure 5.16 shows the time Vs load, time Vs losses and time Vs temperature characteristics of S2 duty type motors.

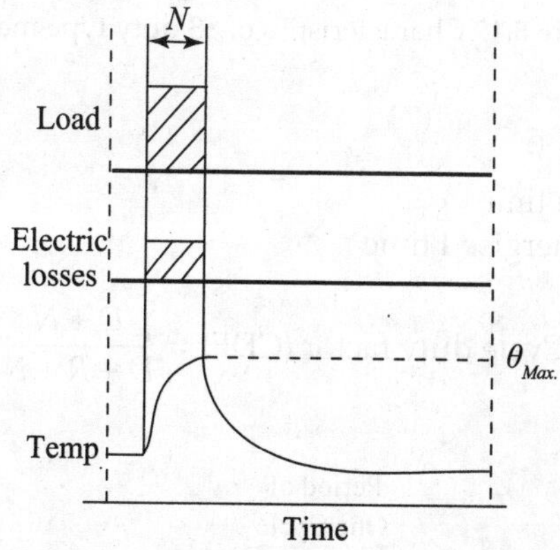

Figure 5.16 Characteristics of S2 duty type motors

Intermittent periodic duty type S3
Motors designed for operation of a sequence of identical duty cycles come under this type. Each duty cycle consists of operation at a constant load and a rest period. The periods are too short that thermal equilibrium will not be reached. The duration of the duty cycle is normally 10 minutes. Figure 5.17 shows the time Vs load, time Vs losses and time Vs temperature rise characteristics of S3 type motors.

$$\text{Cycle duration factor} = \frac{N}{R+N} \tag{5.4}$$

where, N = operating time and
R = rest time or de energised time.

Intermittent periodic duty with starting-type S4
Motors designed for identical duty cycles, each consisting of a period of starting, a period of constant load and a rest period. Thermal equilibrium is not reached in one duty cycle. Stopping of the motor is either by natural deceleration or by braking by mechanical means. Figure 5.18 shows the time Vs load, time Vs losses and time Vs temperature rise characteristics.

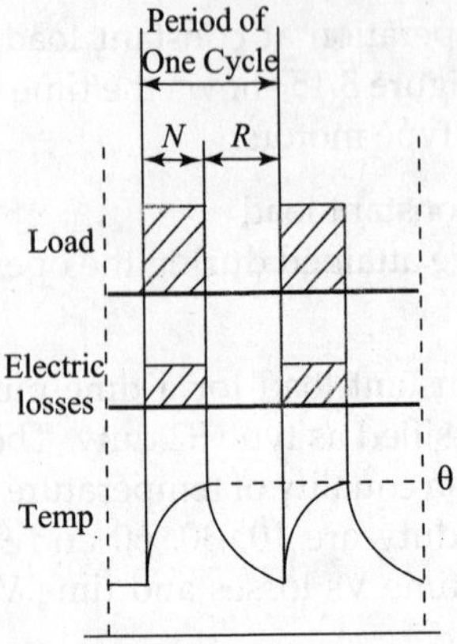

Figure 5.17 Characteristics of S3 duty type motors

D = Starting time
N = Normal operation time
R = Rest time or de-energised time

$$\text{Cycle duty factor (CDF)} = \frac{D+N}{D+R+N} \tag{5.5}$$

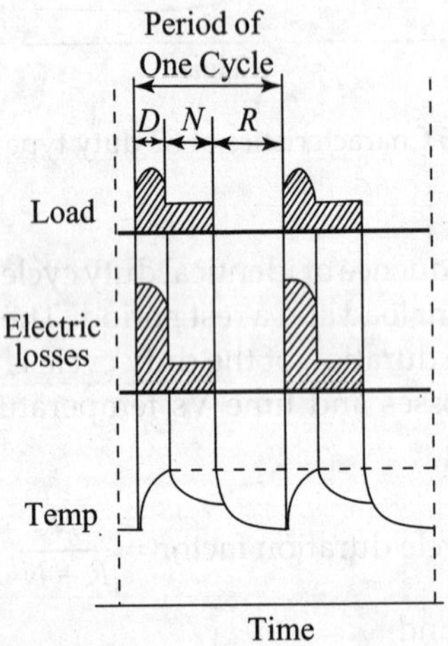

Figure 5.18 Characteristics of S4 duty type motors

Additional details required when purchasing these motors are, number of start/stop per hour, percentage cycle duty factor and load GD^2 (factor of inertia-FI), e.g., 50 start/stop, 40% CDF, FI=2.

Intermittent periodic duty with starting and braking-duty type S5

In this case motors are designed for identical duty cycles each consisting of starting, operation at constant load, a braking period and a rest or de energised period being too small to attain thermal equilibrium. Braking is by electrical means and is very rapid. Figure 5.19 shows the time Vs load, time Vs losses and time Vs temperature rise characteristics of S5 type motors.

$$\text{CDF} = \frac{D+N+F}{D+N+F+R} \tag{5.6}$$

where F = braking period

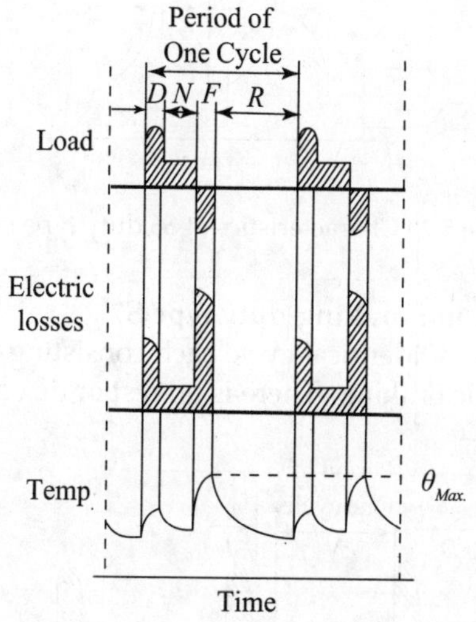

Figure 5.19 Characteristics of S5 duty type motors.

Additional details to be specified are number of start/stop per hour, percentage cycle duty factor and factor of inertia, e.g., S5, 100 start/stop, 40%, FI=2.

Continuous duty with intermittent periodic loading-duty type S6

These motors are designed for identical duty cycles consisting of operation at constant load followed by operation at no load. The operating and no load periods are too short to attain thermal equilibrium during a duty cycle. The standard duty cycle is 10 minutes. Figure 5.20 shows the time Vs load, time Vs losses and time Vs temperature characteristics of S6 type motors.

$$\text{Cycle duration factor} = \frac{N}{N+R} \tag{5.7}$$

where, R = no load running period.

The details required are, number of start/stop per hour and percentage CDF. E.g. S6, 50 start/stop, CDF = 40%.

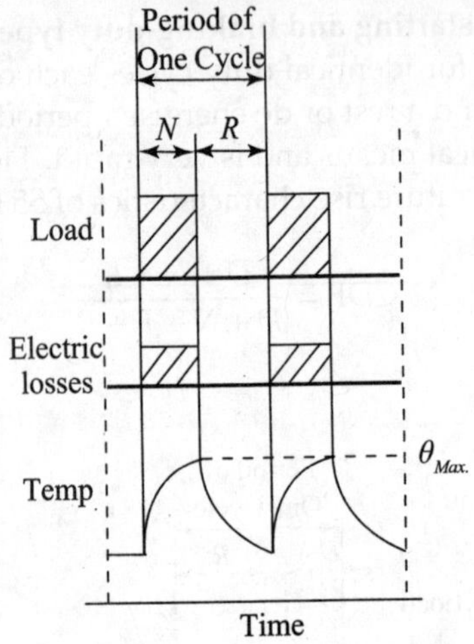

Figure 5.20 Characteristics of S6 duty type motors

Continuous duty with starting and braking-duty type S7

Motors with a loading sequence of identical cycle each consisting of a starting time, constant load operation and a period of electric braking. There is no rest or de energised period.

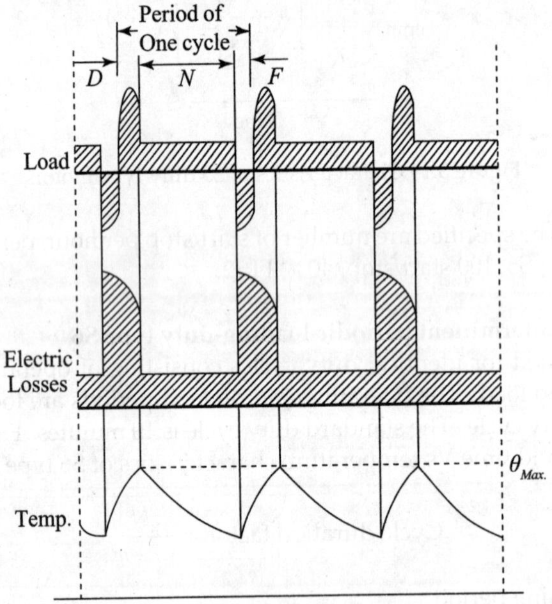

Figure 5.21 Characteristics of S7duty type motors

Figure 5.21 shows the time Vs load, time Vs losses and time Vs temperature rise characteristics of type S7 motors. Additional details required are number of start/stop per hour, percentage CDF and factor of inertia.

Continuous duty with periodic speed changes-duty type S8
These motors are designed for identical duty cycles each consisting of a starting time, operation at constant load at a specific speed, followed by operation at another load with a different speed.

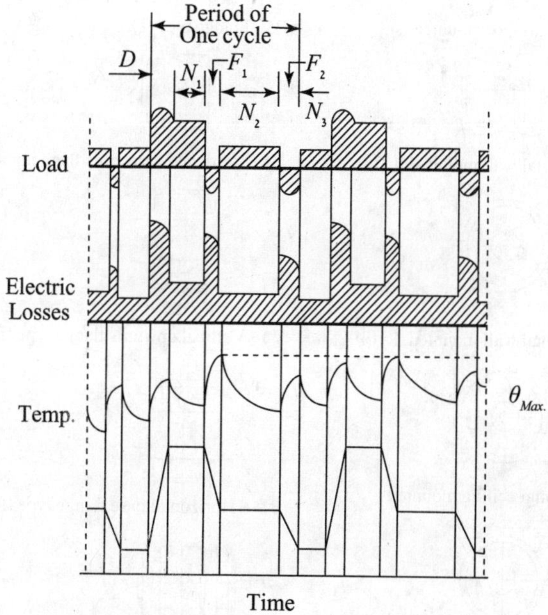

Figure 5.22 Characteristics of S8 duty type motors

There is no rest or deenergised condition. The operating cycle is too short to attain thermal equilibrium. Figure 5.22 shows the time Vs load, time Vs losses, time Vs temperature rise and time Vs speed characteristics of type S8 motors. There is no rest or de energised period. For S8 duty type motors the specifications are, duty cycles per hour, factor of inertia, load, rpm and the cyclic duration factor, e.g., S8, 30 c/h, FI = 30, 20 kW, 1460 rpm, 40%.

$$\text{Cycle duration factor} = \frac{D+N_1}{D+N_1+F_1+N_2+F_2+N_3} \tag{5.8}$$

$$= \frac{F_1+N_2}{D+N_1+F_1+N_2+F_2+N_3} \tag{5.9}$$

$$= \frac{F_2+N_3}{D+N_1+F_1+N_2+F_2+N_3} \tag{5.10}$$

5.6.2 Mechanical features

Apart from specifying the electrical characteristics, mechanical features of the machine also will have to be specified. The main mechanical features are:

- Type of mounting
- Position of terminal box
- Details of shaft extension
- Type of drive and particulars of the driven machine and
- Frame size.

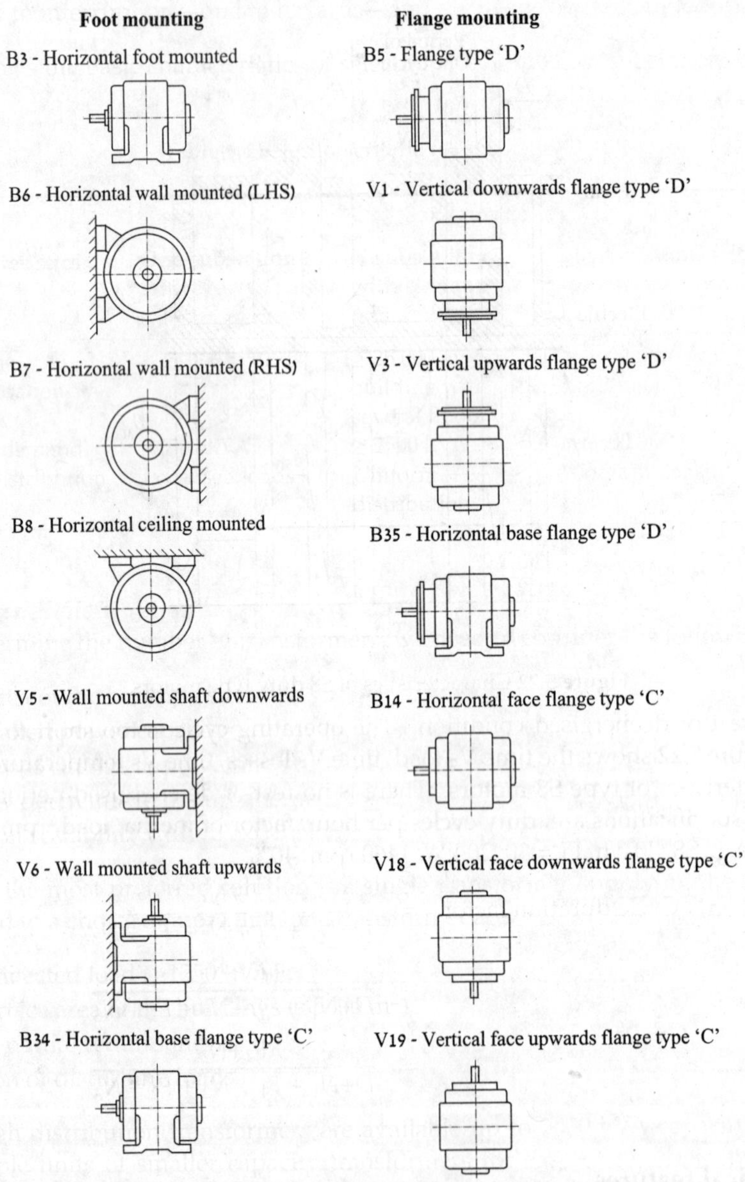

Figure 5.23 Standard types of mounting of motors (Reference: IS-2253)

5.6.2.1 Types of mounting

Motors are generally mounted horizontally with feet attached to the floor, but other types of mounting are also common. IS-2253 specifies the details of different types of mounting possible. Figure 5.23 lists some standard types of mounting of motor.

5.6.2.2 Position of terminal box

From practical point of view, the positioning of the terminal box is very important. The right location of the terminal box makes it easy for the cable connection and the regular maintenance. The location of the terminal box can be specified as on top or on the left hand side or on the right hand side when looking from the non driving end. Normally, the manufacturer's catalogues provide the details of the shaft extensions and selection can be made accordingly. But, some industrial applications may require non standard shaft extensions. The motor manufacturers have different options to meet the customer requirements. For example: double end shafts for use in lift machines, tapered shaft, threaded shaft, etc., are some of the non-standard shaft extensions.

5.6.2.3 Details of shaft extension

Normally, the manufacturers' catalogues provide the details of the shaft extensions and selection can be made accordingly. But, some industrial application may require non-standard shaft extensions. The motor manufacturers have different options to meet the customer requirements. For example: double end shafts for use in lift machines, tapered shaft, threaded shaft, etc., are some of the non-standard shaft extensions.

5.6.2.4 Type of drive and particulars of driven machine

Motors are coupled to the driven machine either directly, through V-belt drive, through gears with/without clutch, etc. The type of connection is important and should be taken into consideration while choosing the motor. The particulars of the driven machine include:

i) Speed torque characteristics of the driven machine
ii) Duty cycle
iii) Peak torque and power
iv) Load inertia
v) Constant/variable or adjustable speed
vi) Frequency of start/stop
vii) Acceleration restrictions and
viii) Braking requirements.

5.6.2.5 Frame size

The frame sizes are specified for motors of general purpose applications with standard nominal output and continuous rating. The distance of the centre line of the shaft to the bottom of the feet of the motor in millimeter is referred to as the frame number. This is followed by the diameter of the shaft extension in millimeter. An English alphabet "S", "M" or "L" indicates the frame length, S-stands for small, M-for medium and L-for long.

5.6.3 Protection and cooling classification

5.6.3.1 Protection classification, IP-XX

It is not just sufficient that a motor fulfils the functional requirements assigned to it. It also has to be protected against possible external influences, and likewise to ensure that the motor is not harmful to the user or to the environment. Protection is required against certain influences such as:

- The ingress of foreign bodies which could disturb the mechanical or electrical operation. (e.g., solid particles, small creatures, flying/creeping insects, etc.)

- The ingress of water or other liquids which could damage the insulation and generate degradation.
- Mechanical impact which might deform or break brittle or sensitive components.
- Corrosive gases from the environment.

First numeral : Protection against contact and ingress of foreign bodies	No special protection	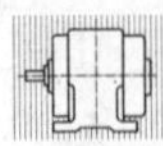 Protection against vertically falling water drops	 Protection against dropping water when inclined by up to 15 degrees	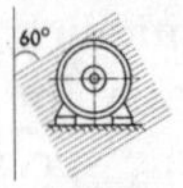 Protection against water spray when inclined by up to 60 degrees from vertical	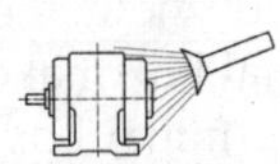 Protection against water splashed from any direction	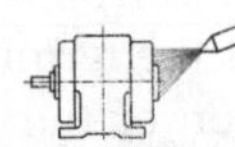 Protection against water projected by nozzle from any direction	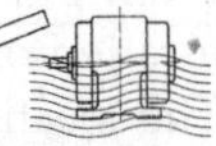 Protection against heavy seas
Second Numeral : Protection against ingress of water	0	1	2	3	4	5	6
0 No special protection							
1 Protection against solid foreign bodies >50 mm (Example: inadvertent contact with hand)							
2 Protection against solid foreign bodies >12 mm (Example: inadvertent contact with the fingers)		IP 21	IP 22	IP 23			
3 Protection against solid foreign bodies >12 mm (Example: inadvertent contact with wire and tools)							
4 Protection against solid foreign bodies >1 mm (Example: inadvertent contact with wire, bands)					IP 44		
5 Protection against dust (harmful deposits of dust)					IP 54	IP 55	IP 56

Figure 5.24 IP classification of motors (Reference: IS-4691).

Protection against ingress of solid bodies, particles or liquids is specified in the Indian standards in the form of IP classification of the motors. IS 4691 outlines the details of IP classification. Each motor is specified by the letters IP followed by two numerals (IP-XX). The first characteristic numeral specifies the protection against solid particles and the second characteristic numeral specifies the protection against the ingress of liquids. Figure 5.24 shows the most frequently used types of protection.

The atmosphere in chemical industries is generally laden with corrosive gases which can deteriorate the life of the motor. Protection against such conditions is ensured by choosing special materials that can withstand the deleterious effect of the corrosive atmosphere. Locations where explosive gases can be present, calls for explosion proof motors.

Degree of protection

Degrees of protection for mechanical machines are designated in accordance with IS 4691/IEC 60034-5 by the letters IP and two characteristic numerals.

5.6.3.2 Cooling classification:IC-XX

While providing the motor with the required amount of protection against external influences, it is likely to affect the cooling of the motor.

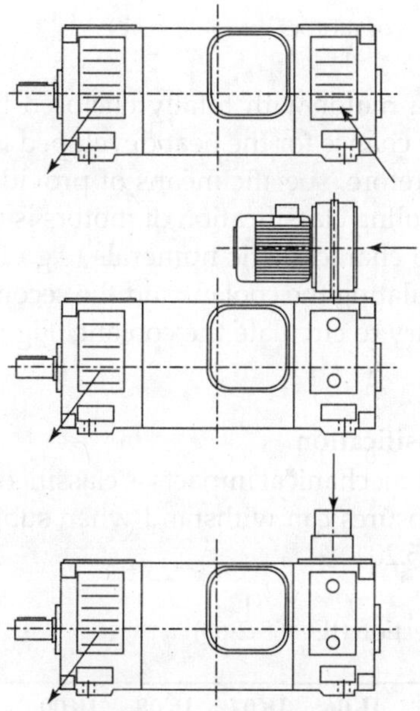

IC 01 — Enclosure IP 21 - IP 23
Self - ventilated with integral fan cooling (DP)
Cooling air is blown through the motor by a fan mounted on the shaft.

IC 06 — Enclosure IP 21 - IP 23
Separate ventilation with radial fitted fan unit (FV)
Cooling air is blown through the motor by a separately excited fan motor. The inlet side may be equipped with an air filter.

IC 17 — Enclosure IP 21 - IP 23
Single pipe ventilated (FV)
Cooling air is blown across the motor through the pipe connection with a separate customer provided external blower fan and discharges on the other side to open space.

IC 410 — Enclosure IP 44 - IP 55
Totally-enclosed non-ventilated (TENV)
Cooling without using a fan, only by natural ventilation and radiation on the totally enclosed motor surface.

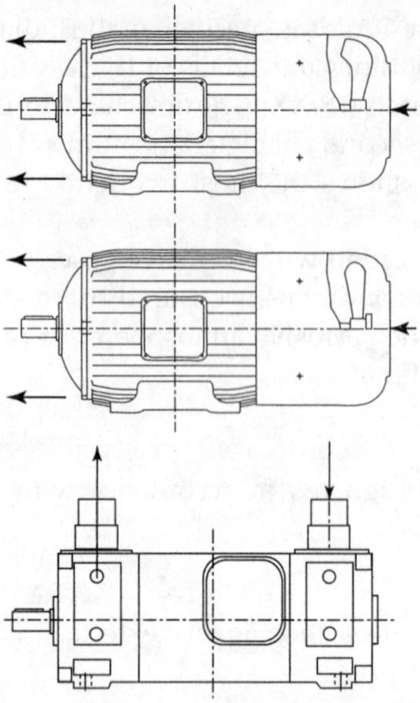

IC 411 — Enclosure IP 44 - IP 55
Totally-enclosed fan-cooled (TEFC)
Cooling air is blown over the totally enclosed motor surface by a fan mounted on the shaft.

IC 416 — Enclosure IP 44 - IP 55
External surface cooling (TEFV)
Cooling air is blown over the totally enclosed motor surface by an separately excited fan motor.

IC 37 — Enclosure IP 44 - IP 55
Double pipe ventilated (TEPV)
Cooling air is blown across the motor through a pipe connecting by means of a separate customer provided external blower fan and discharges on the other side's pipe connecting.

Figure 5.25 IC classification of motors.

A higher degree of mechanical protection means a motor with totally enclosed body of the motor. When the motor is totally enclosed, there is little chance for the heat developed in the windings to dissipate to the surrounding atmosphere. Therefore, specific means of providing cooling designations are highlighted in the IS-6362. And the cooling classification of motors is represented by IC classification. The format is IC-followed by two characteristic numerals (e.g. IC-XX). The first characteristic numeral designates the mode of circulating the coolant and the second numeral, designates the method of supplying the power necessary to circulate the coolant. Figure 5.25 lists some of the frequently used type of IC classification.

5.6.3.3 Protection against mechanical impact, IK classification

The degree of protection provided by enclosures against mechanical impacts is classified as IK code. This corresponds to the impact energy levels, the enclosures can withstand when subjected to IK impact tests. The standard IK coded are listed in Table 5.20.

Table 5.20 IK-XX degree of protection (IEC-68-2-63)

IK Code	IK01	IK02	IK03	IK04	IK05	IK06	IK07	IK08	IK09	IK10
Energy-joules	0.15	0.20	0.35	0.50	0.70	1.00	2.00	5.00	10.00	20.00

5.6.4 Environmental factors

The satisfactory working of the motors depends to a great extent on the site conditions where the motor is installed. The temperature of the cooling air in which the motor is working, the altitude of the location where the motor is installed and the nature of the atmosphere-dust laden, excessively humid or presence of corrosive gases, etc., will need careful consideration in selecting the motor. The derating for the ambient temperature and altitude are listed in Tables 5.21 and 5.22.

Table 5.21 Derating of motor capacity for ambient temperature

Cooling air temperature °C	Permissible output %
30	107
35	104
40	100
45	95
50	90
55	83

Table 5.22 Derating of motor capacity for altitude

Altitude–metres	Permissible output%
1000	100
2000	95
3000	90
4000	85

5.6.4.1 Permissible temperature rise

The windings of the motor are insulated with different types of insulation. Each type of insulation permits a maximum temperature rise of the motors. Insulation class is defined according to the ability of the motors to resist aging and failure due to heat. The permissible values of the temperature rise and the maximum temperature at the hottest point on the motor are specified in IS-325. Table 5.23 lists the permissible temperature rise and hot spot temperature.

Table 5.23 Permissible temperatures of motors (Reference:IS-325)

Class of insulation	Permissible temperature rise from 40 °C	Hot spot temperature T_{max}	Average insulation life at/below rated temperature—hours
Class B	70 °C	130°	20000
Class F	95 °C	155°	20000
Class H	115 °C	180°	20000

The maximum temperature at which the motors are operated will have drastic impact on the average insulation life of the motor. If the maximum operating temperature is 10 °C lower than the maximum permissible temperature of the class of insulation, the average insulation life will be double the normal average life. On the other hand if the maximum operating temperature is 10 °C higher than the permissible maximum temperature, the average insulation life will be half of the normal life time.

5.6.4.2 Permissible number of starts per hour

The maximum number of cold starts per hour for motors of different frame sizes is given in Table 5.24. This is subject to the condition that load inertia is less than or equal to the motor inertia, the load torque is proportional to the square of the speed up to the normal torque and the motors are started at even intervals.

Table 5.24 Permissible numbers of starts per hour

Frame size	2 pole	4 pole	6 and 8 pole
56–71	100	250	350
80–100	60	140	160
112–132	30	60	80
160–180	15	30	50
200–225	8	15	30
250–315	4	8	12

5.6.4.3 Voltage flicker

Starting of induction motors or other inductive loads causes a large voltage drop in the supply system which may be a nuisance to the other consumers. This voltage drop appears like a flicker in lighting circuits and becomes objectionable if the frequency of the flicker and the magnitude of the voltage drop exceed a certain threshold. This threshold of objection is shown as a voltage flicker curve (Figure 5.26).

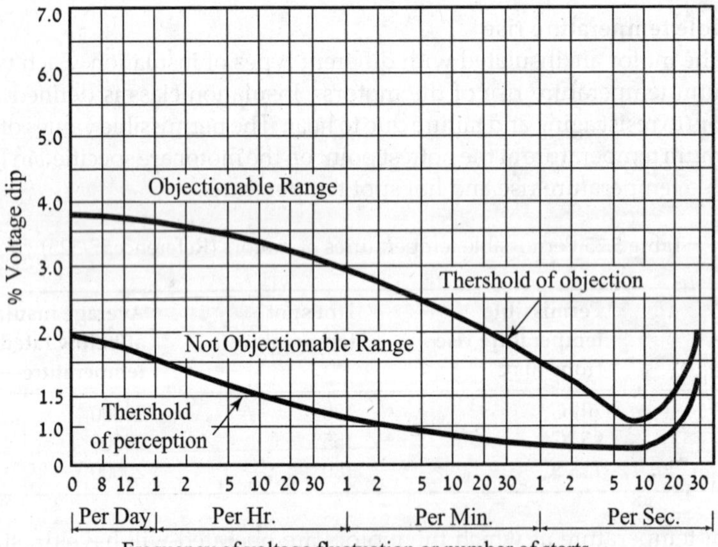

Figure 5.26 Voltage flicker curve (Source: IEEE standard:141-1993)

If the magnitude of the voltage drop and the frequency of occurrence lies below the threshold of objection, but above the threshold of perception, people notice the presence of light flicker, but generally do not find it irritating. If the magnitude of voltage drop and frequency of occurrence are below the threshold of perception, people may not notice the presence of flicker. If the magnitude of the voltage drop and frequency of occurrence are above the threshold of objection, there will be complaints from the public and such a situation should be avoided to the extent possible.

EXAMPLE 5.3

A three-phase, 415 V, 7.5 kW squirrel cage induction motor takes 14 A at rated full load conditions and 77 A when started directly on line. The impedance of the feeder supplying the motor is 0.155 ohms. Check whether the starting of this motor will cause objectionable flicker. If yes, what is the possible alternative?

Solution:

Full load current = 14 A

Starting current = 77 A

Feeder impedance = 0.155 ohm

Voltage dip in the feeder $= \sqrt{3} \times \text{starting current} \times \text{feeder impedance}$

$$= \sqrt{3} \times 77 \times 0.155 = 20.67 \text{ V}$$

Percentage voltage dip $= \dfrac{20.67}{415} \times 100 = 5.00\%$

Supposing the motor is started once every hour, the point is marked as A on the voltage flicker curve (Figure 5.27) and lies above the threshold of objection. Hence, the starting of this motor using DOL starter is objectionable.

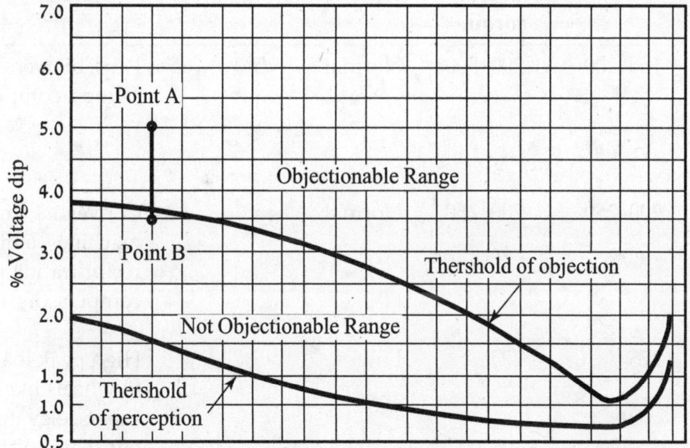

Figure 5.27 Voltage flicker points of example 5.3

To reslove this problem, the solutions available are either to reduce the impedance value of the feeder or to use a reduced voltage starter. The first solution may not be always feasible and economic. But if the starting current can be reduced to 70% of the original value, the voltage dip at starting will come down to 3.5% (0.7 5%) and the operation of the motor will not be a public nuisance. This point is shown on the voltage flicker curve as point B in Figure 5.27. A still better solution will be to use a star delta starter, whereby the starting current can be reduced to 57.7% of the original starting current and the voltage drop comes down to 2.89% (0.577 × 5%).

5.6.4.4 Efficiency and economics

When selecting a motor for a specific application, the initial cost and technical specifications should not be the only criterion. A life cycle costing should preferably be done to assess the economic viability of the selection. The running cost of a motor for one year can easily exceed the original cost of a motor by several times. Because of this, a small increase in efficiency of the motor can yield significant savings. An energy efficient motor, even though costlier than conventional motor can definitely be a better investment over the life period of the motor. It must be remembered that the efficiency of an induction motor varies with load and the efficiency is the maximum between 75% and 100% of full load depending on the design. The efficiency will be very low for loads below 30% of full load. The power factor of the induction motor also will be very poor at low loads. Therefore, induction motors should never be over sized (load < 50% of the motor capacity) as the initial cost as well as the energy costs will be higher and the power factor will be lower. A good engineering practice is to size the motor for operation between 75% and 100% of the full load.

Table 5.25 Induction motor quick selection chart

Classification	Starting torque	(% of FLT) Break-down torque	(% of FLT) Starting current	Slip	Typical applications
Normal starting torque and normal starting current	100–200	200–500	Normal	<5%	Fans, blowers, centrifugal pumps, compressors, etc.
High starting torque and normal starting current	200–250	200–250	Normal	<5%	Conveyors, crushers, stirring machines, agitators, reciprocating pumps, compressors, etc.
High starting torque and high slip	275	275	Low	>5%	High peak loads, load with flywheels like punch wheels, shears, elevators, extractors, winches, hoists, wire drawing machines, etc.
Wound rotor	Any torque up to break-down torque	225–275	Depends on starting torque	Depends on rotor resistance	Where high starting torque with low inrush current, frequent start, or limited to 2:1 speed control are required, and where high inertia must be accelerated

5.6.4.5 Quick selection chart

A quick reference may be made to Table 5.25 to decide upon which design is to be selected to meet a particular requirement. Finer details can be worked out based on the initial selection and the exact requirement.

5.6.5 Selection of motor capacity for intermittent loading

When the motors are subjected to varying loading conditions for varying periods of time, computation of the rating of the motors will be done using the equation

$$kW_{rms} = \frac{\sqrt{\sum_{1}^{n}\left(kW_i^2 \times t_i\right)}}{\sqrt{\left(\frac{t_0}{k_0} + \sum_{1}^{n} t_i\right)}} \tag{5.11}$$

where, kW_{rms} = rms rating of the motor in kilowatts
kW_i = loading of the motor at the ith instant
t_i = time duration of the ith instant
k_0 = is a constant to take care of the rest period
= 3 for open type motors
= 4 for totally enclosed type motors
t_0 = time period of rest for the motor

EXAMPLE 5.4

An industrial application requires a totally enclosed type drive motor to drive a load with the following loading cycle:

1.2 kW	for 60 seconds	No load	for 90 seconds
3 kW	for 90 seconds	Off	for 180 seconds

The motor is to be used for driving a machine at 1440 rpm on a supply system with three phase, 415 volts, 50 Hz. Determine the rating of a suitable motor.

Solution:

Total time for the load cycle = 60 + 90 + 90 + 180
= 300 seconds

Total period of ON time = 60 + 90 + 90
= 240 seconds

$$\text{Cycle duration factor} = \frac{\text{Total ON time}}{\text{Total cycle time}} = \frac{240}{300}$$

$$= 0.8 = 80\%$$

From equation (5.11),

$$\text{The rating of the motor} = kW_{rms} = \frac{\sqrt{\sum_{1}^{n}(kW_i^2 \times t_i)}}{\sqrt{\left(\frac{t_0}{k_0} + \sum_{1}^{n} t_i\right)}}$$

$$= \frac{\sqrt{(1.2^2 \times 60 + 0^2 \times 90 + 3^2 \times 90)}}{\sqrt{\left(\frac{60}{4} + 60 + 90 + 90\right)}}$$

$$= 1.875 \text{ kW}$$

The nearest standard size of a three-phase totally enclosed motor is 2.2 kW. As a precaution, we have to check whether this motor will be able to deliver 3 kW loading for 90 seconds without overheating. This can be verified from the torque speed characteristics of the motor. Normally the short time overload torque should not be greater than 75% of the pull out torque for stable operation.

EXAMPLE 5.5

Determine the capacity of an open type squirrel cage induction motor for an industrial application with the following loading pattern.

No load	for 30 seconds
2.5 kW	for 90 seconds
1.0 kW	for 60 seconds
4.0 kW	for 120 seconds
Off	for 60 seconds

Solution:

$$\text{Cycle duration factor} = \frac{\text{Total ON time}}{\text{Total cycle time}}$$

$$= \frac{300}{360}$$

$$= 0.833 = 83.33\%$$

From equation (5.11),

k_0 for an open type motor $= 3$

$$\text{The rating of the motor} = kW_{rms} = \frac{\sqrt{\sum_{1}^{n}\left(kW_i^2 \times t_i\right)}}{\sqrt{\left(\frac{t_0}{k_0} + \sum_{1}^{n} t_i\right)}}$$

$$= \frac{\sqrt{\left(0^2 \times 30 + 2.5^2 \times 90 + 1^2 \times 60 + 4^2 \times 120\right)}}{\sqrt{\left(\frac{60}{3} + 30 + 90 + 60 + 120\right)}}$$

$$= 2.818 \text{ kW}$$

The next standard size of squirrel cage induction motor is 3.7 kW. Therefore the selection will be a 3.7 kW squirrel cage induction motor with the other specifications like voltage, speed and frequency of operation matching with the actual demand. The cycle duration factor of the machine is 83.33%. Since each cycle takes 300 seconds or 5 minutes, the number of duty cycles per hour will be 12 which can be met by any motor manufactured in the frame size of 80 to 132 (refer Table 5.24).

EXAMPLE 5.6

The following specification pertains to an industrial belt conveyor system (Figure 5.28)

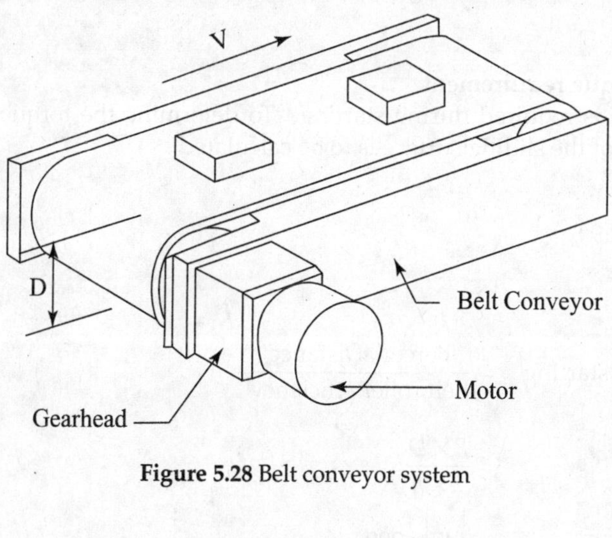

Figure 5.28 Belt conveyor system

Total mass of belt and charge (load) on the belt (m_1) = 50 kg
Friction coefficient of sliding surface (μ) = 0.3
Roller diameter (D) = 200 mm
Mass of roller (m_2) = 5 kg
Belt roller efficiency (η) = 0.9

Belt speed (V) = 200 m/sec ±10%
Working time = 8 hours/day

Choose a suitable electrical drive motor to meet this specification.

Solution:

A belt speed of 200 ±10% mm/sec will correspond to a roller revolution per second of

$$N_r = \frac{V}{\pi D}$$

$$= \frac{200 \pm 10\%}{\pi \times 200}$$

$$= 0.32 \pm 0.032 \text{ rps}$$
$$= 19.2 \pm 1.92 \text{ rpm}$$

Rated synchronous speeds of standard induction motors are in the range of 750 to 3000 rpm. In order to get the required speed range it will be necessary to use a speed reduction gear system. If we choose a gear ratio of 50, the motor speed corresponding to the roller speed will be

$$N_m = 19.2 \pm 1.92 \times 50$$
$$= 960 \pm 9.62 \text{ rpm.}$$

This rpm can be easily obtained by using a six pole induction motor which provides a synchronous speed of 1000 rpm.

Calculation of the torque requirement

The maximum torque is required during starting. To determine the torque during starting, the friction coefficient (F) of the sliding surface is to be calculated.

$$F = m_1 g$$
$$= 0.3 \times 50 \times 9.8$$
$$= 147 \text{ N}$$

Therefore

$$\text{Load torque at starting} = \frac{\text{Force} \times \text{Distance}}{\text{Belt roller efficiency}}$$

$$= \frac{F \times D}{2 \times \eta}$$

$$= \frac{147 \times 200}{2 \times 0.9} = 16333.33 \text{ m Nm}$$

$$T_L = 16.333 \text{ Nm}$$

This torque is required at the gear drive shaft. Therefore, the torque required at the motor shaft

$$T_m = \frac{\text{Torque at the gear drive shaft}}{\text{gear ratio} \times \text{transmission } \eta}$$

$$= \frac{16.333}{50 \times 0.66} = 0.4949 \text{ Nm}$$

It is always better to use a safety margin to take care of the fluctuation in loading pattern. Assuming a factor of safety of 2,

$$T_m = 0.9898 \text{ Nm}$$
$$= 0.101 \text{ kg m}$$

Check for inertia

Moment of inertia of the roller (J_{m2}) $= \frac{m_2 \times D^2}{8} \times 2$

$$= \frac{5 \times 0.2^2}{8} \times 2 = 0.05 \text{ kg.m}^2$$

(for the two rollers on either end of the belt)

Moment of inertia of belt and workpiece (J_{m1}) $= \frac{m_1 \times (\pi D)^2}{4\pi^2}$

$$= \frac{50 \times (\pi \times 0.2)^2}{4\pi^2} = 0.5 \text{ kg.m}^2$$

Therefore total gear head shaft load inertia (J) $= J_{m1} + J_{m2}$
$= 0.55 \text{ kg.m}^2$

The corresponding motor shaft inertia $= \frac{J}{\text{gear ratio}^2}$

$$= \frac{0.55}{50^2} = 2.2 \times 10^{-4} \text{ kg.m}^2$$

Referring to Table 5.26, we see that a 0.18 kW, 3-phase, 415 volt squirrel cage induction motor will provide a full load torque of 0.19 kg.m and a rotor GD^2 of 0.004 kg.m^2.

EXAMPLE 5.7

A floor charger machine with machine inertia of 0.5 kg.m^2 needs a full load torque of 5 kg.m at 750 rpm. The charging cycle per hour of the machine is 15. The charging cycle is:

Start and run	= 45 sec
Stop time with 2-3 inching operation	= 60 sec
Start and run	= 45 sec
Stop time	= 90 sec

Assuming each inching operation as equivalent to 0.5 starting operation, determine the size of a suitable motor for the purpose.

Solution:

Total starts per cycle $= 2 + 3 \times 0.5$
$= 3.5$

Total starts per hour $= 15 \times 3.5$
$= 52.5$

Total ON time for starting in one charging cycle $= 45 + 45$
$= 90$ sec

Assuming 1 sec for each inching operation,

Total ON time including inching $= 90 + 1 \times 3$
$= 93$ sec

Total cycle time $= 45 + 60 + 45 + 90$
$= 240$ sec

Cycle duty factor (CDF) $= \dfrac{\text{Total ON time}}{\text{Total cycle time}} = \dfrac{93}{240}$

$= 0.3875$

Therefore a CDF of 0.4 (40%) is chosen.

Maximum output at rated rpm $= \dfrac{2\pi NT}{4500}$

$$= \frac{2 \times \pi \times 750 \times 5}{4500}$$

$= 5.236$ kW

We can choose a 5.5 kW, 6-pole, squirrel cage induction motor with a frame size of 160M which provides a rated full load torque of 7.55 kg.m and is having rotor inertia of 0.46 kg m^2. (Refer Table 5.26)

$$\text{The inertia factor} = \frac{\text{Machine GD}^2 + \text{Motor GD}^2}{\text{Motor GD}^2}$$

$$= \frac{0.5+0.46}{0.46} = 2.087$$

The motor specifications are:

Rated output = 5.5 kW
Supply voltage = 415 V three-phase, 50 Hz
Type of motor = squirrel cage motor
CDF = 40%
Factor of inertia = 2

Table 5.26 Performance data of squirrel cage induction motors (Courtesy: M/s.Crompton Greaves Ltd.)

PERFORMANCE DATA
3-PHASE TEFC SCR INDUCTION MOTOR

VOLTAGE	: 415 ± 10% V	TYPE	: SQUIRREL CAGE (SCR)	AMBIENT	: 50 DEG. C
FREQUENCY	: 50 ± 5% Hz	RATING	: CONTINUOUS	TEMP. RISE (R)	: 70 DEG. C
COMBINED VARIATION	: ±10% (ABSOLUTE SUM)	INSULATION	: CLASS 'P' WITH CLASS 'B' TEMP. RISE	DEGREE OF PROTECTION	: IP55

OUTPUT kW	OUTPUT HP	POLE	FRAME SIZE	FL RPM	FLC AMPS.	FLT Kg-m	EFFICIENCY (%) FL	EFFICIENCY (%) 3/4 LOAD	EFFICIENCY (%) 1/2 LOAD	POWER FACTOR FL	POWER FACTOR 3/4 LOAD	POWER FACTOR 1/2 LOAD	DOL STG. STG. T % FLT	DOL STG. STG. C % FLC	POT % FLT	GD² KGM²	NET WT. kG.
0.18	0.25	2	AD63	2700	0.50	0.06	64.0	62.0	58.0	0.79	0.68	0.59	300	500	350	0.001	5.6
		4	AD63	1330	0.50	0.13	64.0	62.0	58.0	0.76	0.73	0.60	200	400	250	0.003	5.6
		6	AD71	900	0.55	0.19	65.0	64.0	60.0	0.68	0.60	0.50	200	400	250	0.004	7.0
0.25	0.33	2	AD63	2700	0.65	0.09	68.0	67.0	63.0	0.79	0.72	0.60	275	500	325	0.001	5.6
		4	AD71	1350	0.75	0.18	62.0	61.0	55.0	0.76	0.73	0.60	225	400	275	0.004	7.0
0.37	0.50	2	AD71	2820	0.95	0.13	68.0	67.0	63.0	0.79	0.72	0.60	275	500	325	0.002	7.0
		4	AD71	1400	0.90	0.26	77.0	76.0	74.0	0.76	0.73	0.60	225	600	275	0.004	7.0
		6	AD/D80	910	1.05	0.40	69.0	68.0	66.0	0.71	0.63	0.52	210	400	260	0.011	10/17
		8	ADND90S	680	1.40	0.53	65.0	64.0	60.0	0.57	0.50	0.40	170	400	220	0.015	13/22
0.55	0.75	2	AD/D80	2820	1.30	0.19	73.0	72.0	68.0	0.81	0.73	0.62	250	600	300	0.003	10/17
		4	AD/D80	1410	1.25	0.38	77.0	76.0	74.0	0.78	0.75	0.64	200	500	250	0.007	10/17
		6	AD/D80	910	1.55	0.59	69.0	68.0	66.0	0.71	0.63	0.52	210	400	260	0.011	10/17
		8	AD/ND90L	680	1.75	0.79	69.0	68.0	64.0	0.63	0.54	0.45	150	400	200	0.021	13/22
0.75	1.00	2	AD/D80	2820	1.65	0.26	77.0	76.0	74.0	0.81	0.73	0.62	300	600	350	0.003	10/17
		4	AD/D80	1410	1.75	0.52	77.0	76.0	74.0	0.78	0.75	0.64	200	500	250	0.007	10/17
		6	AD/ND90S	935	2.00	0.78	73.0	71.0	69.0	0.72	0.65	0.58	200	400	250	0.015	13/22
		8	AD/ND100L	700	2.55	1.04	70.0	69.0	64.0	0.58	0.51	0.41	200	400	250	0.030	19/32
1.10	1.50	2	AD/D80	2820	2.35	0.38	79.0	78.0	76.0	0.82	0.77	0.70	225	600	275	0.004	10/17
		4	AD/ND90S	1415	2.50	0.76	78.0	76.0	74.0	0.78	0.75	0.64	200	500	250	0.014	13/22
		6	AD/ND90L	935	2.85	1.15	74.0	73.0	70.0	0.72	0.65	0.58	200	400	250	0.021	16/25
		8	AD100L/ND100L	700	3.30	1.53	72.5	71.0	68.0	0.63	0.54	0.44	175	400	225	0.034	19/35
1.50	2.00	2	AD/ND90S	2830	3.20	0.52	80.0	79.0	77.0	0.82	0.77	0.70	225	600	275	0.006	13/22
		4	AD/ND90L	1415	3.20	1.03	80.0	79.0	77.0	0.81	0.78	0.71	200	500	250	0.019	16/25
		6	AD/ND100L	935	3.80	1.56	76.0	75.0	73.0	0.72	0.65	0.58	200	400	250	0.030	19/32
		8	AD112M/ND112M	700	4.00	2.09	77.0	77.0	75.0	0.68	0.60	0.52	190	400	240	0.057	29/45
2.20	3.00	2	AD/ND90L	2830	4.60	0.76	81.0	80.0	78.0	0.82	0.77	0.70	300	600	350	0.008	16/25
		4	AD/ND100L	1440	4.60	1.49	82.0	81.0	79.0	0.82	0.78	0.72	200	600	250	0.030	19/32
		6	AD/ND112M	935	5.30	2.29	77.0	76.0	73.0	0.75	0.70	0.60	200	500	250	0.048	29/42
		8	ND132S	710	5.40	3.02	77.5	76.5	73.0	0.73	0.68	0.61	180	450	230	0.174	68.0
3.70	5.00	2	AD/ND100L	2875	7.60	1.25	83.0	82.0	80.0	0.82	0.77	0.70	250	650	300	0.022	19/36
		4	AD/ND112M	1440	7.60	2.5	84.0	83.0	81.0	0.81	0.76	0.69	200	600	250	0.052	29/42
		6	ND132S	950	7.70	3.79	85.0	84.0	82.0	0.79	0.73	0.63	200	600	250	0.174	68.0
		8	ND132M	710	8.60	5.08	80.0	80.0	78.0	0.75	0.70	0.60	180	600	230	0.214	79.0
5.50	7.50	2	AD/ND112M	2880	10.10	1.86	85.0	84.0	82.0	0.89	0.85	0.80	250	650	300	0.034	29/45
		4	ND132S	1450	10.10	3.69	85.0	85.0	83.0	0.88	0.85	0.75	250	600	300	0.131	68.0
		6	ND132M	950	11.30	5.64	85.0	84.0	82.0	0.80	0.75	0.68	200	600	250	0.214	79.0
7.50	10.00	2	ND132S	2890	13.80	2.53	85.0	84.0	82.0	0.89	0.85	0.80	275	650	325	0.062	68.0
		4	ND132M	1455	13.60	5.02	87.0	86.0	84.0	0.88	0.85	0.75	275	600	325	0.161	79.0
9.30	12.50	2	# ND132M	2890	16.70	3.13	87.0	86.0	84.0	0.89	0.85	0.80	275	600	325	0.076	79.0

PERFORMANCE DATA
3-PHASE TEFC SCR INDUCTION MOTOR

VOLTAGE	: 415 ± 10% V	TYPE	: SQUIRREL CAGE (SCR)	AMBIENT	: 50 DEG. C
FREQUENCY	: 50 ± 5% Hz	RATING	: CONTINUOUS	TEMP. RISE (R)	: 70 DEG. C
COMBINED VARIATION	: ±10% (ABSOLUTE SUM)	INSULATION	: CLASS 'P' WITH CLASS 'B' TEMP. RISE	DEGREE OF PROTECTION	: IP55

OUTPUT		POLE	FRAME SIZE	FL	FLC	FLT	EFFICIENCY (%)			POWER FACTOR			DOL STG.		POT	GD^2	NET WT.
kW	HP			RPM	AMPS.	Kg-m	FL	3/4 LOAD	1/2 LOAD	FL	3/4 LOAD	1/2 LOAD	STG. T % FLT	STG. C % FLC	% FLT	KGM^2	kG.
3.7	5.0	8	ND160M	710	8.0	5.08	83.0	83.0	81.0	0.74	0.70	0.62	150	500	200	0.46	125
5.5	7.5	6	ND160M	970	11.0	5.52	86.0	86.0	84.0	0.80	0.76	0.68	200	550	250	0.46	125
		8	ND160M	710	12.0	7.55	85.0	85.0	83.0	0.74	0.70	0.62	150	500	200	0.46	125
7.5	10.0	2	ND160M	2910	14.0	2.51	85.0	84.0	82.0	0.88	0.86	0.78	225	600	275	0.13	125
		4	ND160M	1460	14.0	5.00	87.0	87.0	85.0	0.85	0.81	0.73	225	600	275	0.31	125
		6	ND160M	975	15.0	7.49	87.5	87.0	85.0	0.80	0.76	0.68	200	550	250	0.46	125
		8	ND160L	710	16.0	10.29	85.0	85.0	83.0	0.76	0.72	0.64	150	500	200	0.64	148
9.3	12.5	2	ND160M	2920	17.0	3.10	87.0	87.0	85.0	0.88	0.86	0.78	250	650	300	0.13	125
		4	ND160M	1460	17.0	6.20	88.5	88.5	86.5	0.84	0.81	0.73	175	600	225	0.31	125
		6	ND160L	975	18.0	9.29	87.5	87.0	84.0	0.80	0.76	0.68	200	550	250	0.59	148
		8	ND180M	720	20.0	12.58	86.0	86.0	84.0	0.74	0.70	0.60	175	500	225	0.99	174
11	15	2	ND160M	2920	20.0	3.67	88.0	88.0	86.0	0.88	0.86	0.78	250	650	300	0.13	125
		4	ND160M	1460	21.0	7.34	89.0	89.0	86.0	0.82	0.79	0.70	200	600	250	0.36	125
		6	ND160L	975	22.0	10.99	88.0	87.5	86.0	0.80	0.76	0.68	200	550	250	0.64	148
		8	ND180L	720	24.0	14.88	87.0	87.0	85.0	0.74	0.70	0.60	175	500	225	1.16	210
15	20	2	ND160M	2920	26.0	5.00	89.5	89.5	87.5	0.88	0.86	0.79	250	650	300	0.17	125
		4	# ND160L	1460	27.0	10.01	90.0	90.0	88.0	0.85	0.83	0.75	200	600	250	0.47	148
		6	ND180L	975	29.0	14.98	90.0	90.0	88.0	0.79	0.73	0.66	225	600	275	1.16	210
		8	ND200L	725	33.0	20.15	88.5	88.5	86.5	0.71	0.65	0.55	225	500	275	2.14	282
18.5	25	2	ND160L	2920	32.0	6.17	90.0	90.0	88.0	0.88	0.86	0.79	275	650	325	0.21	148
		4	ND180M	1475	33.0	12.22	92.0	92.0	90.0	0.84	0.80	0.72	200	600	250	0.81	174
		6	ND200L	975	34.0	18.48	91.1	91.1	89.9	0.84	0.82	0.73	200	550	250	1.69	282
		8	ND225S	725	39.0	24.85	89.0	89.0	87.0	0.75	0.71	0.63	175	500	225	3.24	345
22	30	2	ND180M	2930	41.0	7.31	91.0	91.0	89.0	0.83	0.80	0.72	225	600	275	0.44	164
		4	ND180L	1475	40.0	14.53	92.0	92.0	90.0	0.84	0.80	0.72	200	600	250	0.95	210
		6	ND200L	975	40.0	21.98	91.5	91.5	90.1	0.84	0.82	0.78	200	550	250	2.04	282
		8	ND225M	725	46.0	29.56	89.0	89.0	87.0	0.75	0.71	0.63	175	550	225	3.61	375
30	40	2	ND200L	2950	51.0	9.91	91.5	91.5	90.0	0.90	0.89	0.87	200	600	250	0.80	282
		4	ND200L	1475	51.0	19.81	92.0	92.0	90.2	0.89	0.86	0.78	250	600	300	1.62	282
		6	ND225M	980	53.0	29.82	92.0	92.0	90.5	0.85	0.82	0.73	200	550	250	3.61	375
		8	ND250M	735	61.0	39.76	90.5	90.5	88.5	0.75	0.71	0.63	175	550	225	4.82	473
37	50	2	ND200L	2950	62.0	12.22	92.0	92.0	90.0	0.90	0.89	0.87	200	600	250	0.89	282
		4	ND225S	1480	63.0	24.35	92.5	92.5	91.6	0.89	0.86	0.78	250	600	300	2.64	345
		6	ND250M	980	66.0	36.77	93.0	93.0	92.0	0.84	0.81	0.72	250	600	300	4.82	473
		8	ND280S	735	75.0	49.03	91.5	91.5	89.5	0.75	0.71	0.63	200	500	250	8.01	600
45	60	2	ND225M	2955	72.0	14.83	92.0	92.0	90.0	0.94	0.92	0.88	225	650	275	1.87	375
		4	ND225M	1480	76.0	29.61	93.0	93.0	91.5	0.89	0.86	0.78	250	600	300	3.13	375
		6	ND280S	980	79.0	44.72	93.0	93.0	91.9	0.85	0.82	0.73	250	550	300	8.01	600
		8	ND280M	725	91.0	60.46	92.0	92.0	90.5	0.75	0.71	0.63	175	500	225	9.89	670
55	75	2	ND250M	2955	88.0	18.13	92.5	92.5	90.5	0.94	0.92	0.88	175	600	225	2.70	473
		4	ND250M	1475	92.0	36.32	93.5	93.5	92.0	0.89	0.85	0.82	200	600	250	3.45	473
		6	ND280M	980	95.0	54.66	93.5	93.5	92.5	0.86	0.82	0.74	175	600	225	9.89	670
		8	ND315S	740	113.0	72.39	93.0	93.0	91.5	0.73	0.66	0.56	250	550	300	14.12	900
75	100	2	NO280S	2975	124.0	24.55	93.5	93.5	92.0	0.90	0.86	0.78	225	600	275	7.14	600
		4	ND280S	1480	123.0	49.36	94.0	94.0	92.5	0.90	0.88	0.84	250	600	300	7.21	600
		6	ND315S	987	129.0	74.01	94.0	94.0	92.5	0.86	0.82	0.74	250	600	300	14.12	900
		8	ND315M	740	153.0	98.72	93.5	93.5	91.5	0.73	0.66	0.56	200	600	250	18.98	950

***Note**: All performance figures are subject to tolerances as per IS325 # for 45 °C/75 °C only.*

5.7 SHORT CIRCUIT STUDIES

The initial current that flows in a power system due to a short circuit, will not be purely sinusoidal in nature. Rather, the current waveform will have a DC offset portion which decays to zero after some time. The offset depends on the inductance L and resistance R of the system and the larger the value of inductance when compared to the resistance, the larger will be the offset. The actual current in a power system during a fault condition is referred to as asymmetrical fault current. The analysis of power systems in time domain to determine the asymmetrical fault current is more complex and time consuming. Hence the analysis is carried out under steady state condition (when the transients have died out) and multiplying factors are used to determine the asymmetrical fault currents.

A power system in its simplest form can be represented by a voltage source in series with impedance. Figure 5.29 shows such a power distribution system with a fault simulated on the system by closing the switch at time $t = 0.0$ seconds.

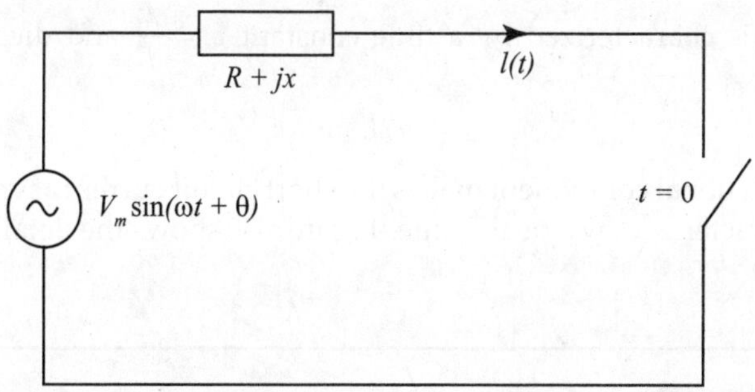

Figure 5.29 A typical power system with a fault simulation

The expression for the instantaneous value of current

$$i_{(t)} = \sqrt{2} \times I_k \left[\sin(\omega t - \theta_z) + \sin\theta_z e^{-\left(\frac{\omega R}{X}\right)t}\right] \tag{5.12}$$

where:

$I_k = \dfrac{V_m}{\sqrt{2} Z_s}$ = rms symmetrical short circuit current

$Z_s = \sqrt{(R^2 + X^2)}$ = system impedance

$\theta_s = \tan^{-1} \dfrac{X}{R}$

V_m = Maximum phase to neutral voltage of the system

The term $\sqrt{2}\left[\sin(\omega t - \theta_z) + \sin\theta_z e^{-\left(\frac{\omega R}{X}\right)t}\right]$ represents a multiplying factor to the rms value of the current I_k and can be used to determine the instantaneous values of fault currents at any instant.

The maximum peak value of the fault current is more important when specifying the short circuit withstand capacity. The multiplying factor that is used to determine the maximum peak instantaneous fault current can be obtained by taking the derivative of the bracketed term in equation (5.12) with respect to time and equating it to zero. The result is then solved to find the time of occurrence of maximum peak current (referred to as t_p). This value of time is then substituted in the bracketed term to determine the appropriate multiplying factor.

The short circuit current may be considered to be formed by two components:

- A symmetrical component i_s with sinusoidal wave form and precisely symmetrical with respect to the x-axis, representing the time axis.

$$i_s = \sqrt{2}I_k \sin(\omega t - \theta_z) \tag{5.13}$$

- An unidirectional component i_u with exponential curve due to the presence of inductance. This component is characterized by a time constant $\tau = \frac{L}{R}$ and dies down in a few time constants. (3 to 6 τ)

$$i_u = \sqrt{2}I_k \sin\theta_z e^{-\left(\frac{\omega R}{X}\right)t} \tag{5.14}$$

The presence of unidirectional component makes the short circuit current asymmetrical with respect to the x-axis and is characterized by a peak value. Figure 5.30 shows the details of this asymmetrical current.

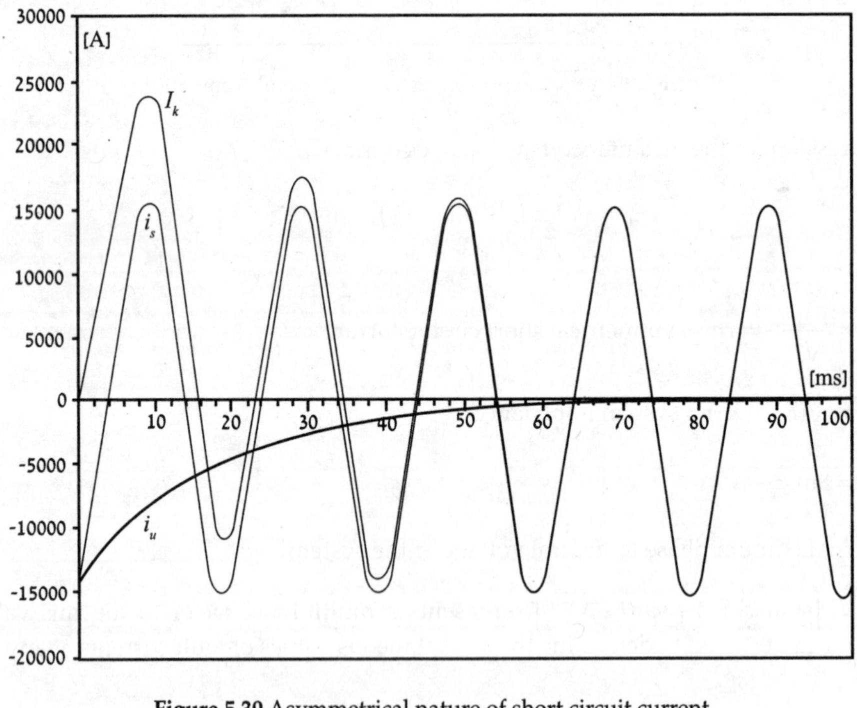

Figure 5.30 Asymmetrical nature of short circuit current

It can be proved that the first current peak after the occurrence of the fault may vary from $\sqrt{2}I_k$ to $2\sqrt{2}I_k$. Table 5.27 shows the values of this first peak for different $\frac{X}{R}$ ratios of the power system. After the transient period, the current settles down to a symmetrical value.

Table 5.27 Instantaneous and first half cycle factors of a power system (IEC 60947-2)

$\frac{X}{R}$ **ratios**	**Instantaneous peak factor**	**Half cycle factor**	**Time to peak t_p (ms)**
0.0	1.4142	1.00	5.00
0.5	1.4250	1.00	6.48
1.0	1.5122	1.002	7.32
2.0	1.7560	1.042	8.16
3.0	1.9495	1.115	8.52
4.0	2.0892	1.191	8.88
5.0	2.1924	1.263	9.00
6.0	2.2708	1.304	9.12
7.0	2.3323	1.347	9.24
8.0	2.3817	1.381	9.36
9.0	2.4222	1..412	9.36
10.0	2.4561	1.438	9.48
20.0	2.6250	1.570	9.72
30.0	2.6890	1.618	9.84
40.0	2.7224	1.643	9.84
50.0	2.7427	1.697	9.84
100.0	2.7848	1.697	9.96
∞	2.8284	1.732	10.00

The rms value of the current waveform over the first half cycle is also important. This value is determined by multiplying the rms symmetrical short circuit current by first half cycle rms multiplying factor. This first half rms multiplying factor is given by:

$$\text{First half multiplying factor} = \frac{\sqrt{\frac{1}{T}\int_0^t i_t^2 dt}}{I_k} \tag{5.15}$$

This factor also is a function of $\frac{X}{R}$ ratio of the system. Table 5.27 lists the instantaneous peak factor and first half cycle factor for various values of $\frac{X}{R}$ ratios. This table will allow us to calculate the instantaneous peak value and the first half cycle value of the fault currents from the knowledge of the symmetrical rms fault current. The direct determination of the instantaneous peak value and the first half cycle value is time consuming since the necessary parameters like short circuit power factor, ratio between inductive reactance and resistance, etc., are not always readily available. Whereas, calculation of the symmetrical (rms) faults current is more direct and less time consuming.

EXAMPLE 5.8

The source impedance of a 11 kV distribution substation is 0.2 + j1.2 ohm per phase. Calculate:

A) rms fault current
B) maximum peak instantaneous vale of fault current and
C) rms value of half cycle fault current,

if a balanced three-phase fault occurs on the 11 kV side of the transformer

Solution:

$$\text{The source impedance} = \sqrt{0.2^2 + 1.2^2} = 1.2166\Omega$$

$$\text{The rms symmetrical fault current} = \frac{V_{\text{ln}}}{Z_s} = \frac{11000/\sqrt{3}}{1.2166}$$

$$= 5220.16 \text{ A}$$

$$\text{Power system } \frac{X}{R} \text{ ratio} = \frac{1.2}{0.2} = 6$$

From Table 5.27, the instantaneous peak factor for an X/R ratio of 6 is 2.708

$$\text{Therefore, the peak instantaneous value of fault current} = 2.708 \times I_k$$
$$= 2.708 \times 5220.16$$
$$= 14136.21 \text{ A}$$

and the half cycle multiplying factor for an X/R ratio of 6 is 1.304.

$$\text{Therefore, rms value of first half cycle asymmetrical fault current} = 1.304 \times 5220.16$$
$$= 6807.09 \text{ A}$$

The Bureau of Indian Standards specification IS-13234-92 and IEC standard 60909-1988 gives very useful information for the calculation of the peak current.

$$\text{Initial symmetrical short circuit current} = I_k'' = \frac{cU_n}{\sqrt{3} \times Z_k} A \quad (5.16)$$

(5.17)

$$\text{And, peak short circuit current} = i_p = k\sqrt{2} \times I_k'' \quad (5.18)$$

Where,

U_n = Nominal line to line system voltage
Z_k = System impedance up to the fault point 'k'
c = Voltage factor
= 1 for low voltage (240/415)
= 1.05 for other low voltages
= 1.1 for voltages above 1kV up to 230 kV

$$k = 1.02 + 0.98e^{\left(\frac{-3R_k}{X_k}\right)} \tag{5.19}$$

The value of 'k' can also be evaluated using the Figure 5.31 shown below.

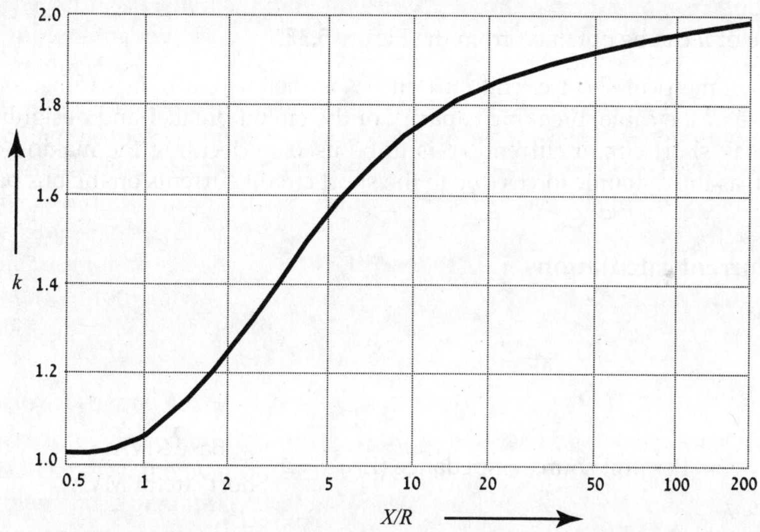

Figure 5.31 Graph showing the relation between $\frac{X}{R}$ and k

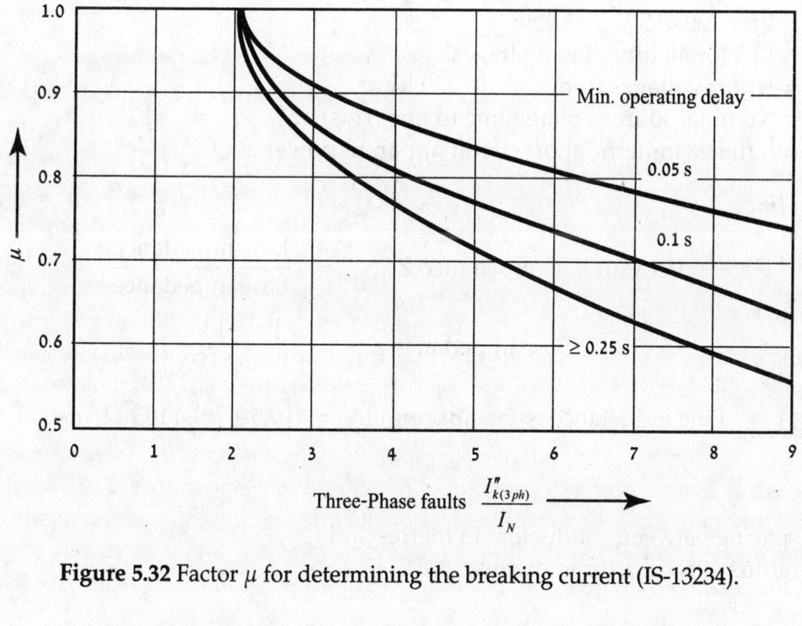

Figure 5.32 Factor μ for determining the breaking current (IS-13234).

The breaking current or the rms value of the alternating current flowing through the first phase at the instant of contact separation, during the opening of the circuit breaker in the event of a short circuit is given by

$$I_{cu} = \mu I_k'' \tag{5.20}$$

Where, the value of μ depends on the minimum operating delay of the circuit breaker and the ratio of $\frac{I_k''}{I_n}$. The value of μ can be obtained from the Figure 5.32.

The initial symmetrical short circuit current 'I_k''' is the current that is to be considered while selecting the size of the cable, breaking capacity of the circuit breaker and designing the earthing system. The peak short circuit current 'i_p' is to be used in deciding the making capacity of the breakers as well as the dynamic forces due to the short circuit currents on the bus bars.

5.7.1 Fault current calculations

Basic equations:

Source

$$\text{Per unit source impedance } (Z_Q) = \frac{\text{Base MVA}}{\text{Short circuit MVA}} \tag{5.21}$$

$$\text{Source impedance in ohms } (Z_Q) = \frac{cU_{nQ}^2}{S_{kQ}''} \tag{5.22}$$

where,

c = Voltage factor
= 1 for 230/415 V
= 1.05 for all other low voltages
= 1.1 for voltages from 1 kV to 230 kV

U_{nQ} = Nominal source voltage-line to line (rms) at Q

S_{kQ}'' = Initial symmetric short circuit apparent power at Q

Transmission line

$$\text{Per unit line impedance } Z_{l(pu)} = \frac{\text{Line impedance}}{\text{Base impedance}} \tag{5.23}$$

$$\text{Base impedance } Z_b = \frac{\text{Base kV}^2}{\text{Base MVA}} \tag{5.24}$$

$$\text{Line inductance (sym. spacing) } L = \left(0.5 + ln\frac{d}{r}\right)10^{-7}\text{H/m} \tag{5.25}$$

Where,

d = spacing between conductors in metres and

r = radius of conductor in metres.

$$\text{Inductive reactance of the overhead line } X_l = 0.0628\left(\frac{0.25}{n} + l_n\frac{d}{r}\right)\Omega/km \tag{5.26}$$

Where,

n = number of conductors per phase
d = spacing of overhead line in metres

$$r = \text{mean radius of the conductor} = \frac{1.14}{1000}\sqrt{\frac{q_n}{\pi}}\text{m} \tag{5.27}$$

q_n = area of cross-section of the conductor in mm^2

Transformer

S_{rT} = Rated apparent power of transformers - VA
U_{rT} = Rated voltage of transformer on HV or MV side - V
u_{kr} = Percentage impedance (impedance voltage) of the transformer

$$Z_T = \text{Transformer impedance} - \Omega = \frac{u_{kr}}{100}\frac{U_{rT}^2}{S_{rT}} \tag{5.28}$$

P_{krT} = Full load loss of transformer - W

$$R_T = \text{Resistance of the transformer} = \frac{P_{krT}}{3I_{rT}^2} \tag{5.29}$$

$$X_T = \text{Reactance of the transformer} = \sqrt{(Z_T^2 - R_T^2)}\Omega \tag{5.30}$$

I_{rT} = Full load current of the transformer on HV or MV side-A

$$\text{Soil resistivity } \rho = 2\pi \text{DR ohm} \cdot \text{m} \tag{5.31}$$

Where,

D = Spacing of the spikes in meter and
R = reading of the earth tester in ohms.

$$\text{Permissible current density of the earth electrode } I_d = \frac{7.57X10^3}{\sqrt{\rho t}}\text{A/m}^2 \tag{5.32}$$

Where,

t = Duration of fault in seconds (3 sec)
ρ= Soil resistivity in ohm $\cdot$ m

Data required:

1. Supply voltage of the system (U_n)
2. Fault level on the nearest substation- (S''_{kQ})
3. Length of the transmission line from the substation to the consumers transformer (l)
4. Spacing of the conductors of the O/H transmission line (d)
5. Radius of the conductor of the O/H line-r (cross-section of the conductor-mm^2 q_n)
6. Resistance of the line (R_l)
7. kVA capacity of the consumer transformer (S_{rT})
8. Percentage impedance (impedance voltage) of the consumer transformer (u_{kT})
9. Full load loss of the transformer (P_{krT})

EXAMPLE 5.9

Electric supply to an industry is provided from the utility substation through an overhead transmission line. The length of the 11 kV line is 2.5 km and the conductor used is MINK with an equilateral spacing of 900 mm. The industry proposes to have a transformer substation with a 750 kVA, 11 kV/433V, delta/star connected transformer. The percentage reactance of the transformer is 4% and the full load copper loss of the transformer is 1.5%. The three-phase short circuit power at the utility substation is 350MVA. The resistance of the line conductor = 0.454 ohm per km. Calculate:

1. Initial symmetrical short circuit current and
2. Peak short circuit current

on the primary and secondary terminals of the proposed transformer.

Solution:

The schematic diagram of the power distribution arrangement is shown in Figure 5.33
The equivalent single diagram of the power distribution system is shown below (Figure 5.34)

$$\text{The source impedance } Z_Q = \frac{cU_{nQ}^2}{S''_{kQ}}$$

$$= \frac{1.1 \times 11000^2}{350 \times 10^6} = 0.3803\ \Omega$$

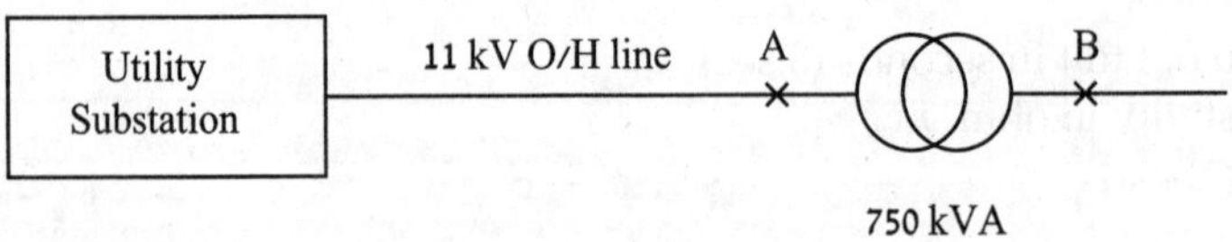

Figure 5.33 Schematic arrangement of the distribution system

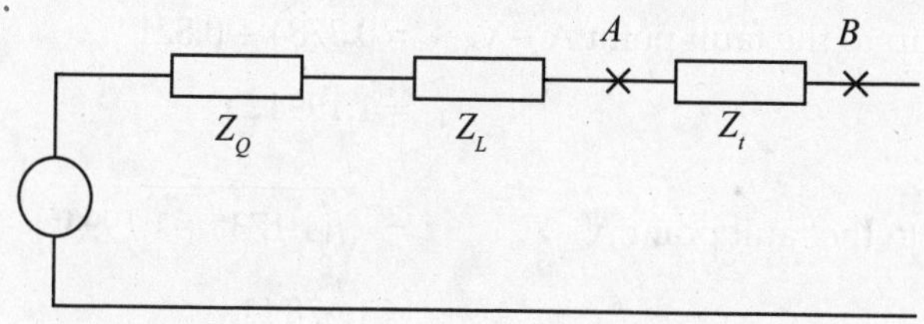

Figure 5.34 Equivalent single line diagram of example.

The source resistance R_Q $= 0.1\ Z_Q$

$= 0.03803\ \Omega$

The source reactance X_Q $= 0.995\ Z$

$= 0.3784\ \Omega$

The resistance of the line conductor $= 0.454\ \Omega/\text{km}$

Total resistance of the line R_l $= 2.5 \times 0.454$

$= 1.135\ \Omega$

Nominal area of cross section of MINK conductor, $q_n = 93.3\ \text{mm}^2$

Mean radius of the conductor (r) $= \dfrac{1.14}{1000}\sqrt{\dfrac{q_n}{\pi}}$

$= 0.0062\ \text{m}$

Number of conductors per phase $(n) = 1$

Inductive reactance of the line x_l $= 0.0628\left(\dfrac{0.25}{n} + \ln\dfrac{d}{r}\right)$

$$= 0.0628\left(\frac{0.25}{1} + \ln\frac{0.9}{0.0062}\right)$$

$= 0.3283\ \Omega/\text{km}$

Total inductive reactance X_l $= 2.5 \times 0.3283$

$= 0.821\ \Omega$

Fault location A (primary side of the transformer)

Total resistance up to the fault point A $-R_A$ $= R_Q + R_l$

$= 0.03803 + 1.135$

$= 1.173\ \Omega$

Total reactance up to the fault point A $-X_A$ $= 0.3784 + 0.821$

$= 1.1994\ \Omega$

Short circuit up to the fault point A $-Z_A$ $= \sqrt{(1.173^2 + 1.1994^2)}$

$= 1.678\ \Omega$

Therefore,

Initial symmetrical short circuit current at point A $I''_A = \dfrac{cU_n}{\sqrt{3} \times Z_k}$

$$= \frac{1.1 \times 11{,}000}{\sqrt{3} \times 1.678}$$

$= 4163$ A

Peak short circuit current at point A i_{pA} $= k\sqrt{2} \times I''_k$

And $k = 1.02 + 0.98e^{\left(\frac{-3R_k}{X_k}\right)}$

$$= 1.02 + 0.98e^{\frac{-3R_k}{X_k}}$$

$$= 1.02 + 0.98e^{\frac{-3 \times 1.135}{1.173}}$$

$= 1.0738$

Therefore, $i_{pA} = 1.0738 \times \sqrt{2} \times 4163$

$= 6322$ A

Fault location B (secondary side of the transformer)

In order to calculate the total impedance up to the fault location, first the impedance value of the supply system and the line are converted to the value referred to the secondary side.

Resistance up to point A referred to the secondary side of the transformer

$$R'_A = 1.173\left(\frac{433}{11{,}000}\right)^2$$

$= 0.00182\ \Omega$

Similarly $X'_A = 0.464\left(\dfrac{433}{11{,}000}\right)^2$

$= 0.00071\ \Omega$

Impedance of the transformer Z_T $= \dfrac{u_{kr}}{100} \dfrac{U_{rT}^2}{S_{rT}}$

$$= \frac{4 \times 433^2}{100 \times 750000}$$

$$= 0.01\ \Omega$$

Power loss in the transformer P_T $= \frac{1.5 \times 750000}{100}$

$$= 11{,}250\ \text{W}$$

Full load current on the secondary side $= \frac{750000}{\sqrt{3} \times 433}$

$$= 1000\ \text{A}$$

Resistance of the transformer R_T $= \frac{P_{krT}}{3I_{rT}^2}$

$$= \frac{112500}{3 \times 1000^2}$$

$$= 0.00375\ \Omega$$

Reactance of the transformer X_T $= \sqrt{(0.01^2 - 0.00375^2)}$

$$= 0.00927\ \Omega$$

Total resistance up to the fault point B R_B $= 0.00182 + 0.00375$

$$= 0.00557\ \Omega$$

Total reactance up to the fault point B X_B $= 0.0018 + 0.00927$

$$= 0.01107\ \Omega$$

Total impedance up to the fault point B Z_B $= \sqrt{(0.00557^2 + 0.01107^2)}$

$$= 0.0124\ \Omega$$

Therefore

Initial symmetrical fault current at B $I_B'' = \frac{cU_n}{\sqrt{3} \times Z_B}$

$$= \frac{1.05 \times 433}{\sqrt{3} \times 0.0124}$$

(Voltage factor $c = 1.05$ for a voltage of 433 V)

$$= 21.167\ \text{kA}$$

Peak short circuit current at point A $i_{pB} = k\sqrt{2} \times I_k''$

And $k = 1.02 + 0.98e^{\left(\frac{-3R_k}{X_k}\right)}$

$= 1.02 + 0.98e^{\left(\frac{-3\times0.00557}{0.01107}\right)}$

$= 1.237$

$i_{pB} = 1.237 \times \sqrt{2 \times 21.167}$

$= 37.03$ kA

Summary of short circuit currents

Fault location	Initial symmetrical short circuit current (I_k'')	Peak short circuit current i_p
Point A	4.163 kA	6.322 kA
Point B	21.167 kA	37.03 kA

5.7.2 Earthing design

Earth electrode resistances are given by the equations:

Plate electrodes $R = \frac{\rho}{4}\sqrt{\frac{\pi}{A}}$ ohm

Pipe or rod electrodes $R = \frac{100\rho}{2\pi l}\ln\frac{2l}{d}$ ohm

Strip or conductor $R = \frac{100\rho}{2\pi l}\ln\frac{4l}{t}$ ohm

Where, ρ = soil resistivity in ohm.m

A = area of both sides of the plate electrode-m^2

l = length in m

d = diameter of pipe or rod in m

t = thickness of strip conductor or twice the diameter of circular conductor

The number of earth electrodes can be calculated based on the fault level on the H.T side of the consumer transformer.

Permissible current density of an electrode $= \dfrac{7.57 \times 10^3}{\sqrt{\rho t}} = I_d$

Where t = duration of fault (3s)

Assuming $\rho = 100$ ohm.m

$$I_d = \frac{7.57 \times 10^3}{\sqrt{100 \times 3}}$$

$$= 437 \text{ A/m}^2$$

Area of earth plates required $= \dfrac{I''_A}{I_d} \times 1000 \text{ m}^2$

$$= \frac{4.163}{437} \times 1000$$

$$= 9.526 \text{ mm}^2$$

Standard size of earth electrodes is $= 1.2 \times 1.2 \times 0.012$ m

Hence, the area of a plate electrode $= 2\,(1.2 \times 1.2)\text{ m}^2$

$$= 2.88 \text{ m}^2$$

Hence, number of earth electrodes required $= \dfrac{9.526}{2.88}$

$$= 3.36 \text{ say } 4$$

5.7.3 Determination of earthing conductor size

(Assume that copper is used as earthing conductor)

Permissible current density of copper = 118 A/mm^2

Area of earth conductor on H.T side $= \dfrac{I''_A}{118}$ mm^2

$$= \frac{4163}{118}$$

$$= 35.28 \text{ mm}^2$$

From Table 5.28, choose the nearest standard size on the higher side.
(Choose the nearest standard section, say 25 × 3 mm on the HT side)

$$\text{Area of earth conductor on the LT side} = \frac{I''_B}{118}\ \text{mm}^2$$

$$= \frac{21167}{118}$$

$$= 180\ \text{mm}^2$$

(Choose the nearest standard section such as 32 × 6 mm on the LT side)

Table 5.28 Standard sizes of rectangular copper conductors for earthing

Standard sizes of rectangular copper conductors for earthing	
25 × 3 mm	copper strip tinned at the points of contacts
25 × 6 mm	-do-
32 × 3 mm	-do-
32 × 6 mm	-do-
40 × 3 mm	-do-
40 × 6 mm	-do-
50 × 3 mm	-do-
50 × 6 mm	-do-
65 × 3 mm	-do-
65 × 6 mm	-do-

5.8 SELECTION OF SWITCHGEARS

The basic functions of switchgears are:

- Electrical protection
- Electrical isolation of a section of the installation and
- Local or remote switching

These functions are summarised in Table 5.29.

Table 5.29 Basic function of switchgears

Electrical protection against	**Isolation**	**Control**
Overload current Short circuits Insulation failure	Isolation clearly indicated by a fail proof mechanical indicator. A gap interposed insulating barrier between the open contacts, clearly visible.	Functional switching Emergency switching Emergency stopping Switching off for maintenance.

5.8.1 Electrical protection

The most important function of switchgear is the electrical protection. The important issues to be addressed while providing the electrical protection are:

- To provide protection of elements against the thermal and mechanical stresses of short circuit currents
- To provide protection to persons and livestock in the event of insulation failure and
- To provide protection of apparatus and appliances.

The protective function of a switchgear aims to avoid or to limit the dangerous or destructive consequences of excessive currents (short circuit), or those due to overloading and insulation failure. It also aims at separating the defective circuit from the rest of the installation. A clear distinction is made in defining the protective functions to the following.

- The circuit elements of the installation like cables, transformers etc.
- The persons and livestock and
- The equipment and appliances connected to the installation.

5.8.2 Protection of circuit elements

Circuit elements need to be protected from two different issues. One is the protection against the possibility of overloading. Overloading is a condition of drawing slightly more than the rated current from a healthy installation. If this situation is continued, it will generate excessive heat in the equipment or circuit elements and may finally cause damage to the equipment or circuit elements. The second possibility is a short circuit due to complete failure of the insulation between all the phases, between two phases or between a phase and the ground. This situation may give rise to a very large current to exist through a section of the installation. Protection in such cases is provided either by fuses or circuit breakers in the switchboards at the origin of the final circuit to which the load is connected.

5.8.3 Protection of persons and livestock

Protection of persons or livestock is essential when there is a possibility of insulation failure or a chance for direct contact with live parts of the installation. Protection is provided by fuses, circuit-breakers, residual current devices and/or by permanent monitoring of the insulation resistance, depending upon the type of system earthing of the installation (TN, TT or IT).

5.8.4 Protection of equipment

Protection of equipment is needed to protect them against overheating due to prolonged overloading, single phasing, stalling of the rotors etc. Thermal relays designed to match the characteristics of the equipment can be used for this purpose. Such a relay, if required, can also be used to protect the motor circuit cables against over loading. Short circuit protection is invariably provided by a fuse or circuit breaker.

5.8.5 Electrical isolation

Isolation is another important function of switchgear. The main purpose of an isolator is to separate a circuit or equipment from the remainder of the installation which is energised, in order that personnel may carry out work on the isolated circuit or equipment with perfect safety. All medium voltage installations must have means of isolation of its different components. In order to maintain continuity of service, it is preferable to provide isolation at the origin of each circuit and within a distance not exceeding 3 metres from the apparatus and in direct line of sight. The requirements to be fulfilled by an isolating device are:

- All poles of the circuit including the neutral pole must open. An exception to this case is when the neutral is also the protective earth conductor (as in the case of TN system with PEN conductor). In the case of neutral being the PEN conductor, the neutral pole should not be open when the isolator is opened. It is also necessary that when all the poles are opened, the neutral open after the phase poles are opened and the neutral pole closes before the phase poles are closed.
- All isolators must be provided with a mechanical locking system in the OPEN position with a key and padlock, in order to avoid unauthorised reclosure by inadvertent operations.
- Isolators must comply with National Standards concerning the clearances between contacts, creepage distance and over voltage withstand capability etc.
- Isolators must have fail proof visual indications showing that the contacts of the isolators are open. These indications may be visual when the device is designed to allow the contacts to be seen. It can also be a mechanical arrangement with the help of a OPEN/CLOSE indicator solidly connected to the operating mechanism. In this case, the construction of the isolator must be such that, in the eventuality that the contacts become welded together in the closed position, the indicator should never indicate that the isolator is OPEN.
- The leakage current with the isolator contacts in the open position shall not exceed 0.5 mA for a new device and 6.0 mA at the end of its useful life.
- Isolators shall withstand voltage surge of 1.2/50 μs impulse voltage of appropriate magnitude in the open position.

5.8.6 Switchgear control

The control function of switchgear allows system operating personnel to modify a loaded power distribution system at any point of the time during their operation, and at all levels of the installation. These include:

- Functional control
- Emergency switching operation and
- Maintenance operation.

5.8.6.1 Functional control

This includes all switching operations during normal conditions for energizing and de-energizing of a part of a system or installation, or an individual equipment. All switchgear for functional control must be located at the final load circuit or circuits. Making (closing) of the circuits must be clear and unambiguous. In order to provide maximum flexibility and continuity of operation, particularly when the switching device also includes the protective function (e.g. circuit breaker, switch fuse unit), it is preferable to include a switch at every level of the installation. In other words, there should be a switch at every outgoing circuit on all switchboards and sub-switchboards. The control operations may be manual or by means of a push button on the switch. They are designed for instantaneous operation (with no deliberate delay) and are basically omnipolar in nature, i.e., one break in each pole and one break in the neutral. The break in the neutral depends on the type of system earthing provided. All devices used as the main controller for the installation (main circuit breaker) as well as any device used for changeover to another source of supply, must be unipolar devices (with break on all poles including the neutral).

5.8.6.2 Emergency control

Emergency switching function is a built in facility to de-energise a live circuit or system which is or which could become dangerous to persons and property (e.g., electric shock hazard or fire). The important factors to be considered while providing emergency control functions are:

- The emergency control device or its means of operation such as a large red mushroom headed emergency stop button, must be located in a clearly recognisable and easily accessible location and in proximity to the location at which the danger could arise or be seen.
- A single action must guarantee complete switching off of all the live conductors. (A PEN conductor must not be opened in a power distribution system where the earthing system flowed is TN system).
- A "break glass" switching initiation device authorises to switch off the system in an emergency. But in unmanned locations, re-energising of the circuit must only be achieved by means of a key held by an authorised person. In systems where the control devices are energised through an auxiliary supply, it is essential that the auxiliary supply should be available until all final stoppage of the machines and equipment in the system.

5.8.6.3 Maintenance operations

This aspect of control function assures the stopping or disconnecting a piece of equipment or circuit for enabling maintenance work to be carried out, with the assurance that there is no possibility of inadvertently switching on the circuit or equipment, while the maintenance work is in progress. The shut down is carried out at the functional switching device with the help of a suitable safety lock and with a warning notice at the switching device to this effect.

5.8.7 Switching devices

There are different types of switching device in switchgears. These devices independently, in general, do not fulfil all the basic functions of switchgears, viz: protection, control and Isolation. Therefore, a combination of these components may be required to be used to form complete switchgear. An exception to this case is circuit-breaker which satisfies all the basic functions of the switchgear. The basic switching devices are:

- Isolators (switch disconnectors)
- Load break switch
- Contactors and discontactors
- Changeover switches (off load and load) and
- Automatic transfer switch (ATS).

5.8.7.1 Isolators (switch disconnectors)

This is the simplest form of a switching device. It consists of a manually operated ON/OFF switch, with provision for locking and provides a safe isolation of a circuit or equipment when kept locked in OFF position. An isolator is not designed to make or break a load current and hence no values are specified for breaking or making capacity of isolators. They are not supposed to be operated when the circuit is loaded. However, it must be capable of withstanding the passage of short circuit currents and is assigned a rated short circuit time (usually 1 second) withstand capacity, unless otherwise specified by the manufacturer. This short time rating is more than sufficient for sustaining periods of operation under overloaded conditions, as in the case of motor starting operation etc. The basic specifications include, the rated current, mechanical switching rate, over-voltage withstand capacity and leakage current. Figure 5.35 shows the schematic representation of an isolator.

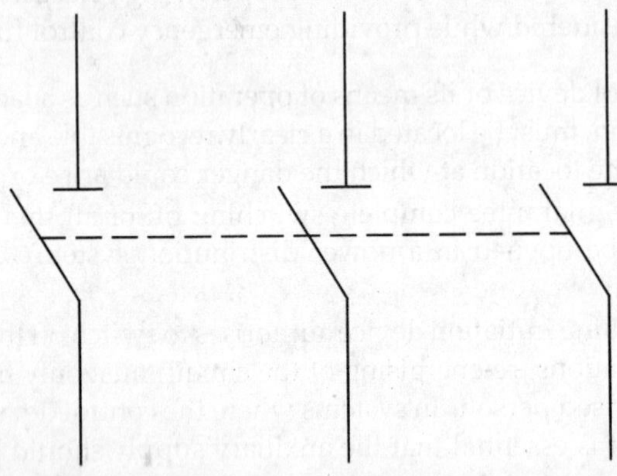

Figure 5.35 Symbol of an isolator (switch disconnector)

5.8.7.2 Load break switches

The load break switch is functionally identical to isolators and is designed for operation manually. Sometimes it is provided with electrically operated tripping mechanism for the convenience of operation. It is again a non-automatic ON/OFF switching device. The basic difference is that load break switches are designed for operation on loaded circuits, under normal operating conditions. However, load break switches do not provide any form of protection to the circuits or to the equipment. The basic specifications (as per IEC standard IEC-60947-3) include:

- The frequency of switching operations (600 open/close operations per hour)
- Mechanical and electrical endurance capacity
- Making and breaking current ratings for normal and infrequent operations.

Table 5.30 Utilisation categories of load break switches (IEC-60947-3)

Utilisation category		Typical applications	cos φ	Making current	Breaking current
Frequent operation	**Infrequent operation**				
AC-20A	AC-20B	Connecting and disconnecting under no load conditions	–	–	–
AC-21A	AC-21B	Switching of resistive loads with moderate overloads	0.95	$1.5I_n$	$1.5I_n$
AC-22A	AC-22B	Switching of mixed resistive and inductive loads with moderate overloads	0.65	$3I_n$	$3I_n$
AC-23A	AC-23B	Switching of motor loads or other highly inductive loads	0.45 for I<100 A 0.35 for I>100 A	$10I_n$	$8I_n$

The problem associated with closing of a load break switch to energise a circuit is that there is a possibility of an unsuspected short circuit in the circuit. For this reason, the load break switch is assigned a fault current making rating. In other words, the device will be guaranteed for successful closing against the electromechanical forces due to a short circuit in the circuit. Such a load break switch is referred to as "Fault make load break switch". The protective function however is provided by an upstream device built into the system.

The load break switches are categorised according to the type of utilisation. Table 5.29 shows the different categories of load break switches.

The graphic symbol of a load break switch is shown in Figure 5.36.

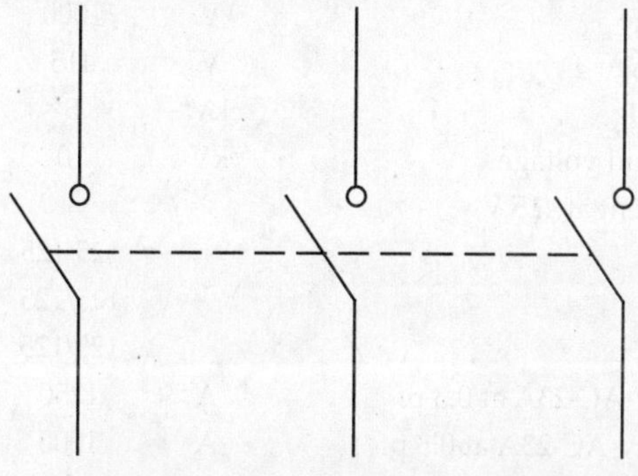

Figure 5.36 Symbol of a load break switch

EXAMPLE 5.10

Calculate the breaking and making current capacities of a load break switch rated for 63 A and coming under AC-23A utilisation category.

Solution:
From Table 5.30,

Breaking current $= 8I_n$

$= 8 \times 63 = 504$ A

Making current $= 10I_n$

$= 10 \times 63 = 603$ A

Typical specifications for load break switches are shown in Table 5.31.

5.8.7.3 Contactors

A contactor is a solenoid operated switch which is held closed by a current through a solenoid coil. Contactors are designed to carry out numerous open/close operations are commonly controlled locally or remotely byON/OFF push buttons. The basic properties of contactors as per IEC standard 60947-4-1 are:

- The operating duration: (e.g., 8 hours, uninterrupted, intermittent, temporary of 3, 10, 30, 60 and 90 minutes).
- Utilisation category: (e.g., AC3 for starting and stopping of induction motors)
- Start-stop cycle: (e.g., 1 to 1200 cycle per hours)
- Mechanical endurance: number of OFF load operations.
- Electrical endurance: number of ON load
- Rated making and breaking performance.

Table 5.31 Typical specifications for load break switches

Description	Unit	Ratings		
Rated current at 40 °C	A	125	160	200
Rated insulation voltage	V	1000	1000	1000
Rated operational voltage	V	415	415	415
Dielectric strength 50 Hz	kV	5	5	5
Rated impulse withstand voltage	kV	10	10	10
Rated operational current at 415 V				
AC-21A/B		125/125	125/125	125/125
AC-22A/B		125/125	125/125	125/125
AC-23A/B		125/125	125/125	125/125
Rated making capacity-AC-23A at 0.3 pf	A	1250	1600	1600
Rated breaking capacity-AC-23A at 0.3 pf	A	1000	1280	1280
Rated motor power	kW	55	55	100
Fuse operated short circuit withstand				
Rated conditional short circuit current	kA_{rms}	80	80	80
Max allowed cut-off current	kA_{peak}	17	18	22
Rated short time withstand current (1sec)	kA_{rms}	7.5	7.5	7.5
Electrical durability-Operating cycles-AC-23A/B		3000/600	3000/600	3000/600
Mechanical durability-operating cycles		30000	30000	30000
Temperature range	°C	–5 to 50	–5 to 50	–5 to 50

(Courtsey: Havell's India Ltd.)

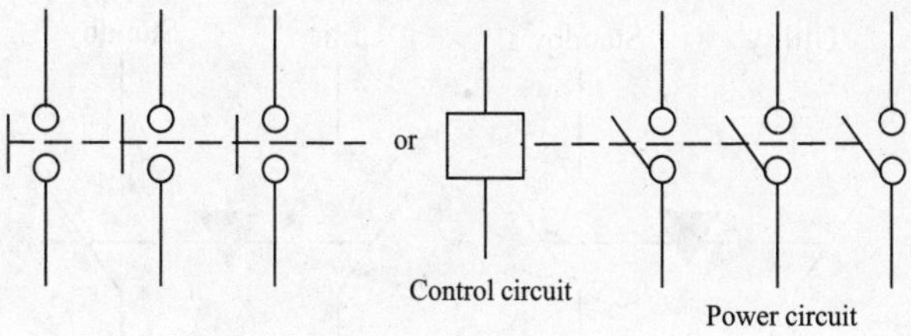

Figure 5.37 Symbol of a contactor

The graphic symbol of a contactor is shown in Figure 5.37.

Discontactors

A contactor equipped with a thermal relay for protection against over load current is defined as a discontactor. Discontactors are essential elements of direct on line starters (DOL). A discontactor is not a circuit breaker as the short circuit current breaking capacity is limited to 8 to 10 times I_n only. Therefore, in order to assure short circuit protection, a fuse or circuit breaker must be used in series with a discontactor, on the upstream side. The graphic symbol of a discontactor is shown in Figure 5.38.

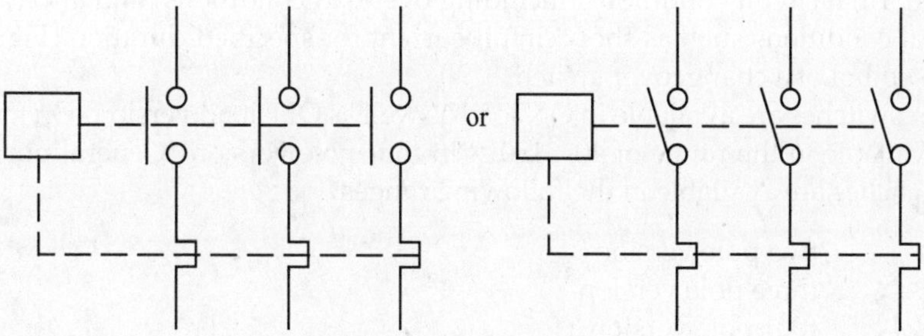

Figure 5.38 Symbol of a discontactor

5.8.7.4 Changeover switches

The availability of continuous power supply and its reliability is the life of any industry. Modern systems are power dependent and complex in operation. Therefore continuous information and communications are needed to control automatic processes, be in industries, commercial complexes, hospitals or other places. Therefore the need for independent standby power system has increased manifold. Standby generator systems are required to cater to:

- **Sensitive loads:** e.g., computer loads, hospital equipments, microprocessor controlled machine etc. These loads are generally supplied by UPS systems. The period of non-availability of utility supply, before the standby generator takes over is bridged by battery backup.
- **Critical loads:** mostly inverter standby generator system to supply power to lightings, air-conditioning, lifts etc., in hospitals, air ports, commercial complexes etc.
- **Essential loads:** Mostly used in process industries as they relate to high restarting time in the event of power failure. They also include common service providing devices in high rise buildings, hospitals, hotels, etc.

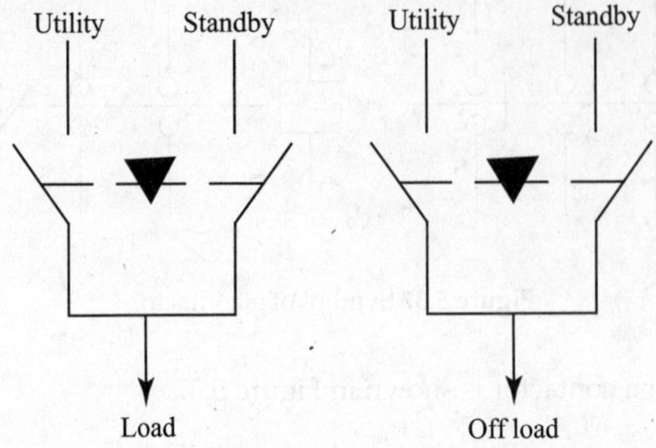

Figure 5.39 Symbol of changeover switches

Changeover switches find a wide application in situations when the reliability of electrical supply from the utility is low and are used in lighting/industrial applications wherever continuity of supply is necessary, for switching on from utility to an alternate source and vice-versa. They are switch disconnectors with independent manual operation capable of making, carrying and breaking current under normal circuit conditions (including overload conditions) and also carrying current under abnormal conditions such as short circuit current for specified duration. Figure 5.39 shows the schematic symbol of a changeover switch.

Changeover switches are available in ON load as well as OFF load version. On load changeover switches are available in the range of 40–3150 A in four-pole version. General purpose OFF load change over switches are available in the following ranges:

32 A	two pole version
32–100 A	three pole version
16–2000 A	four pole version

Switches up to 100 A can also be used as ON load change over switches. Table 5.32 shows typical specifications of load changeover switches.

5.8.7.5 Automatic transfer switches (ATS)

Automatic transfer of power from main supply to standby supply is vital for sensitive, critical or essential loads. In the event of power failure, standby power is expected to take over automatically. In order to achieve the automatic transfer, electrical starting equipment and battery backup are required in addition to the standby generator. The integration of standby generator with electrical starting and battery backup is achieved by using automatic mains failure (AMF) system. ATC is self acting equipment containing the transfer switching devices and other accessories for monitoring the supply sources and for transferring load circuit from one supply to another. The operating sequence consists of an automatic transfer of a load from the normal supply to an alternate supply in the event of a monitored supply deviation and automatically returning the load to the normal supply when it is restored. This transfer may be with or without a pre-determined time delay and may include an OFF position. In case both normal supply and alternate supply are available; the ATS will assume the normal supply position when in the preferred supply. The utilisation categories of ATC as per IEC 9476-1 are shown in Table 5.33.

Table 5.32 Typical specifications for changeover switches

Current rating at 40 °C	**A**	**125**	**160**	**200**
Number of poles		4	4	4
Rated insulation voltage	V	1000	1000	1000
Rated operational voltage	V	415	415	415
Rated impulse withstand voltage	kV	10	10	10
Rated operational current at 415V				
AC-21A/B	A	125/125	160/160	200/200
AC-22A/B	A	125/125	160/160	200/200
AC-23A/B	A	125/125	160/160	200/200
Rated making capacity	A	1250	1600	1600
Rated breaking capacity	A	1000	1280	1280
Fuse protected short circuit withstand				
Rated max current of gG fuse	A	125	160	200
Rated conditional short circuit current	kA	80	80	80
Max allowed cut off current	kA	17	18	22
Electrical durability				
No. of operating cycles AC-23A/B		3000/600	3000/600	3000/600
Mechanical durability				
No. of no load operating cycles		30000	30000	30000
Terminal connection cable size (Al)	mm^2	70	95	150
Temperature withstand range	°C	–5 to 50	–5 to 50	–5 to 50

(Courtesy: Havell's India Ltd.)

Table 5.33 Utilisation categories of automatic transfer switches (IEC 947-6-1)

Supply	**Utilisation category**		**Applications**
	Frequent operation	**Infrequent operation**	
AC	AC-31A	AC-31B	Non-inductive or slightly ind. loads
	AC-33A	AC-33B	Motor loads or mixed loads
	AC-35A	AC-35B	Electric discharge lamps
	AC-36A	AC-36B	Incandescent lamps.
DC	DC-31A	DC-31B	Resistive load,
	DC-33A	DC-33B	Motor loads or mixed loads,
	DC-36A	DC-36B	Incandescent lamp load.

The schematic symbol of automatic switch is shown in Figure 5.40.

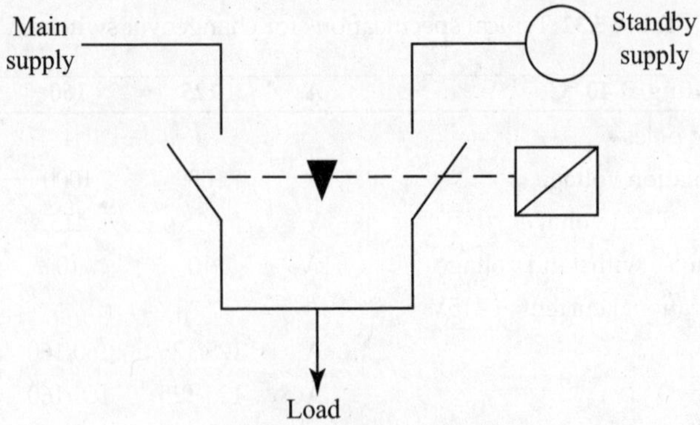

Figure 5.40 Symbol of ATC

5.8.8 Fuses

Fuses are the simplest form of a protective element in an electrical system. Fuses break a circuit by controlled melting of the fuse element when the current exceeds a given value for a corresponding period of time. Each type of fuse has a current/time relationship presented in the form of a performance characteristic. Figure 5.41 shows the current-time characteristics of fuses.

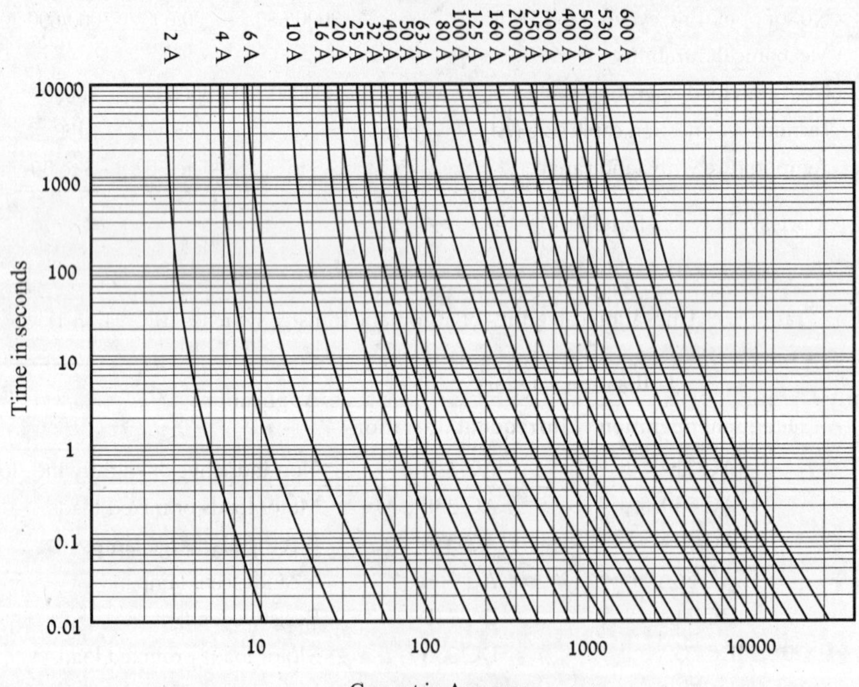

Figure 5.41 Time-current characteristics in fuses

There are two types of fuses, e.g., rewirable fuses and high rupturing capacity fuse (HRC). The rewirable fuses were intended for use only on domestic application, and are being phased out with the modern gadgets like miniature circuit breakers (MCB). The advantages of the rewirable fuses are that they are very cheap and easy to replace. The main disadvantages are that they have a low breaking capacity and are likely to cause fire hazard. Rewirable fuses are not recommended for industrial applications.

HRC fuses are manufactured with and without "fuse blown" mechanical indicators. They have very high breaking capacity (80 kA at 415 V) and are suitable for both AC/DC operations. They have very low let through energy resulting in reduced electromagnetic stresses and they provide very reliable short circuit protection. They are available in the following ranges:

- 2-630 A in bolted type (BS type)
- 6-630 A in blade contact (BIN type)
- 4-63 A in round head designs.

They conform to IEC standard 601 269 and IS-13703 part I and II.

Standards define the classes of fuses as:

- Those intended for installations in the form of a cartridge for rated current up to 100 A and designated by the symbol gG.
- Those for industrial application, with cartridge type designated as gG (general use), gM and aM (for motor circuits).

The main difference between the domestic and industrial fuses is that terminal voltage, current loads and fault current breaking capacities are higher in industrial fuses. Type gG fuse can also be used for motor circuits provided their characteristic are capable to withstand the motor starting current without deterioration in performance.

5.8.8.1 Fuse characteristics

The conditions of fusing of a fuse are defined by the standards. (Type gG fuses are designed to provide protection against overloads and short circuits).

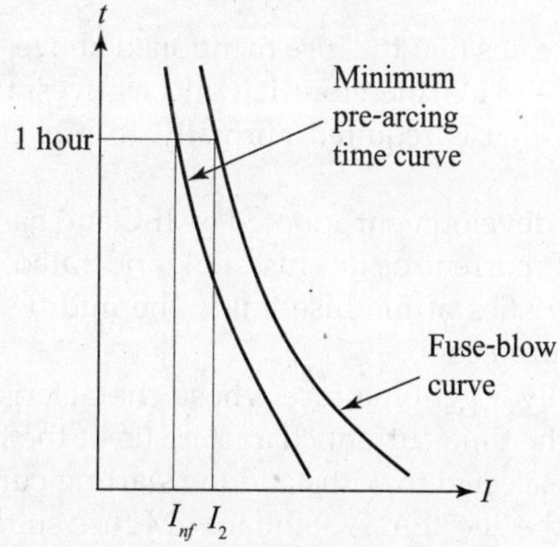

Figure 5.42 Fusing characteristics

The fusing characteristics are shown in Figure 5.42. The conventional non-fusing current I_{nf} is the value of current that the fuse element can carry for a specified time without melting. The conventional fusing current I_f is the value of current which will cause melting of the fuse element before the expiry of the specified time limit. Table 5.34 shows the fusing zones of gG and gM class fuses.

Table 5.34 Fusing and non-fusing zones for medium voltage gG and gM fuses (IEC 60269)

Rated current I_n	Non-fusing current I_{nf}	Fusing current I_f	Conventional time (hours)
$I_n \leq 4A$	1.5 I_n	2.1 I_n	1
$4<I_n \leq 16$ A	1.5 I_n	1.9 I_n	1
$16<I_n 63$ A	1.25 I_n	1.6 I_n	1
$63<I_n 160$ A	1.25 I_n	1.6 I_n	2
$160<I_n 400$ A	1.25 I_n	1.6 I_n	3
$400<I_n$	1.25 I_n	1.6 I_n	4

EXAMPLE 5.11

Calculate the non-fusing current and fusing current of a 63A HRC fuse.

Solution:

From table

Non-fusing current $= 1.25 I_n$

$= 1.25 \times 63 = 78.75$ A

The fuse must not melt in less than one hour when carrying 78.75 A

Fusing current $= 1.6\ I_n$

$= 1.6 \times 63$ A $= 100.8$ A

The fuse must melt within one hour when carrying 100.8 A.

The above example clearly indicates that the fuse mentioned above, provides very poor protection in the low overload range (63–78 A) therefore it will be necessary to install a cable of larger current carrying capacity than that is required normally, to prevent possible damage from long term overloading.

The gM type fuse is a recent development adopted by IEC and has a dual rating. It is represented by (I_n M I_{ch}). I_n denotes rated current of the fuse link and rated current of the fuse holder, I_{ch} corresponds to the G characteristics of the fuse link. The middle letter represents that they are intended to be used for motor protection.

A gM type fuse is essentially a gG type fuse whose characteristics correspond to the current value I_{ch}. (e.g. 63 A). So that the time current characteristic of the fuse is identical to that of 63 A gM fuse. This current value is selected to withstand the starting current of the motor; whereas the normal operating current may be 20–30 A. A standard gM fuse suitable for this purpose would be designated as 32M63, the first current rating I_n corresponds to steady state thermal performance of the fuse link (allows smaller size in construction) and second current I_ch relates to its (short time) starting current performance. Even through short circuit protection is provided, overload protection is not provided and so a separate thermal relay is always necessary along with gM fuse. The only advantage of a gM fuse is the reduced physical size and lower cost.

The aM class fuses provide protection against short circuit currents only and therefore must be used in conjunction with other switchgear to ensure overload protection. They will not provide problem for currents less than 4 I_n. Therefore conventional fusing and non-fusing currents are not specified for this class of fuse.

5.8.8.2 Rated short circuit breaking current

The modern cartridge fuse offers very rapid fusion in case of high short circuit current levels. These fuses are basically current limiting in design and hence short circuit currents cannot rise to the full prospective value owing to this very short clearing time. This limiting of the current significantly reduces the thermal and dynamic stresses in the switchgear, thereby minimising the danger and damage at the fault position. The rated short circuit breaking current is therefore based on the rms value of AC component of the prospective fault current. No making current specification is provided for this type of fuses. Table 5.35 shows the typical specifications for the type gG fuses in different configurations.

Table 5.35 Typical specifications of type gG fuses

Specifications	Type of fuse		
	Bolted connection (BS)	**Blade contact (DIN)**	**Cylindrical cap (RH)**
Rated voltage	415 V	415 V	415 V
Rated current	2 A to 630 A	6 A to 630 A	4 A to 63 A
Rated frequency	50 Hz	50 Hz	50 Hz
Breaking capacity	80 kA	80 kA	80 kA
Utilization category	gG	gG	gG
Ambient temperature range:			
70 °C	2 A to 32 A	6 A to 32 A	4 A to 63 A
65 °C	35 A to 80 A	35 A to 80 A	
60 °C		100A to 630A	
Non-fusing current	$1.25I_n$	$1.25I_n$	$1.25I_n$
Fusing current	$1.6I_n$	$1.6I_n$	$1.6I_n$
Material of body	Steatite ceramic	Steatite ceramic	Steatite ceramic
Material of fusing element	Copper with tin solder	Copper with tin solder	Copper with tin solder
Material of filler	Silica quartz	Silica quartz	Silica quartz
Fuse blown indicator	-	Provided	-

(Source: Havells India Ltd.)

5.8.8.3 Current limiting property of fuses

Modern electrical distribution systems have very high short circuit levels. However, many components used in the distribution systems have relatively low short circuit withstand capability. If the components are not capable of handling the very high short circuit levels, they could easily be damaged. A characteristic feature of the modern cartridge fuses is that, owing to the rapidity of fusion, in the case of high short circuit current levels, a current cut-off begins before the occurrence of the first major peak current. This ensures that fault current never reaches its prospective peak value. Figure 5.43 shows the current limiting action of a fuse.

This feature of current limiting action reduces significantly the thermal and dynamic stresses which would otherwise exist, thereby minimising the danger and damage at the fault position. The rated short circuit breaking current of the fuse is therefore based on the rms value of the ac component of the prospective fault current. No short circuit making capacity rating is specified for fuses.

The degree of current limiting of a given type and size of fuse depends upon the available short circuit current that can be delivered by the electrical distribution system. The current limiting feature is best described by fuse let through charts. Fuse let through charts are prepared from actual test data. Each type or class of fuse has its own family of let through charts. The let through data is generated by actual short circuit tests on current limiting fuses. The circuit parameters that affect the let through characteristics are:

- Short circuit power factor
- Short circuit closing angle
- Applied voltage

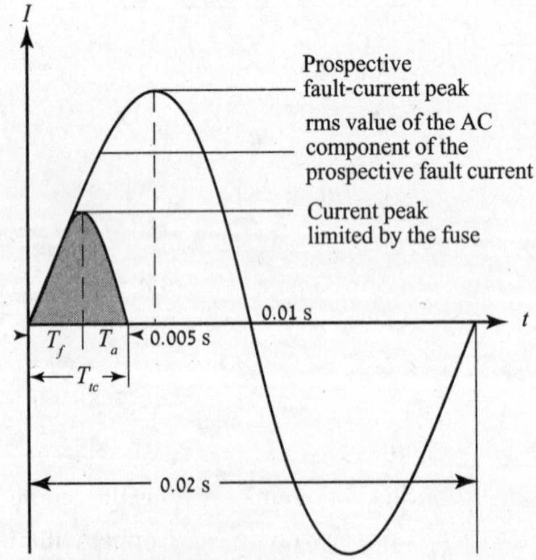

T_f = Fuse pre-arc fusing time
T_a = Arcing time
T_{tc} = Total fault-clearance time

Figure 5.43 Current limiting feature of fuses

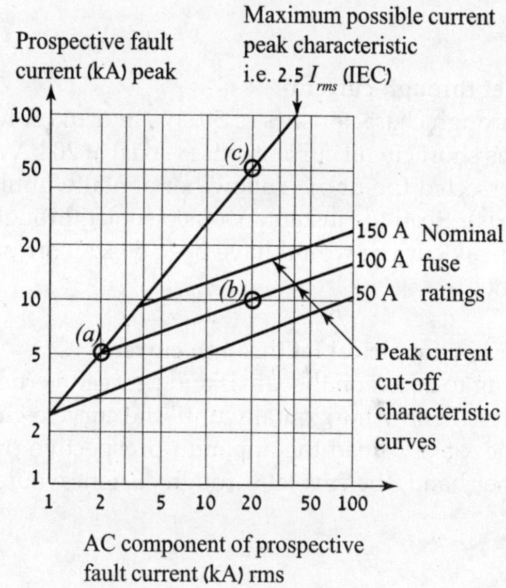

Figure 5.44 Fuse let through curves

Let through curves of current limiting fuses are generated under worst case conditions, based on the three variable parameters mentioned above. Under actual field conditions, changing any one or combinations of the three will result in lower let through currents. A typical let through characteristic is shown in Figure 5.44.

The current limiting fuse let through characteristics are plotted between ac component of prospective symmetrical rms fault current and prospective instantaneous peak value of the fault current. The use of let through chart is explained with the help of a numerical example.

EXAMPLE 5.12

Consider a power distribution system as shown in Figure 5.45. If the rms symmetrical prospective short circuit current at the supply end is 20 kA, find

1. Peak value of the let through current
2. Apparent prospective rms symmetrical let through current
3. Clearing time

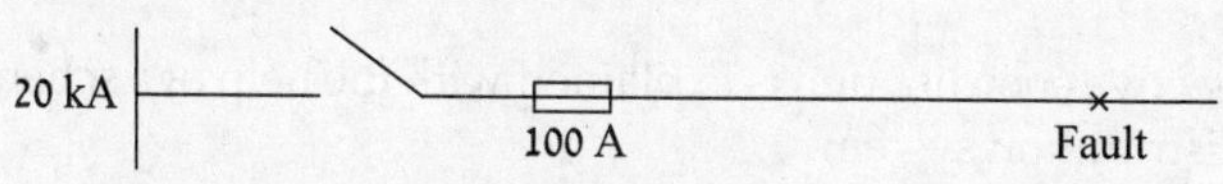

Figure 5.45 Power distribution system for example 5.12

Solution:

1) Determination of peak let through current

Figure 5.44 shows the let through characteristics of 50, 100 and 160 A fuses. Enter the chart on the prospective symmetrical rms short circuit current axis (x-axis) at 20 kA and proceed vertically until the 100 A fuse curve is intersected (point b). Follow horizontally until the prospective peak fault current axis is reached (y-axis). Read the instantaneous peak let through current as 10 kA. If a fuse has not been installed, the peak current would have been 50 kA corresponding to the intersection of the characteristics at point c.

2) Apparent prospective rms symmetrical let through current

Enter the chart corresponding to 20 kA on the x-axis and proceed vertically until the line intersects the 100 A fuse characteristics. Follow horizontally until the curve a-c is intersected. Proceed vertically downwards on to the x-axis. Read the apparent prospective rms symmetrical let through current as 4 kA. On the other hand, the rms let through current would be 20 kA if there were no fuse in the circuit.

3) Clearing time

If the rms symmetrical short circuit current is greater than the point at which the fuse characteristics intersect the diagonal line a-c, then the fuse clearing time will be half cycle or less. In this case, the intersection is at 2 kA. So for short circuit currents above 2 kA, the fuse is current limiting in nature.

5.8.9 Selective coordination (discrimination)

One of the important features of any industry is the electrical distribution systems. A power failure can stop all activities, paralyze production, create inconvenience and discomfort people and possibly cause a panic. Selective Coordination is an aspect of electrical system design critical for maintaining the reliability of the distribution network.

Selective coordination by definition is 'the localisation of an over current condition to restrict the outage of the circuit or equipment affected, accomplished by the choice of over current protective devices and their ratings or settings.'

It may be noted that the type of over current protective devices and their settings determine whether a system is selectively coordinated or not. A properly designed electrical system will allow only the nearest up stream over current protective device to open for both over loads and all types of short circuits, leaving the remainder of the system unaffected and thereby preserving the continuity of service. Isolation of a faulted part of a distribution system is very critical in modern electrical systems.

The principle of selective coordination is explained with the help of a schematic diagram. Figure 5.46 shows a typical distribution system.

If a fault on the load side opens all the over current devices, A, B, C and D, the system is not selectively coordinated. (Figure 5.46a) and there will be unnecessary power loss to the remaining parts of the system. Such a situation is also referred to as cascading effect or lack of coordination. The same system is shown in figure 5.46b. If only the over current device at D opens for a fault on the load side and the remaining over current devices are unaffected, the system is said to be selectively coordinated. Only the circuit with the fault is removed from the system and the remaining parts of the installations are unaffected.

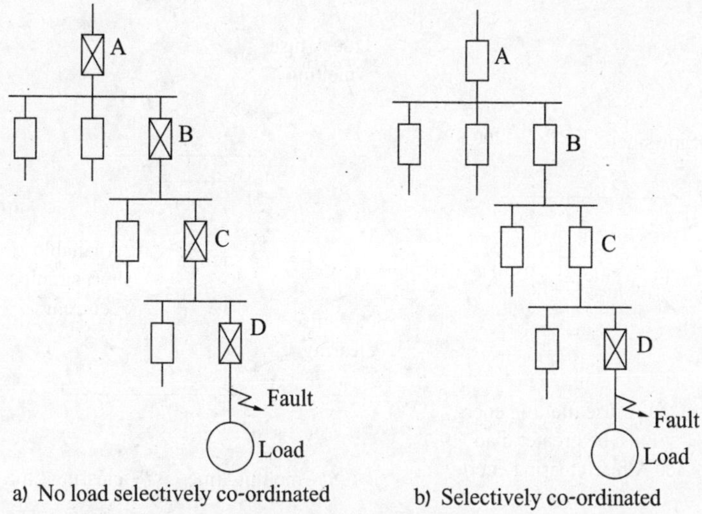

Figure 5.46 Selective coordination of fuses.

Selective coordination can be performed by any one of the following methods.

1. For systems using fuses (medium voltage systems below 600), use of the fuse selectivity ratio charts published by the manufacturers. These ratios apply to all over current conditions including short circuits. Using the fuse selectivity ratio is easy and quick and there is no need to refer to time-current characteristics of the fuses.
2. Use of computer software allowing us to select the time-current characteristics published by the manufacturers and placing it on a single graph. These graphs are then analysed and interpreted properly in relation to the possible fault currents at various points in the installation.
3. Overlays of the time-current characteristics with the manufacturer's data are hand traced on a log-log sheet and analysed and interpreted for coordination.

Most of the circuit breaker manufacturers provide tested coordination tables that may be used instead of methods referred to in 2 and 3.

5.8.9.1 Selective coordination with fuses

Determination of selectivity between fuses is relatively simple. Selectivity between fuses exists when the total clearing energy of the load side fuse (downstream) is less than the melting energy of the line side (upstream) fuse. Figure 5.47 shows the principle of selective coordination between fuses.

When a fuse is in its current limiting range, the fuse will clear the fault in approximately half cycle or less. If T_m refers the melting time and T_c, the total clearing time, the area under the time current characteristics is indicative of the energy let through by the fuse. The thermal energy is proportional to I^2t. The amount of energy released during the melting of the fuse is called melting energy and the energy produced during the entire interruption process is called the total clearing

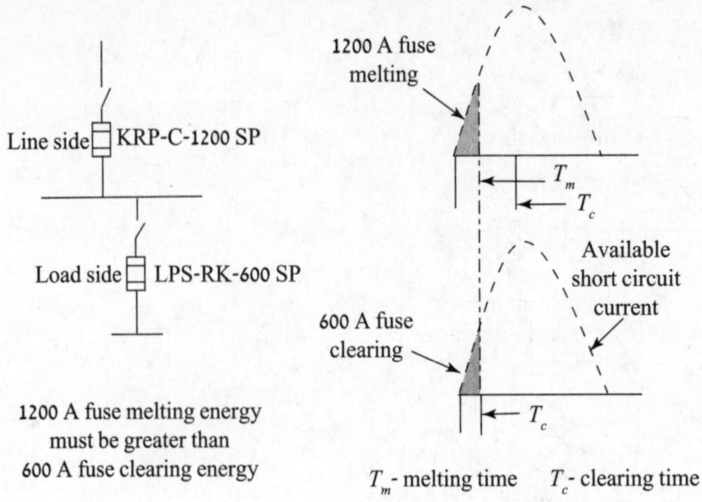

Figure 5.47 Example of selective coordination.

energy. For proper coordination, the T_c and clearing energy of the down stream fuse must be lower than the T_m and melting energy of the upstream fuse.

Figure 5.48 shows the energy let through vs current for various fuse ratings. The bottom of the shaded box represents the pre-arc energy and the top of the shaded box represents the total clearing energy. Selective coordination can be obtained by checking whether the total let through energy (clearing) of the down stream fuse is less than the pre arcing energy of the upstream fuse.

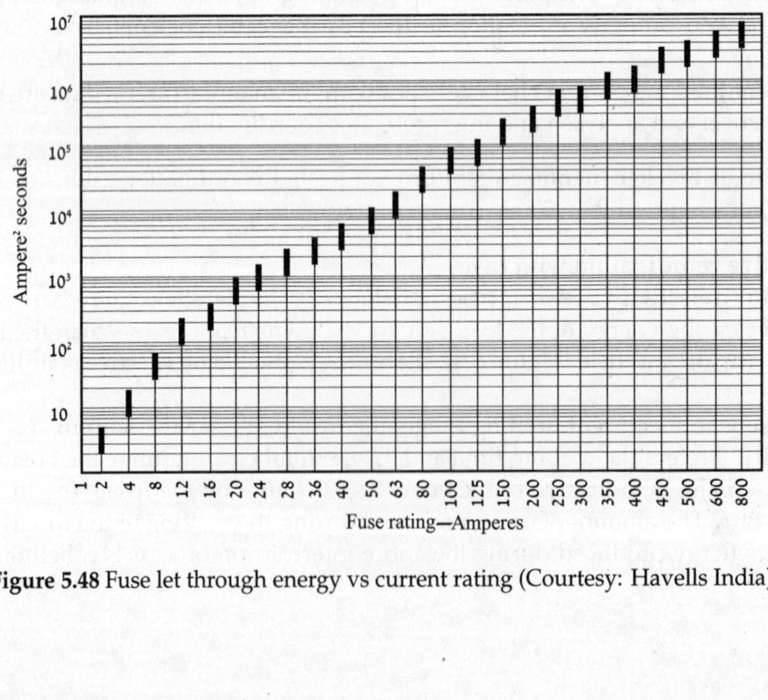

Figure 5.48 Fuse let through energy vs current rating (Courtesy: Havells India).

EXAMPLE 5.13

Examine whether selective coordination exists between the following fuses
1. 250 A and 200 A
2. 100 A and 80 A
3. 400 A and 250 A

Solution:

With reference to Figure 5.48,

1. The total clearing energy of the 200 A fuse is not less than the pre-arc energy of the 250 A fuse. Hence these fuses are not selectively coordinated.
2. The total clearing energy of the 80 A fuse is not less than the pre-arc energy of the 100 A fuse. Hence these fuses are not selectively coordinated.
3. The total clearing energy of the 250 A fuse is less than the pre-arc energy of the 400 A fuse. Hence these fuses are selectively coordinated.

With modern current limiting fuses, selective coordination can be achieved by adhering to fuse selectivity ratio charts provided by the manufacturers. It is neither necessary to plot the time-current characteristics nor to compute the short circuit current levels. Maintaining the ratios as provided in the fuse selectivity guide will guarantee selective coordination.

5.8.10 Switchgear elements in combinations

Switchgear elements like isolator load break switch contactors or fuse do not provide all the basic protection functions like protection, control and isolation. Circuit breaker is the only switchgear element that can offer all the above functions. When the use of a circuit breaker is inappropriate due to the high rate of switching or switching over extended period, it is recommended to use a combination of base switching gear elements. The most commonly used combinations are:

- Switch and fuse combinations
- Fuse disconnector + discontactor
- Fuse switch disconnector + discontactor.

5.8.10.1 Switch and fuse combinations

There are two basic types of combinations of switch and fuse. In one case, a non-automatic switch is associated with a set of fuse in a common enclosure. Such a combination can be either a 'switch-fuse' or 'fuse-switch' unit. A switch fuse unit generally consists of a switch with double break per pole on the upstream side of the fixed fuse bases. The fuse carriers are inserted into the fixed fuse bases. A fuse switch consists of switch blades for each pole with a double break per pole. The switch blade carries the fuse unit and when the switch unit is in the OFF position the switch blade carrying the fuse are isolated from the upstream as well as the downstream connections. The symbols of switch fuse and fuse switch units are shown in Figure 5.49.

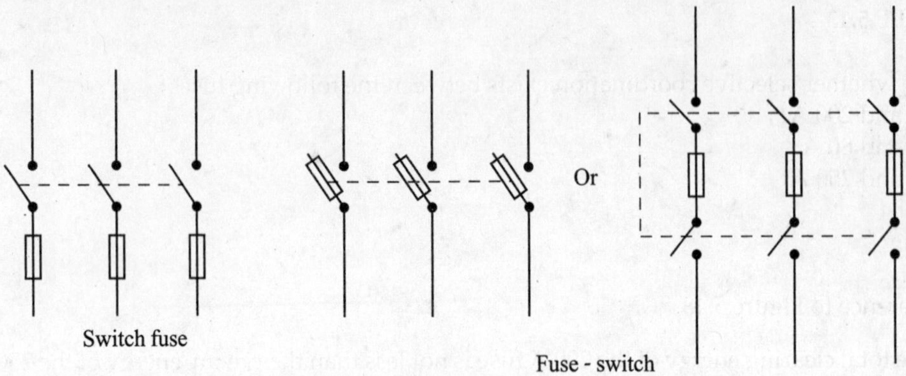

Figure 5.49 Symbols of switch fuse and fuse-switch unit

The maximum ratings of fuse devices are generally limited to 100 A for the three-phase 415 V systems. Their principal use is in domestic and low power applications.

5.8.10.2 Fuse – disconnector + discontactor

Since a discontactor does not provide short circuit protection, it is necessary to add short circuit protection by adding a set of fuse along with discontactor. This combination is mainly used in motor circuits, where the disconnecter allows safe operations for changing the fuse and to work on the down stream side of the disconnecter. The fuse-disconnecter must be interlocked with the discontactor so that no operations (opening or closing) of the disconnector are possible unless the discontactor is in the OFF position. This is becomes, the fuse disconnector has no load switching capacity. Figure 5.50 shows the graphic symbols of fuse disconnecter + discontactor.

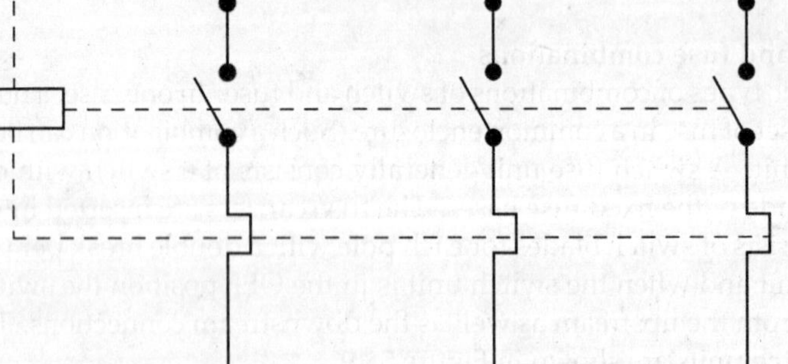

Figure 5.50 Fuse disconnector + discontactor

5.8.10.3 Fuse switch + disconnector + discontactor

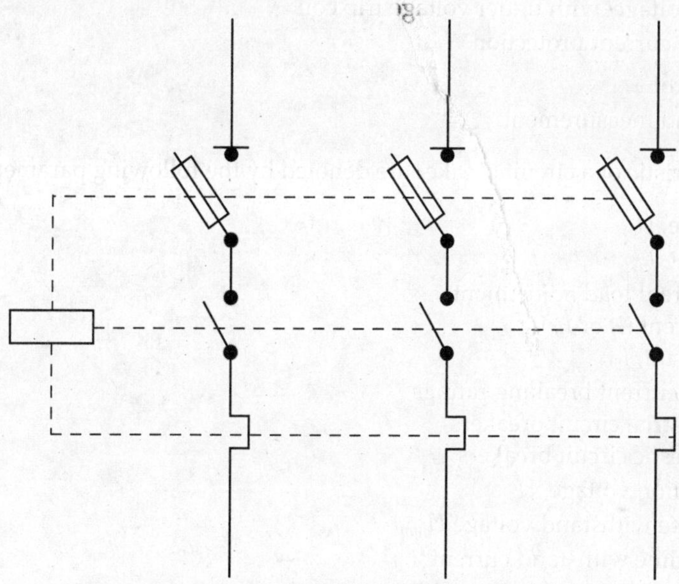

Figure 5.51 Fuse switch disconnector + discontactor

The main difference between fuse disconnector + discontactor and fuse switch disconnector + discontactor is that the second combination does not require an interlocking between the fuse switch disconnector and discontactor since the fuse switch disconnectors are capable of switching operations, under loaded conditions. The only requirement is that the switches shall be of AC-22 or AC-23 duty type. If it is used in motor applications. Figure 5.51 shows the graphic symbol.

5.8.11 Circuit breakers

Circuit breaker is the only switchgear component capable of simultaneously satisfying all the basic function necessary in all electrical installations. It can also, by means of auxiliary units, provide a wide range of functions such as, indications (ON-OFF-TRIPPED), under voltage tripping, remote operations, diagnosis, communication, monitoring etc. These features make a circuit breaker the most essential unit of any electrical installation. The function performed by a circuit breaker can be summarised as follows:

- Isolation
- Control
    1. functional
    2. emergency switching
    3. switching for maintenance
- Protection
    1. overload
    2. short circuit

3. insulation failure–with differential relay
4. under voltage–with under voltage trip coil
5. residual current protection

- Remote control
- Indication and measurement.

The basic characteristic of a circuit breaker are denoted by the following parameters.

- Rated voltage, U_e
- Rated current, I_n
- Tripping current load adjustment
  For over current (I_r or I_{th})
  For short circuit, I_m
- Short circuit current breaking ratings
  I_{cu} - for industrial circuit breakers
  I_{cn} - for domestic circuit breakers
- Rated insulation voltage, U_i
- Rated impulse withstand voltage U_{imp}
- Rated short time withstand current, I_{cw}
- Rated making capacity, I_{cm}
- Rated service short circuit-breaking capacity, I_{cs}.

5.8.11.1 Rated operational voltage, U_e
This is the voltage at which the circuit-breaker is designed to operate under normal conditions.

5.8.11.2 Rated current, I_n
This is the maximum value of current that a circuit-breaker can carry indefinitely at ambient temperature as stated by the manufacturer, without exceeding the specified temperature limits of the current carrying parts. It is to be understood that the CB is provided with a specified over current tripping relay mechanisms. I_n specified for the circuit-breakers is equal to the uninterrupted current rating of the switchgears, I_u. The rated current I_n for a circuit-breaker is specified for an ambient temperature of 40 °C. The same circuit-breaker can be used at higher values of ambient temperature if its rating is suitably 'derated'. The use of electronic thermal tripping unit permits the circuit-breaker to be de rated over wide range. For example a 250 A circuit-breaker with electronic tripping relay can be de rated to operate in the range of 0.64 I_n to I_n. If the circuit-breaker is called upon to function at an ambient temperature of 50 °C, the circuit-breaker can be de rated to 93.6% of its original rating, i.e., 234 A, without exceeding the specified temperature limits. If a circuit-breaker is fitted with over current unit with different current level settings, the rating assigned to the circuit-breaker will be the highest current setting of the tripping unit. This rating also is called the frame size rating of the circuit-breaker.

5.8.11.3 Overload trip-current setting (I_r or I_{rth})
Industrial circuit-breakers in general are equipped with exchangeable over current trip relays. This is essential, because of the need to adopt a circuit-breaker to the exact requirements of the circuit which the circuit-breaker is controlling and to avoid the need to use oversized cables. The trip current setting of the circuit-breaker is the current above which the circuit-breaker will trip. It

also represents the maximum current the circuit-breaker can carry without tripping. Then value of current must be greater than the maximum load current, but it should also be less than the maximum current permitted in the circuit.

The thermal trip ratings are usually adjustable from 0.7 I_n to I_n. But with the use of electronic trip relay, the adjustment range can be increased from 0.4 I_n to I_n.

EXAMPLE 5.14

A circuit-breaker rated for 630 A is fitted with a 400 A electronic trip relay. If the relay is set at 0.9 find

a. the overload relay trip current setting I_r
b. Adjustment range of the overload trip current setting and
c. Circuit-breaker frame size rating.

Solution :

Rated current of the electronic tripping and I_m = 400 A
Since the electronic trip relay setting is 0.9,

a. the overload trip current setting = 0.9 × 400 = 360 A
b. The adjustment range of the electronic trip relay = $0.4I_n$ to I_n
= 0.4 × 400 to 400 = 160 to 400 A
c. Since the rating of the circuit-breaker is 630 A, the frame size setting of the CB is 630 A.

5.8.11.4 Short circuit relay trip current setting, I_m

The second important function of a circuit-breaker is to disconnect the circuit in the event of a short circuit. Short circuit tripping relays (instantaneous or with slight time delays) are used for the purpose. The short circuit tripping threshold current I_m is normally specified by the standards (IEC standard 60898 or by the manufacturers for industrial circuit-breakers, according to IEC standard 60947-2). In the case of industrial circuit-breakers, there are a wide variety of tripping devices which allows matching the protection performance of the CB to the exact requirement of the load. Typical values of tripping current range of over current and short circuit protective devices for medium voltage circuit-breakers are shown in Table 5.36.

5.8.11.5 Rated short circuit breaking capacity (I_{cu} or I_{cn})

The short circuit breaking capacity of a circuit-breaker is the highest prospector value of current that the circuit-breaker is capable of breaking without being damage. Thus value of current is the RMS value of AC component of the fault current (DC offset is assumed to be zero even though they are present). The value of current, I_{cu} for industrial circuit-breakers and I_{cm} for domestic circuit-breakers are expressed in kV (rms).

The short current breaking capacity of a circuit-breaker is related to the power factor of the fault current path. In practice, the power system fault current are always at lagging power factor, the values are low because of the predominantly inductive nature of the fault path. In general greater the magnitude of the fault current (closer to the source) lower will be the power factor. IEC standard 60947-2 details the standard value of power factor for industrial circuit-breakers according to their rated breaking capacity. Then power factors are representative of the majority of power system in operation. Table 5.37 gives the relation between power factor and I_{cu}.

Table 5.36 Tripping current ranges of overload and short circuit protective devices

	Type of protective relay	Overload protection	Short circuit protection		
Domestic breakers-IEC 60898	Thermal magnetic	$I_r = I_n$	Low setting type B $3I_n \leq I_m \leq 5I_n$	Standard setting type C $5I_n \leq I_m \leq 10I_n$	High setting type D $10I_n \leq I_m \leq 20I_n$
Modular industrial breakers	Thermal magnetic	$I_r = I_n$ fixed	Low setting type B or Z $3.2I_n \leq$ fixed $\leq 4.8I_n$	Standard setting type C $7I_n \leq$ fixed $\leq 10I_n$	High setting type D or K $10I_n \leq$ fixed $\leq 14I_n$
Industrial circuit-breakers IEC 60947-2	Thermal magnetic	$I_r = I_n$ fixed Adjustable $0.7I_n \leq I_r \leq I_n$	Fixed, $I_m = 7$ to $10\ I_n$ Adjustable Low setting : 2 to 5 I_n Standard setting : 5 to 10 I_n		
	Electronic	Long delay $0.4I_n \leq I_r \leq I_n$	Short delay - adjustable $1.5I_r \leq I_m \leq 10I_r$ Instantaneous(I) fixed $I = 12$ to $15\ I_n$		

Table 5.37 Relation between power factor and I_{cu}

I_{cu}	$\cos\phi$
6 kA $< I_{cu} \leq$ 10 kA	0.5
10 kA $< I_{cu} \leq$ 20 kA	0.3
20 kA $< I_{cu} \leq$ 50 kA	0.25
$I_{cu} >$ 50 kA	0.2

5.8.11.6 Rated insulation voltage, U_i

This is the voltage to which the dielectric test voltages and creepage distances are referred to. The dielectric test voltage generally will be greater than $2U_i$.

5.8.11.7 Rated impulse withstand voltage, U_{imp}

This is the peak value of the voltage with a prescribed polarity, the circuit-breaker can withstand without failure under test conditions. The standard values for medium voltage circuit-breakers are 8 kV for industrial circuit-breakers and 6 kV for domestic circuit-breakers.

5.8.11.8 Rated making capacity, I_{cm}

Rated making capacity of a circuit-breaker is the instantaneous value of current the circuit-breaker can establish at rated voltage. In alternating current system, this value is related to the rated breaking current I_{cu} by a factor 'k' which depends upon the power factor of the short circuit current.

Table 5.38 shows the relation between I_{cu} and I_{cm} at different power factors.

Table 5.38 Relation between I_{cu} and I_{cm}

I_{cu}	$\cos\varphi$	$I_{cm} = kI_{cu}$
6 kA $< I_{cu} \leq$ 10 kA	0.5	$1.7I_{cu}$
10 kA $< I_{cu} \leq$ 20 kA	0.3	$2.0I_{cu}$
20 kA $< I_{cu} \leq$ 50 kA	0.25	$2.1I_{cu}$
50 kA $\leq I_{cu}$	0.2	$2.2I_{cu}$

5.8.11.9 Rated service short circuit breaking capacity I_{cs}

The rated breaking capacity I_{cu} (I_{cn}) is the maximum fault current, the breaker can successfully interrupt without being damaged. However, under normal circumstances, the fault currents are considerably lower than the breaking capacity and the probability of a circuit-breaker being subjected to its rated breaking capacity is very low. In a properly designed installation, the circuit-breaker will never be required to operate at its rated ultimate breaking capacity (I_{cu}). Therefore, a new parameter, rated service short circuit breaking capacity (I_{cs}) is introduced and this is normally expressed as a percentage of rated ultimate breaking capacity (I_{cu}). E.g. I_{cs} = 75% I_{cu} etc.

5.8.11.10 Categories of utilisation

There are two basic categories of circuit-breakers depending on the type of tripping mechanisms used in the circuit-breakers, Category A and Category B.

Circuit-breakers in which there is no deliberate delay in operation of the short circuit magnetic tripping mechanism is referred to as category A circuit-breakers. These circuit-breakers are also classified as 'instantaneous and are generally of moulded type circuit-breakers (MCCB).

Circuit-breakers in which, in order to discriminate with other circuit-breakers, if it is possible to delay the tripping of the circuit-breaker, when the fault level is lower than that of the short circuit withstand capacity I_{cw} of the circuit-breaker, are classified as category B circuit-breakers.

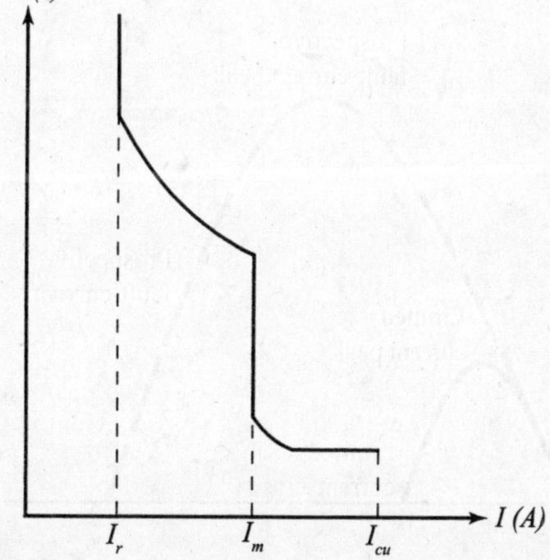

Figure 5.52a Category A circuit-breaker characteristics.

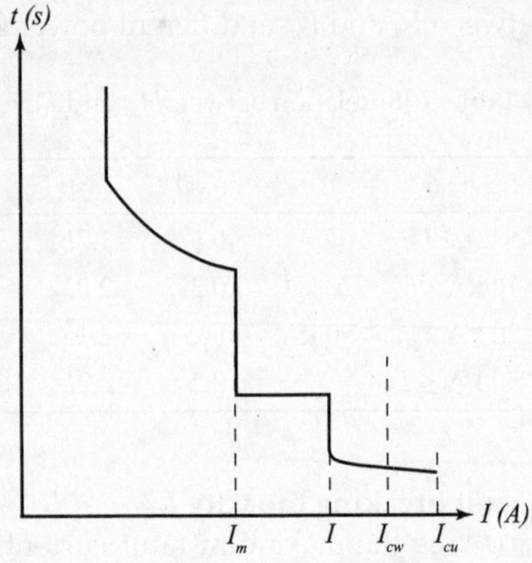

Figure 5.52b Category B circuit-breaker characteristics.

They are generally large open type circuit-breakers or heavy duty moulded case circuit-breakers. I_{cw} is the maximum current that the category B circuit-breakers can withstand, thermally and electromagnetically, without being damaged. The performance characteristics of category A and category B circuit-breakers are shown in Figures 5.52a and 5.52b.

I_r = Overload (thermal or long delay) relay trip setting.
I_m = Short circuit (magnetic or short delay) relay trip setting.
I or I_i = Short circuit instantaneous relay trip setting.
I_{cu} = Breaking capacity.

5.8.11.11 Current limiting circuit-breakers

Current limiting capacity of a circuit-breaker is the capability of the circuit-breaker to prevent the passage of the maximum prospective short circuit current through the circuit-breaker. The current limiting action is illustrated in Figure 5.53.

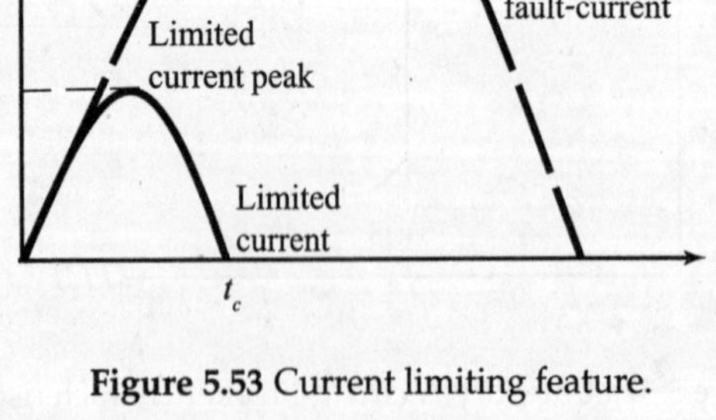

Figure 5.53 Current limiting feature.

The advantages of current limiting property in circuit-breakers are:

- Conductor heating (and thereby insulation heating) is reduced so that life of cables are correspondingly increased.
- Forces due to electromagnetic action are lower
- Reduced electromagnetic interference effects
- Reduces considerably the harmful effects of short circuit currents
- Permits improved exploitation of cables, switchgear and thereby reduced aging of the installation.

5.8.12 Selection of circuit-breakers

Choice of a circuit-breaker for a specific application is determined by the following considerations:

- Electrical characteristics of the installation
- The environment in which the circuit-breaker is intended to operate
- Short circuit breaking and making requirement
- Operational flexibility
- Installation regulations and
- Load characteristics.

The primary consideration is to match the circuit-breaker with the basic system parameters like rated system voltage, system frequency, rated insulation voltage etc. The positioning of the circuit-breaker, whether in an open execution or in a closed environment like in a switch board, greatly influences the current carrying capacity of a circuit-breaker. The rated current carrying capacity of a circuit-breaker is defined for an ambient temperature of 30 °C in the case of domestic circuit-breakers and 40 °C in the case of industrial circuit-breakers. When the circuit-breaker is enclosed in the switchboard, the ambient temperature in which the circuit-breaker is operating will definitely be more than the surrounding atmosphere because of the closed nature of the switchboard as well as the statutory requirement for keeping the switchboards dust and vermin proof.

This necessitates some correction in the rated current of the circuit-breakers when they are being used in different environments. Circuit-breakers normally come with any one of the following tripping mechanisms:

- Uncompensated thermal-magnetic tripping
- Compensated thermal-magnetic tripping and
- Electronic tripping.

Circuit-breakers with uncompensated thermal tripping elements are greatly affected by the surrounding temperature. Therefore, if such a circuit-breaker is used in tightly enclosed switchboards or in a hot environment like a boiler room, the current required to trip the circuit-breaker will be sensibly reduced. If the temperature in which the circuit-breaker is located exceeds the reference temperature, the circuit-breaker is to be derated to account for the increase in ambient temperature. Manufacturers therefore provide tables which indicate the modified rating of the circuit-breakers for different ambient temperatures. A typical data for domestic and industrial circuit-breakers with uncompensated tripping device are given in Tables 5.39 and 5.40.

Table 5.41 shows the basic data of industrial circuit-breakers manufactured by M/s Legrand India Ltd.

In the case of domestic applications, circuit-breakers are generally modular type and are mounted side by side in close contact with each other in a compact metal enclosure. In such situations, because of the mutual heating, while carrying normal current, the circuit-breakers need to be derated by a factor of 0.8. This de rating is over and above the values specified in the tables.

Table 5.39 Re rating factors for domestic circuit-breakers

Current rating (A)	Ambient temperature °C							
	20	**25**	**30**	**35**	**40**	**45**	**50**	**55**
6	6.24	6.12	6.00	5.88	5.76	5.64	5.52	5.4
10	10.6	10.3	10.0	9.7	9.3	9.0	8.6	8.2
16	16.8	16.5	16.0	15.5	15.2	14.7	14.2	13.8
20	21.0	20.6	20.0	19.4	'9.0	18.4	17.8	17.4
25	6.2	25.7	25.0	24.2	23.7	23.0	22.2	21.5
32	33.5	32.9	32.0	31.4	30.4	29.8	28.4	28.2
40	42.0	41.2	40.0	38.8	38.0	36.8	35.6	34.4
50	52.5	51.5	50.0	48.5	47.4	45.5	44.0	42.5
63	66.2	64.9	63.0	61.1	58.0	56.7	54.2	51.7

Table 5.40 Re rating factors for domestic circuit-breakers

Current rating (A)	Ambient temperature °C							
	20	**25**	**30**	**35**	**40**	**45**	**50**	**55**
16	18	18	17	16.6	16	15.6	15.2	14
25	27.5	25	26.3	25.6	25	24.5	24	23.5
32	35.2	34.4	33.6	32.8	32	31.3	30.5	30
40	44	43	42	41	40	39	38	37
50	55	54	52.5	51	50	49	48	47
63	69	68	66	65	63	61.5	60	58
80	88	86	84	82	80	78	76	74
100	110	108	105	103	100	97.5	95	92.5
125	138	134	131	128	125	122	119	116
160	176	172	168	164	160	156	152	148
200	220	215	210	205	200	195	190	185
250	277	269	263	256	250	244	238	231

(Courtesy: Schneider Electric)

Table 5.41 Typical data of moulded case circuit-breakers

Type and frame size		DPX 125			DPX 160		DPX 250 ER		DPX 250			DPX 250 electronic Release	DPX 630		
Number of poles		3-4	3-4	3-4	3-4	3-4	3-4		3-4			3-4	3-4		
Normal rating in (A)		16-125	16-125	16-125	160		100-250		25-250			40-250	400-630		
Electrical characteristic conforming to EN 60947-2 and IS 13947-2															
Rated operating voltage U_e (V dc)		500			500		500		690			690	690		
Rated insulating voltage U_i (V ac)		250			250		250		250			-	250		
Rated impulse withstand voltage		500			500		500		690			690	690		
Rated impulse withstand voltage U_{imp} (kV)		6			6		6		8			8	8		
Category of use		A			A		A		A			A	A		
Ultimate breaking capacity I_{cu} (kA)	400/415 VA	16	25	36	36	50	36	50	36	70	100	70	36	70	100
Standard braking capacity I_{cs} (% I_{cu})		100	50	75	75	50	75	50	100	75	50	75	100	75	50
Rated closing capacity on short circuit		32	52.5	75.6	75.6	100	75.6	105	75.6	154	220	154	75.6	154	220
Endurance (o.c cycle)	mechanical	25,000			20,000		20,000		20,000			20,000	15,000		
	electrical	8000			8000		8000		8000			8000	5000		

(Source: Legrand(India) Pvt. Ltd.)

EXAMPLE 5.15

What current rating should be selected for a domestic application requiring a maximum current of 16 A, if the circuit-breaker is installed side-by-side with other circuit-breakers in a distribution board and if the ambient temperature is 50 °C.

Solution:

From Table 5.39, a 20 A domestic circuit-breaker will be derated to 17.8 A for an ambient temperature of 50 °C. Since the circuit-breaker is used in close contact with other circuit-breakers, additional de-rating of 0.8 also will have to be accounted. The total derated value of 20 A circuit-breaker will be only $17.8 \times 0.8 = 14.24$ A. This will not be suitable for supplying 16 A. The next higher rating is 25 A circuit-breaker. The effective derating of this circuit-breaker will be $22.2 \times 0.8 = 17.76$ A. Hence, a 25 A circuit-breaker will be required to supply the 16 A load for the specified condition.

Circuit-breakers with compensated tripping units are provided with a bimetal compensating strip which allows the over current trip setting (I_r or I_{rth}) to be adjusted, within a specified range, irrespective of the ambient temperature. Medium voltage circuit-breakers rated below 630 A are generally equipped with compensated tripping units for temperature variation of –5 °C to 40 °C.

Circuit-breakers with electronic tripping units provide highly stable performances in varying temperature conditions. However, operational limits are imposed by the manufacturers on the devices and they provide operation charts relating the maximum value of permissible trip current levels with ambient temperature. Electronic trip units can also provide information that can be used in building management systems including energy efficiency and power quality. Table 5.42 shows the temperature dependence of a typical electronics trip unit.

Table 5.42 Typical temperature dependence of electronic trip units

Version	Current	40 °C	45 °C	50 °C	55 °C	60 °C
Withdrawable CB with horizontal plugs	I_n (A) Maximum adjustment	2000 1	2000 1	2000 1	1980 0.99	1890 0.95
Withdrawable CB with edge plugs	I_n (A) Maximum adjustment	2000 1	2000 1	1900 0.95	1850 0.93	1800 0.90

5.8.12.1 Selection based on the short circuit breaking capacity

While selecting a circuit-breaker for medium voltage applications, any one of the following conditions must be fulfilled:

- The circuit-breaker must have a rated short circuit breaking capacity I_{cu} (or I_{cn}) which is equal to or greater than the prospective short circuit current calculated for the point at which the circuit-breaker is being installed.
- The circuit-breaker can be associated with another switchgear device located upstream, and which has the required short circuit breaking capacity.

In the latter case, the characteristics of the two devices must be so coordinated such that the energy permitted to pass through the upstream device must not exceed that, which the downstream device and all the associated cables, wires and other components can withstand, without being damaged. This technique is used in system design with economic benefits. The two basic forms are:

- Association of fuses and circuit-breakers
- Association of current limiting circuit-breakers and standard circuit-breakers.

This is also known as cascading. Association is the technique by which the breaking capacity of a circuit-breaker is increased by coordinating it with another protective device placed upstream. The coordination makes it possible to use a protective device with a breaking capacity which is lower than the maximum prospective short circuit current at its point of installation.

The associations between switchgear components are normally provided by the manufacturers. A typical coordination and association table is given in Table 5.43.

Table 5.43 Association and coordination of MCCBs and MCBs (in kA)

(For 3 phase networks +N 400/415V as per IEC 60947-2)

		DPX-E 125	**DPX 125**	**DPX/DPX-H160**	**DPX 250 ER**				**DPX/H/L 250**		**DPX/H/L 630**
MCBs downstream		16–125 A	16–125 A	25–160 A	63 A	100 A	160 A	250 A	160 A	250 A	250–400 A
	0.5–20 A	16	25	25	25	25	25	25	25	25	25
	25 A	16	25	25	25	25	25	25	25	25	25
Lexic 10	32 A	16	25	25	25	25	25	25	25	25	25
kA MCB	40 A	16	25	25	25	25	25	25	25	20	20
	50 A	16	25	20	25	25	20	20	20	15	15
	63 A	16	25	15		20	15	15	15	15	15

(Source: Legrand (India) Pvt. Ltd.)

5.8.12.2 Selective discrimination of circuit-breakers

As described earlier in section 5.8.9, selective discrimination is said to be achieved by the automatic protective devices, if an over current condition (overload or short circuit), occurring at any point in the installation, is cleared by the protective devices located immediately upstream of the faulted point, while all other protective devices remain unaffected. Consider a schematic diagram as shown in Figure 5.54.

Discrimination between circuit-breakers A and B is said to be total if the maximum value of short circuit current on the circuit-breaker B (I_{scB}) does not exceed the short circuit trip setting of the circuit-breaker A (I_{rA}). This situation is shown in Figure 5.55.

Discrimination will be partial if the short circuit current on circuit-breaker B exceeds the short circuit trip setting of circuit-breaker A as indicated in Figure 5.56. Discrimination is said to be total if it is assured, irrespective of the value of the fault current, up to the maximum value of the fault current likely to exist in the installation. Otherwise, discrimination is only partial. For each type of fault there is a specific protective device (e.g., protection against overload, short circuit, earth fault etc.). Each one of these can cause loss of discrimination, if proper coordination of the devices has not been taken into consideration at the time of design.

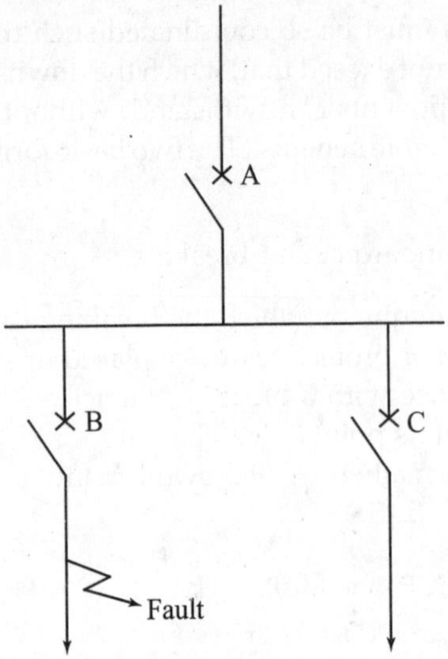

Figure 5.54

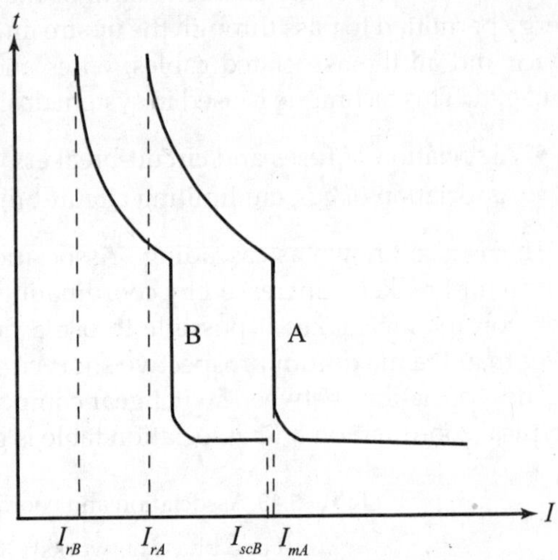

Figure 5.55 Total discrimination between circuit-breakers A and B.

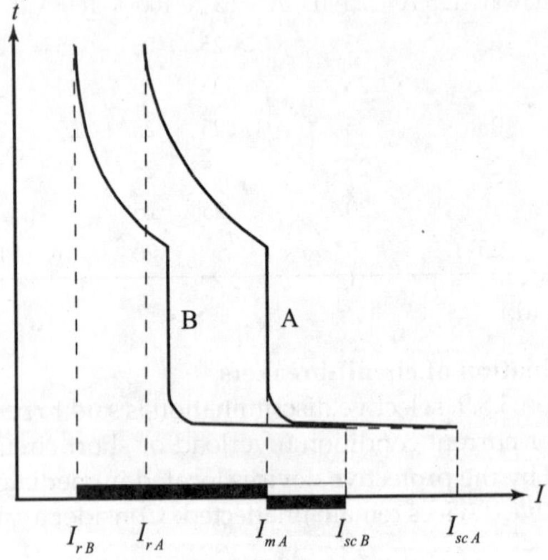

Figure 5.56 Partial discrimination between circuit-breakers A and B

5.8.12.3 Discrimination against overload

This is a situation where the current levels are between 1 to 10 times the normal duty current. Such a situation should be removed within a period which is compatible with the thermal withstand capacity of the piece of equipment/device concerned. The tripping time is generally proportional to the square of the current magnitude and is known as inverse time tripping. Discrimination is obtained by setting the successive tripping thresholds of circuit-breakers in stepped levels starting from the down stream circuit-breaker (Lowest setting) towards the source circuit-breaker (highest setting). Figure 5.57 shows the selective discrimination between circuit-breakers A and B. As a rule of thumb, discrimination will be achieved if $\frac{I_{rA}}{I_{rB}} > 2$.

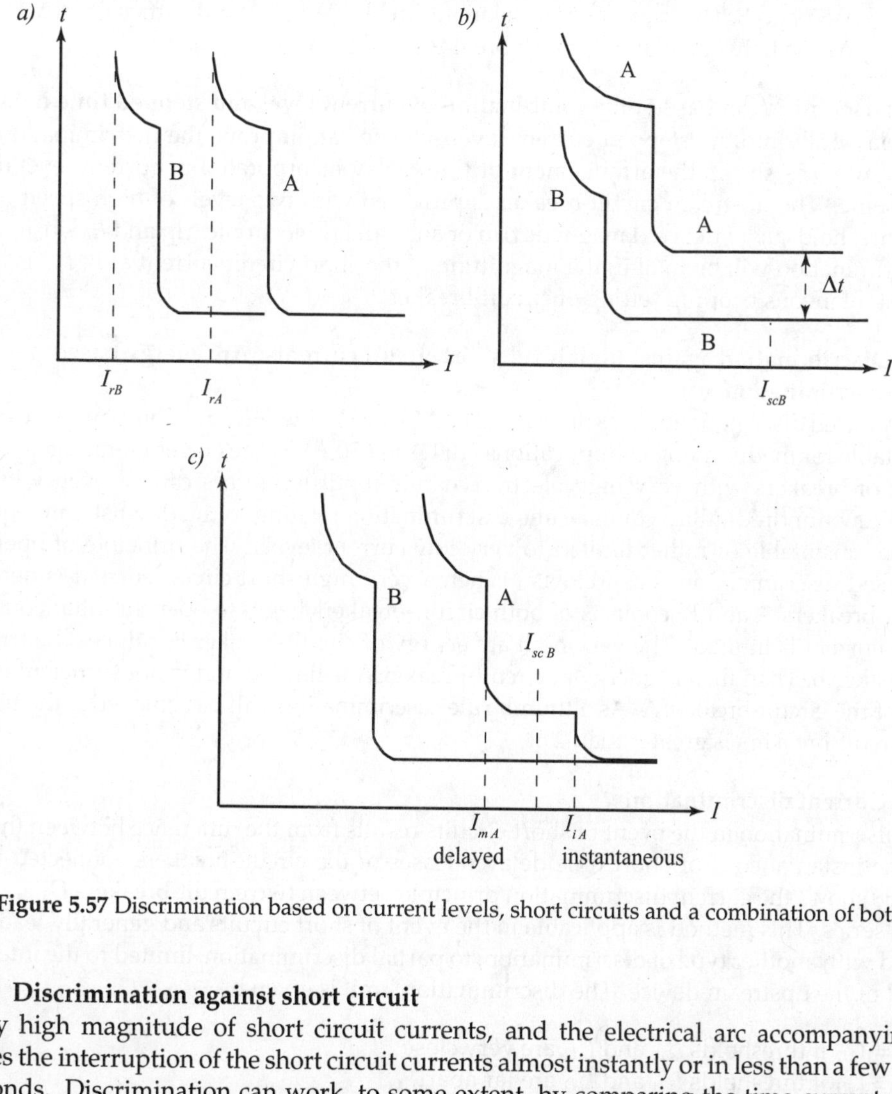

Figure 5.57 Discrimination based on current levels, short circuits and a combination of both

5.8.12.4 Discrimination against short circuit

The very high magnitude of short circuit currents, and the electrical arc accompanying them, mandates the interruption of the short circuit currents almost instantly or in less than a few hundred milliseconds. Discrimination can work, to some extent, by comparing the time-current curves of circuit-breakers, provided that the clearing time is at least 30 to 40 milliseconds. For a clearing time below this, the time-current curves are not sufficiently precise to provide a definitive result with absolute certainty. In such situations, time-current ceases to be the only discriminating criteria. It may also be necessary to take into account the peak value of the current or a combination of time and current ($\int i^2dt$), depending on the situation prevailing. If so, it will be necessary, to refer to the discrimination tables provided by the manufacturers of the circuit-breakers.

5.8.12.5 Discrimination against low level short circuit currents

This is achieved by adjusting the time delayed tripping units, such that the downstream breakers are set to have the shortest operating time with progressively increasing delays towards the upstream circuit-breakers. Figure 5.57b shows the stepped time delay arrangement between two circuit-breakers A and B. The tripping time of circuit-breaker A is delayed sufficiently (Δt) to ensure total discrimination between the two circuit-breakers.

5.8.12.6 Discrimination based on a combination of current level and stepped time delay

A time delay built into the stepped current level scheme can improve the discrimination performance. Figure 5.57 shows the arrangement of time delay incorporated in current level discrimination scheme. The upstream circuit-breaker is provided with two levels of high speed magnetic tripping thresholds, I_{mA} - delayed magnetic trip or short delay electronic trip and I_{iA} -instantaneous trip. Discrimination will be total if, the magnitude of the short circuit current at B, I_{scB} is less than I_{iA}, the instantaneous tripping setting of circuit-breaker A.

5.8.12.7 Discrimination against high level short circuit currents: Arc energy based discrimination

Arc energy based discrimination was developed by M/s. Schneider Electric Company and are particularly suitable for medium power applications (100 A to 630 A) where current limiting is necessary. This type of breakers with very high electrodynamic repulsive forces cannot even withstand a delay of a few hundred milliseconds. Time discrimination relating to the downstream equipment is therefore unsuitable or rather limited to very low current levels. The principle of operation of energy based discrimination is as follows. When a very high short circuit current is detected by the circuit-breakers A and B, contacts of both circuit-breakers begin to open simultaneously. As a result the current is limited. The very high arc energy of circuit-breaker B induces the tripping of circuit-breaker B. Then the arc energy at circuit-breaker A is limited and is not sufficient to induce tripping of the circuit-breaker A. As a thumb rule discrimination will be achieved if the frame size ratio of circuit-breakers is greater than 2.5.

5.8.12.8 Current discrimination

Current discrimination in the event of short circuits results from the difference between the thresholds of the instantaneous or short time delay releases of the circuit-breakers connected in series. Figure 5.58 shows the current discrimination principle between two circuit-breakers D_1 and D_2 connected in series. This method is applicable in the event of short circuits and generally leads, unless associated with another type of discrimination, to partial discrimination, limited to the intervention threshold of the upstream device. The discrimination limit is given by

$I_s = I_{sd2}$ if thresholds I_{sd1} and I_{sd2} are very close
$I_s = I_{sd1}$ if thresholds I_{sd1} and I_{sd2} are far apart.

As a rule discrimination is achieved if

$$\frac{I_{r1}}{I_{r2}} < 2 \text{ and } \frac{I_{sd1}}{I_{sd2}} > 1$$

and the discrimination limit is given by $I_s = I_{sc}$ at D_2, the maximum value of the short circuit current at D_2 or $I_{sd1} > I_{sc}$ at D_2. This implies that the differences between the ratings of the circuit-breakers D_1 and D_2 should be large or in other words a relatively low level of short circuit current at D_2. This type of discrimination is therefore used in final distribution circuits.

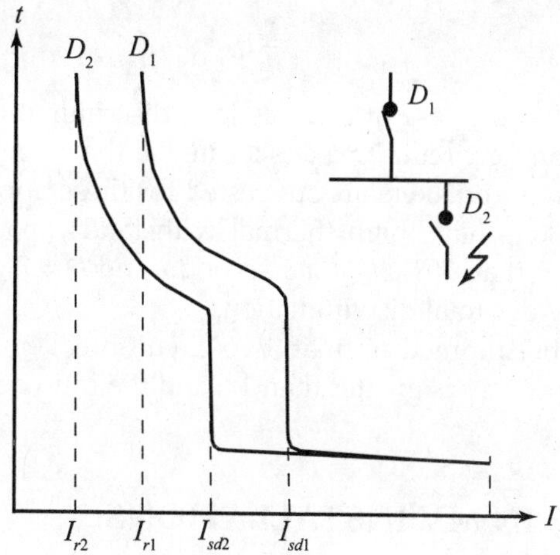

Figure 5.58 Current discrimination

5.8.12.9 Time discrimination

In order to ensure discrimination above the short time threshold I_{sd1} of the upstream circuit-breaker D_1, it is possible to introduce a time delay t which may or may not be adjustable, on the trip unit of upstream circuit-breaker D_1. Figure 5.59 shows this type of discrimination arrangement.

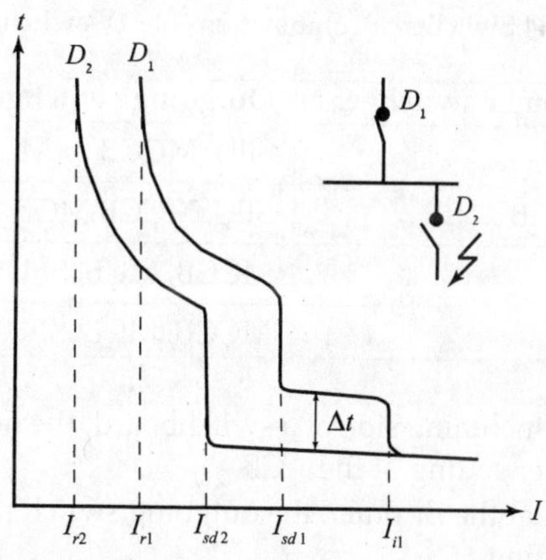

Figure 5.59 Time discrimination between D_1 and D_2

The discrimination limit I_s is at least equal to I_{i1}, the instantaneous threshold setting of D_1. There are two basic applications for this type of discrimination.

- For final or intermediate feeders and
- On the incomers and feeders on the main switchboards.

Circuit-breakers of category A can be used with time delayed tripping of the upstream circuit-breakers. This allows extension of current limiting up to the instantaneous threshold I_{i1} of D_1.

$$ie., I_s = I_{i1}$$

If I_{sd2} is not very high (as in the case of final feeders) discrimination will be total. Continuity of supply is the major criterion for circuit-breakers located at the incomer as well as the feeders on the MSB. Category B type circuit-breakers are suggested for these applications with time delayed tripping. These circuit-breakers have high thermal withstand capacity also. i.e. I_{cw} of these circuit-breakers will be greater than 50%. I_{cn} for 1 second. And $I_s = I_{cw1}$. Even if I_{scD2} is high, time discrimination normally provides total discrimination.

As a general rule, it can be summed up that, two circuit-breakers will be selective as long as their ratio between their frame sizes is greater than 1.6 and the ratio between their current settings is greater than 2.5.

5.9 SWITCHBOARDS IN MV INSTALLATIONS

The design of industrial switchboards is a complex process taking into consideration the magnitude of possibilities that exist with the different switching combinations. Basic guidelines of the design of MV switchboards are listed below.

1. Bus bar section shall be selected based on the current density of 1.2 A/mm^2 for copper and 0.8 A/mm^2 for aluminium. The minimum thickness of bus bar shall be 6 mm.
2. The combinations of switchgear used in switchboards shall be as shown in Table 5.44.

Table 5.44 Switchgear combinations of MV switchboards

Incomer switchgear	Outgoing switchgear
ACB	SFU, MCCB or MCB
MCCB	SFU, MCCB,MCB
SFU	MCCB, MCB, SFU
Isolator	MCCB, MCB, SFU

3. When ACB is used at the incoming side of a switchboard, the outlet switch rating shall not be less than 1/3 of the incomer setting of the ACB.
4. When MCCB is provided at the incomer, the outgoing switch rating shall not be less than 1/5 of the incomer MCCB setting.
5. When CB is provided at the incomer and SFU at the outgoing, outgoing fuse rating shall not exceed 1/3 of the incomer breaker setting.
6. For ACB as incomer and MCCB as outgoing, MCCBs can be set at 80% of incomer breaker setting.

7. If MCCBs are provided at incomer and outgoing, maximum current setting of outgoing MCCBs shall not exceed 50% of incomer MCCB setting.
8. In case of breaker controlled switchboards (3), even though the outlet switch rating is fixed at 1/3 incomer setting, the feeder can be derated by using lower size fuses.
9. For SFU/SFU controlled switchboards, the outgoing SFU can be 1/10 of incomer SFU rating. Max fuse rating at outlet shall not exceed 1/2 the incomer fuse rating.
10. Maximum rating of fuses shall be 630 A and max rating of isolator shall be 800 A.
11. Maximum incomer rating of a LDB shall be 63 A. The maximum number of ways of the DB shall be 12.
12. The max capacity of a motor that can be connected to a DB shall be less than 7.5 kW.
13. VDBs can use switchgears up to 100A as incomer.
14. Separate feeders shall be used for all motors above 7.5 kW and above
15. The minimum clearance in front of a switchboard shall be 1m.
16. Minimum depth of laying of cables shall be 75 cm and at road crossings 1 m and drawn through 100 mm diameter hume pipes.
17. Minimum size of cable for power loads shall be 4 mm^2 Al and for 2.5 mm^2 copper
18. Voltage drop in cable shall not exceed 3% of the rated voltage.
19. Cables shall be loaded only up to 75% of the rated current capacity as specified by the manufacturer.
20. All cables should have adequate short circuit current carrying capacity.

EXAMPLE 5.16

An industry consists of the following connected loads:

(a) 7.5 kW, 3-phase cage induction motor – 1 No.
(b) 9.3 kW, 3-phase cage induction motor – 2 Nos.
(c) 22.2 kW, 3-phase cage induction motor – 1 No.
(d) Power sockets – 15 Nos.
(e) Lighting loads – 40 Nos. of 2 × 28 W LED retrofit lamps.
(f) Exhaust fans 300 mm, 1420 rpm – 4 Nos.
Design the electrical system for the industry, if the industry is located in a village, and also determine:

(a) Type of industry.
(b) Transformer capacity required and type of substation.
(c) Draw the single line electrical schematic diagram showing the details of cable sizes, starters and switch gears. Use a switch board with MCCB incomer and MCCB outgoing and MCB-type distribution boards.

Solution:
Step 1: Selection of the transformer

Total connected load of the motors (CL) in kW= 7.5 + 2 × 9.3 + 22.5 = 48.6 kW

Assume a reasonable efficiency of 0.85 for the industry.

Total power input to the motors = $\frac{48.60}{0.85}$ = 57.18 kW

Therefore, total connected load of the industry = $57.18 + 15 \times 1 + \frac{40 \times 2 \times 28}{1000} + \frac{4 \times 45}{1000}$

= 74.6 kW

Assuming an average power factor of 0.9, connected load in kVA = $\frac{74.6}{0.9}$ = 82.88 kVA

Assuming a diversity factor of 1.5, the maximum demand of the industry = $\frac{82.88}{1.5}$ = 55.26 kVA

a) Since the connected load is >50 kVA, the industry can be classified as Medium Industry. Future expansions or rating changes of motors are to be expected for any industry. Hence 25% spare capacity shall be provided at the design stage itself for taking into account unforeseen load changes.
Thus, the anticipated connected load of the industry in kVA =1.25 × 82.88 = 103.6 kVA
The maximum demand will be definitely less than the connected load, taking into consideration the diversity factor. A reasonable diversity factor of 1.5 may be assumed for an industry of this load.
The anticipated maximum demand = $\frac{\text{anticipated kVA}}{\text{DF}} = \frac{103.6}{1.5}$ = 69.06 kVA

b) From Table 31 of the Data Handbook, the nearest standard size of transformer is 100 kVA, 11 kV/433 V, Delta-Star, Dy11, oil-filled outdoor transformer.
Since the industry is located in a village, we can choose an outdoor-type substation.

Step 2: Primary control devices

Full load current on the primary side = $\frac{100 \times 1000}{\sqrt{3} \times 11 \times 1000}$ = 5.25 A

Referring to Table 34 of the Data Handbook, HT fuse rating is selected as 16 A for taking into account the switching inrush current of the transformer. Therefore, select 11 kV, 630 A, drop out fuse unit with 16 A HT fuse. The other devices required are:

11 kV lightning arrestor set and 11 kV ganged air break switch with locking arrangement.

Considering a fault level of 250 MVA on the primary side, the three-phase short circuit current,

$$I_s = \frac{250 \times 1000 \times 10^6}{\sqrt{3} \times 11 \times 10^3} = 13.12 \text{ kA}$$

The size of the HT cable can be calculated using the equation.

Size of the XLPE cable in mm^2 = $11.1 \times I_s \times \sqrt{t}$

Assuming t = 1 sec

Size of the XLPE cable = 145.63 mm^2

From Table 67 of the Data Handbook, the nearest standard size of 11 kV XLPE cable is 150 mm^2.

HT metering:

For energy metering purpose, an outdoor type 11 kV CT/PT unit is required. The CT and PT specifications are:

CT 10/5 A, 0.5 Accuracy Class, 15 VA
PT 11000/110 V, 0.5 accuracy class, 15 VA
Electronic TOD meter, 3 phase, 5 A, 110 V

Step 3: Secondary control devices

The main switch board (MSB) is assumed to be of MCCB/MCCB-type.

Full load current on the secondary side of the transformer = $\frac{100 \times 1000}{\sqrt{3} \times 433}$ = 133.33 A

From Table 42 and 109 of the Data Handbook, we can choose 160 A MCCB set to operate at 135A(similar to DPX 160 of M/s Legrand) having isolation duty, earth fault release and stand by low set E/F release using CT on neutral earthing conductor. The secondary feeder cable rating can be taken as 160 A for which the cable size required as per Table 31A of the Design Data Handbook is 3.5 core 185 mm^2 AYFY cable. The minimum size of cable for a 100 kVA transformer on the secondary is 3.5 core 90 mm^2 AYFY cable (Table 31).

Step 4: Selection of bus bar size for MSB

Required current rating (same as incomer rating) = 160.0 A
Derating factors:
Temperature rise of 30°C above ambient temperature of 35°C = 0.75
Enclosure factor = 0.60
Impurity factor for 50 S-WP = 0.94
Effective derating = $0.75 \times 0.60 \times 0.94$
= 0.423

Actual current rating of the bus bar section required = $\frac{160.0}{0.423}$ = 378.25 A

From Table 38A, the nearest section of a single bar of 38.1×6.35 mm section with a current rating of 520 A will meet this condition.
Alternatively,

Assuming a current density of 0.8 A/mm^2 for aluminium,

Area of bus bar required = $\frac{160}{0.8}$ = 200 mm^2

From Table 38A, the nearest section available is 38.1×6.35 mm section (241.935 mm^2). For the neutral bus bar, a lower section 25.4×6.35 can be chosen.
Since the control device chosen is MCCB, the maximum and minimum setting of MCCB on the outgoing side will be: (Refer Table 31B)

Maximum setting = 80% of incomer setting = 0.8×135 A= 108 A
Minimum setting = 20% of incomer setting = 0.2×135 = 27 A

Step 5: Calculation of backup fuse setting, cable sizes and starting devices for motors

Since the motor capacities are all 7.5 kW and above, separate feeders are recommended for each motors.

For 7.5 kW motor, from Table 28, assuming Star-Delta starting, backup setting for the motor is 25 A with a TP MCCB rated 40 A and cable size is 3×6.0 sq.mm AYWY.

For 9.3 kW motor, from Table 28 and assuming Star-Delta starting, backup setting for the motor is 35 A with a TP MCCB rated for 63 A and cable size is 3 × 10.0 sq.mm AYWY.

Similarly for 22.5 kW motor, backup setting of 63 A with A TP MCCB rated 100 A with a cable size of 3 × 25 sq.mm AYFY.

Since the minimum setting of the MCCB on the outgoing side of the MSB is 27 A, we cannot connect the 7.5 kW motor directly to the MSB (because, back up setting for 7.5 kW motor is = 25 A). The other motors can be connected to the MSB directly.

MSB details:

Incomer 160 A 4-pole MCCB set at 135 A

Outgoing:

100 A TP MCCB set at 63 A	1 No.	for 22.2 kW motor
63 A TP MCCB set at 35 A	2 Nos.	for 9.3 kW motors
63 A 4P MCCB set at 40 A	1 No.	
Spare feeder	2 Nos.	

Bus bars 38.1 × 6.35 mm Al for phase bars and 25.4 × 6.35 mm Al for neutral bar.

One set LED phase indicating lamps

0-500 V: Volt meter with selector switch and

0-150 A: Ammeters-3 Nos. through 150/5 A CT

The 7.5 kW motor, all the power points (through a PDB) and all the lighting points (through LDB) can be connected to a subswitch board (SSB).

For the lighting and loads, a three-phase MCB-type LDB can be used.

Total number light points = 40 lamps + 4 exhaust fans = 44

Number of points in lighting circuit may be 8 to 10. However, as a better engineering practice, no. of points in a circuit is limited to 6.

Therefore, number of light circuits = $\frac{44}{6}$ = 8.

A three-phase, 4-way MCB type Distribution board (Horizontal) can be used for this with 4-pole, 25 A ELCB with 30 mA sensitivity as incomer and 8 numbers of 6 A B characteristics MCBs as outgoing. The balance four spaces can be dummied.

For the power points,

Total number of power socket outlets = 15

For power socket outlets, 2 Nos. of sockets may be provided in a circuit if ratings of plug-in devices are not known.

Therefore, number of power circuits = $\frac{15}{2}$ = 8

Total number of sub-circuits required from the TPN MCB DB = 8

Hence, a three-phase, 6-way MCB type Vertical Distribution Board (VDB) with 18 sub-circuits may be selected with the following details 4 pole, 40 A 4 P ELCB with 30 mA sensitivity as incomer and 8 numbers of 16 A, C characteristics MCBs as outgoing. The use of vertical DB permits the connection of both three-phase and single loads on the same distribution board.

Step 6: SSB details

Connected load on the SSB $= \frac{7.5}{0.8} + 15 + \frac{40 \times 56}{1000} + \frac{4 \times 45}{1000}$

$$= 26.795 \text{ kW}$$

Maximum demand, assuming a DF of 1.5 $= \dfrac{26.795}{1.5} = 17.86$ kW

Full load current corresponding to MD $= \dfrac{17860}{\sqrt{3} \times 415 \times 0.85} = 29.23$ A (say 30 A)

Incomer can be chosen as 63 A , 4 pole MCB isolator and the outgoing feeders as:
25 A TP MCB- C for 7.5 kW motor feeders,
25 A TPN MCB- B for LDB feeder and
40 A TPN MCB-C for VDB and a spare feeder for future use.
A set of LED phase indicating lamps, 0-500 V, voltmeter with selector switch and 0-60 A, ammeter with selector switch are the other accessories required.

Step 7: Schematic diagram

Figure 5.60 gives the schematic diagram of the substation layout for the industry.

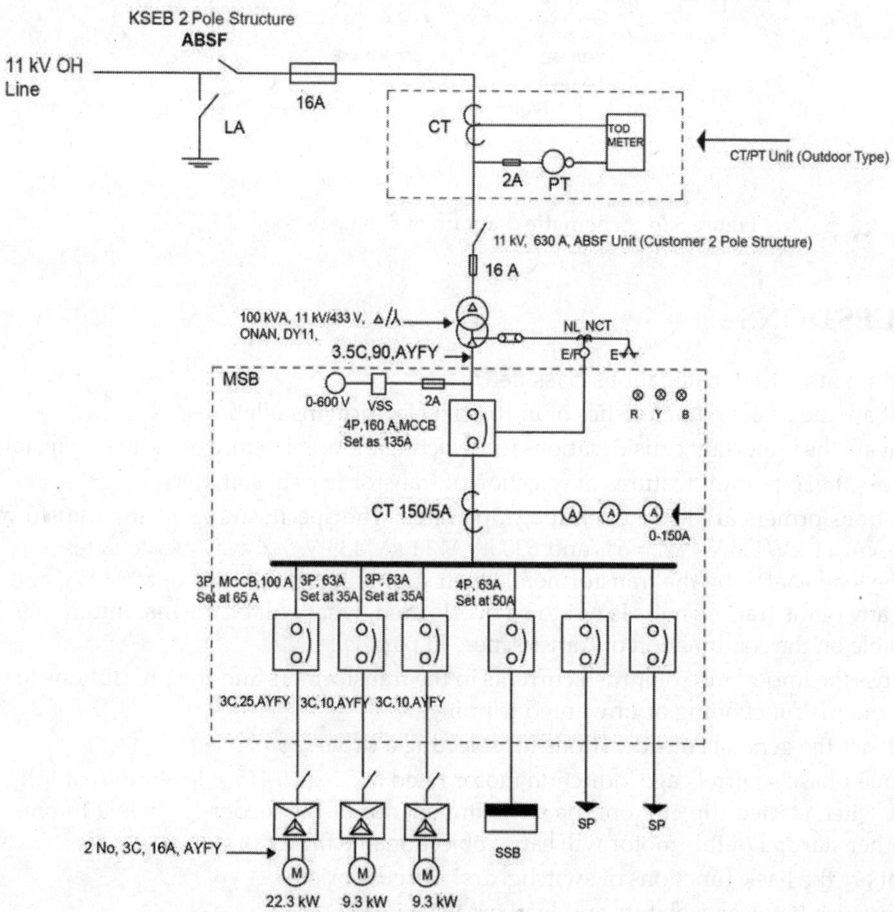

Figure 5.60 Schematic wiring diagram of the industry

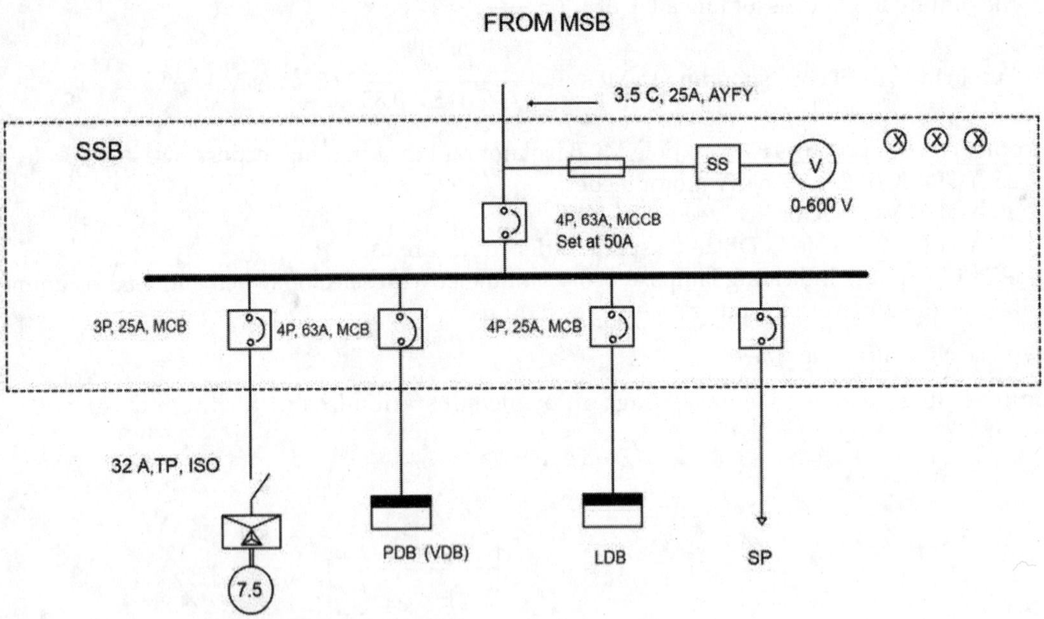

Figure 5.61 Schematic diagram of sub switchboard

5.10 QUESTIONS

1. How are industrial installations classified?
2. What are the basic characteristics of industrial electrical installations?
3. What are the important considerations for selecting the architecture of industrial installations?
4. Discuss the important features of selection of transformer substations.
5. Two transformers are to be operated in parallel. The specifications of the transformers are 400 kVA, 11 kV/433 V, %Z = 6% and 630 kVA, 11 kV/433 V, %Z = 5%. Calculate:
   a) The load shared by the transformers, when supplying a total load of 800 kVA, and
   b) If any of the transformers is not to be overloaded, what will be the maximum total loading possible on the combination of transformers in parallel.
6. Discuss the importance of inrush currents in the transformers and the precautions to be taken to avoid malfunctioning or unwanted tripping.
7. What are the general considerations for selecting a drive motor?
8. A three-phase squirrel cage induction motor rated at 5.5 kW, 415 V takes 10 A at full load and 55 A when started directly on line. The impedance of the feeder cable is 0.15 ohm. Check whether starting of this motor will cause objectionable flicker or not.
9. What are the basic functions of switchgears? Discuss briefly.
10. Distinguish between isolators and load break switches.
11. Discuss the time current characteristics of the fuse.

12. What is meant by current limiting feature of the fuse? Explain.
13. What is the importance of fuse let through curves?
14. What is the difference between switch fuse unit and fuse switch unit.
15. Explain the basic functions of circuit-breakers.
16. What are the basic specifications of a circuit-breaker?
17. Explain the basic considerations for selecting a circuit-breaker.
18. Why is the environment important in deciding the de rating of circuit-breakers?
19. What is meant by selective coordination? How is it implemented in the case of fuses?
20. What is meant by current discrimination? How is it implemented in electrical installations?
21. A cement factory is supplied from a utility substation 5 km away from the factory through an overhead line. The cement factory has two numbers of 630 kVA, 11 kV/ 433 V, 3-phase transformers with a percentage reactance of 6% operating in parallel. If the 3-phase fault level on the 11 kV side of the substation is 500 MVA, determine the number of earthing stations required using plate electrodes for the transformer station. The O/H line conductor (AAC-Grasshopper) has a cross section of 84.1 mm^2 and a spacing of 900 mm. Assume a soil resistivity of 65 Ωm.
22. A small industry 20 m × 10 m in size has the following connected load:

    (a) 5.5 kW, 415 V, 3-phase induction motors – 2 Nos.
    (b) 3.7 kW, 415 V, 3-phase induction motors – 3 Nos.
    (c) 2.2 kW, 415 V, 3-phase induction motors – 3 Nos.
    (d) 0.735 kW, 240 V, single-phase induction motor – 2 Nos.
    (e) Ceiling fans – 8 Nos., 6 A socket outlets – 8 Nos.
    (f) 2 × 28 W LED fluorescent lamps – 18 Nos. exhaust fans – 4 Nos.

    Prepare the single line layout of the electrical installation and control equipment together with the rating of cables, switch fuse units, etc. Assume that the industry is located in a town area.

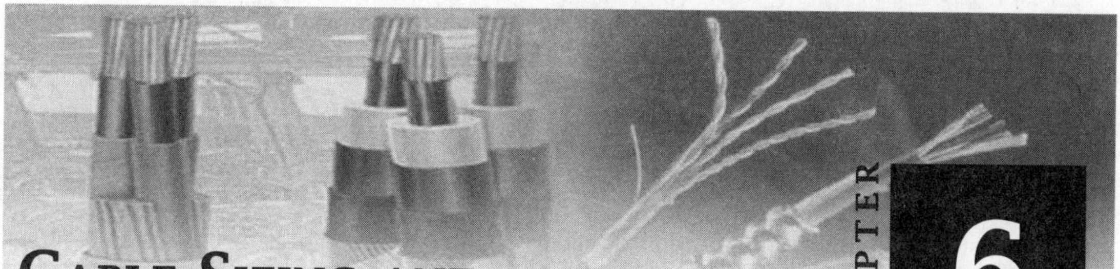

CABLE SIZING AND VOLTAGE DROP CALCULATIONS

CHAPTER 6

OBJECTIVES

At the end of this chapter you will

- Know about the constructional features of cables.
- Know about the current carrying capacities of cables.
- Understand the importance of short circuit capability of cables.
- Understand the significance of voltage drop calculations.
- Will be able to calculate the voltage drop in simple power distribution systems.
- Know the different methods of laying of cables.
- Know how to arrive at the right current carrying capacity of a cable.
- Understand the procedure for derating the cables.

6.1 INTRODUCTION

Economical, technical and practical aspects are to be taken into consideration to establish the optimum solution for transmission, distribution and utilisation of electrical power. The utilisation of electricity in factories, domestic premises and other locations is mainly by cables and they present the most practical means of transporting electrical power to equipment, tools and appliances of all type. To meet these diverse requirements, cable designs vary enormously but there are certain components which are common to all. All types of electric cable consist essentially of a low resistance conductor to carry the current (except in special cases, such as heating cables) and insulation to isolate the conductors from each other and from their surroundings.

The main factors which are to be taken into account for selection of cables are as follows:

1. Voltage designation
2. Choice of materials used in cables

3. Current carrying capacity
4. Short circuit ratings
5. Voltage drop.

6.2 VOLTAGE DESIGNATION

The foremost consideration for selection of cables shall be with respect to the system voltage required. The cables selected shall be suitable for use on AC single-phase or three-phase (earthed or unearthed) systems depending upon the rated voltages. The design voltages for cables are expressed in the form of $U_0 = U$ (formerly $V_0 = V$). U_0 is the power frequency voltage between conductor and earth and U the power frequency voltage between conductors for which the cable is designed, U_0 and U both being r.m.s values. Power cables manufactured in our country are designated as 650/1100 Volts, 1.9/3.3 kV or 6.35/11 kV or 22 kV or 33 kV. The cables are designed and selected largely on the basis of the rated system voltage because the insulation characteristics are designed on mechanical rather than electrical parameters.

6.3 CHOICE OF MATERIALS USED IN CABLES

Taking price into consideration, copper and aluminium are clearly the best choice for conductors.

6.3.1 Copper conductors

Because of the way it can readily be rolled into rod and then drawn into wire, together with its excellent electrical conductivity, copper was virtually unchallenged as a conductor for all types of insulated cable for well over fifty years. Indeed in the electrical world, the International Electrotechnical Commission established an International Annealed Copper Standard (IACS) with copper of resistivity of 1.724 microhm-cm at 20 °C assigned as 100%. The grade and quality of copper is very important and the high conductivity copper used for electrical purposes comfortably exceeds the 100% IACS value. Conductivity is greatly influenced by impurities and by mechanical working. Consequently, the purity is of the order of 99.99%, which nowadays is obtained by final electrolytic refining. The only unsatisfactory feature of copper is the way the price fluctuates widely with the world supply and demand. If its use could always be justified economically it would not have competitors. However, over the last few decades aluminium has become a replacement in the power distribution field solely on cost justification.

6.3.2 Aluminium conductors

Aluminium did not make such serious impact until the price of copper soared in the late 1950s. Unlike copper, the mechanical strength of annealed aluminium is rather soft for single wire or stranded conductors, but fortunately conductivity is little changed by working. Compared with copper, aluminium has a number of technical disadvantages, all of which can be satisfactorily overcome to benefit from its economic attraction. The advantage of a low mass density of one-third that of copper is partly offset by its low conductivity of 61% of copper, but for equal conductance the weight of the expensive conductor metal required is almost exactly halved. However, the cross-sectional conductor area has to be increased by a factor of 1.6 and this means extra usage of insulation, sheathing and armouring material. With conventional paper insulated lead sheathed and armoured constructions, there is no overall saving in cable weight but the situation is quite different if aluminium is also used for sheathing. The above is a generalisation based on cable having equal conductance. In practice there are other factors to be considered such as the cable voltage, current carrying capacity and voltage drop. In general, current ratings of aluminium cables are about 78–80% of those of copper cables of the same conductor size.

As with copper, the addition of alloying metals to provide high tensile strength has no application for insulated cables. Impurities, as distinct from alloying additions, do not greatly affect the conductivity of aluminium but a grade known as EC (electrical conductivity) is commonly accepted in the aluminium industry. A particular disadvantage of aluminium for electrical purposes is the thin, hard protective oxide skin which is so valuable in giving corrosion protection to aluminium installed aboveground. Satisfactory techniques for removing it in making soldered joints have long been developed. Corrosion protection is outside the scope of this book and all that need to be mentioned here is that, whilst claim for aluminium having excellent corrosion resistance are true if made in relation to dry indoor situation, or even outdoors if drying off is rapid after rain, cable engineers must ensure that bare aluminium should not be left exposed anywhere. Cable situation are rarely dry, for example in the base of outdoor pillars, corrosion can be rapid. Protection in a form such as heat shrunk sleeving should always be used. The properties of conductor materials are listed in Table 6.1.

Table 6.1 Properties of conductor material

Description	Annealed copper	Hard drawn copper	Aluminium hard drawn	Aluminium $\frac{3}{4}$ hard
Conductivity percent	100	97	61	61
Density at 20 °C (kg/dm^3)	8.93	8.93	2.7	2.7
Melting point (°C)	1083	1083	660	660
Specific heat (kJ/kg·K)	0.39	0.39	0.92	0.92
Thermal conductivity at (Watts/K·m)	395	395	210	210
Electrical conductivity (MS/m)	56	56	36	36
Thermal conductivity (W/K·m)	395	395	210	210
Tensile strength (N/mm^2)	220	220	70	70

6.3.3 Insulating materials

Paper insulated cables have been extensively used up to 33 kV in the distribution network in our country for over half a century. Till the early 1950's paper insulated cables were being imported for UK, U.S.A., and Germany etc. In the mid 1950's a few cable companies started manufacturing these cables upto and including 33 kV in collaboration with foreign manufacturers. After a decade or so, PVC compound was made available in our country. To check the drainage on foreign exchange, the Government banned the manufacture of paper insulated cables for low voltage application and forced the use of PVC insulated cable. The PVC compound was further improved and its application was extended to higher voltages. The relatively high dissipation factor of PVC cables has restricted its use up to 6.6 kV only. With the availability of XLPE compound, the manufacturers in the country started making XLPE insulated PVC sheathed cables. The first XLPE cable was manufactured in 1978 in the country. Subsequently several companies entered the field for manufacturing XLPE insulated Cable. The voltage range for such cables was gradually extended and at present XLPE cables of 132 kV and above voltage rating are in service in the country.

Materials used as insulation for power cables can be broadly divided as follows

a) Impregnated paper
b) Synthetic dielectrics.

6.3.4 Impregnated paper

For many years, the preferred choice of cable dielectric for any voltage range was impregnated paper. Paper for cable making consists of a felted mat of long cellulose fibres derived by chemical treatment of wood pulp. The thickness of paper normally used is from 65 to 190 micrometer. The density varies from 650 to 1000 kg/m^3. The required electrical properties of the paper are obtained by drying the paper to about 120 °C and impregnating it with suitable oils and compounds. The loss angle and permittivity of impregnated paper are related to the density of the paper. The impregnated paper is to be protected from ingress of moisture by metallic sheaths like lead, aluminium, etc., which are quite heavy and expensive, therefore non-hygroscopic synthetic dielectrics have gradually replaced the oil impregnated paper insulated cables.

6.3.5 Synthetic dielectrics

(i) **Polyvinyl Chloride (PVC)**

PVC dielectric is compound of PVC polymer, plasticizers, hardeners, lubricants, fillers and pigments. These components are mixed together by the cable manufacturers in accordance with specific instructions in order to obtain the required properties.

The advantages of PVC are that it is relatively economical, easy to process and hardly combustible. PVC cables are easy to install. The disadvantages are:

- A thermoplastic material
- Low temperature brittleness
- Electrical properties are only sufficient for medium voltage cables (up to 6.6 kV)
- Softening at higher temperature
- Generation of hydrochloride in case of fire.

(ii) **Polyethylene (PE)**

Polyethylene is thermoplastic insulating material with molecules consisting of carbon and hydrogen atoms only i.e., a non-polar material with excellent electrical characteristics. PE has been used for many years as electrical insulating material for medium and high voltage cables. The main disadvantages are

- Melting at relatively low temperature
- Combustible and
- Life affected by partial discharge and ingress of moisture leading to water treeing, etc.

(iii) **Ethylene Propylene Rubber (EPR)**

The term EPR covers polymers of ethylene and propylene in which peroxides are used for cross linking. The use of peroxides for cross linking provides better long term resistance to aging and electrical properties. The main disadvantages of EPR are

- Formulation and processing is difficult
- Volume cost is more
- For HV applications, EPR is not recommended due to its higher dielectric loss coupled with higher thermal resistivity.

(iv) **Cross linked polyethylene (XLPE)**

The basic material for XLPE is polyethylene (PE). PE has very good electrical properties, like good dielectric strength, very low dissipation factor etc. however, PE has a linear molecular structure of long chains of atoms which can move freely with respect to each other as the temperature is raised. Consequently, the mechanical strength of polyethylene decreases significantly at 75 °C and above 100 °C, the resulting deformation becomes considerable. The improved thermal characteristics of PE are obtained by converting the linear molecular structure of PE into a multi-directional molecular structure having a large number of cross lines between molecular chains. The rigid structure of XLPE greatly increases the resistance to deformation at elevated temperature and significant increase in mechanical strength at temperature well above the melting point of the base polymer, resulting in higher continuous operating temperature of 90 °C and short circuit temperature of 250 °C.

The disadvantages are:

- Process is difficult
- Formation of micro voids during cross linking reactions, which results in reduced life especially in cables operating at high stresses
- Life affected by electrical partial discharges and ingress of water, etc.

6.4 CURRENT CARRYING CAPACITY

During service operation, cables suffer electrical losses which appear as heat in the conductor, insulation and metallic components. The current rating is dependent on the way this heat is transmitted to the cable surface and then dissipated to surrounding. Temperature is clearly an important factor and is expressed as a conductor temperature to establish a datum for the cable itself. A maximum temperature is fixed for a reasonable maximum life, which is commonly the limit for the insulation material, without undue aging. Then by choosing a base ambient temperature for the surrounding, a permissible temperature rise is available from which a maximum cable rating can be calculated for a particular environment.

The economic conditions imposed on cable users by present day costs call for current loadings to be as high as possible, consistent with a satisfactory expectation of cable life. The upper limit to current loading is determined by the maximum permissible power loss which can be tolerated, expressed either in terms of a maximum voltage drop along the cable or maximum conductor temperature rise above the surroundings. The latter limit provides the basis for the normal continuous current rating of a cable. The relation between current loading and conductor temperature rise is determined by the conductor material and the thermal properties of the insulting material of the cable surroundings. Thermoplastic insulation softens with increase in temperature, and PVC loses the plasticising agent at an increasing rate, so that, for a given life expectation, an upper limit can be set to the temperature at which such insulation should be operated continuously. Upper limit can also be set for the overload and short circuit temperatures. The penalty for exceeding these temperatures may be a drastic reduction in cable life and it is important to determine cable temperature accurately under various operating conditions.

Calculation of the continuous current rating for a cable is based on a thermal *Ohm law*. Heat produced in the conductor and armour is dissipated to the environment through the insulation and serving of the cable. Each part of this heat-flow path contributes for the overall temperature rise of the conductor above the cable environment in proportion to its thermal resistance.

6.4.1 Constants employed for current rating calculations

Operating temperatures
The operating temperature of a cable is based on the maximum permissible continuous operating temperature of the conductor. The conductor temperature is determined by the maximum temperature at which the dielectric and other component parts of the cable are able to operate continuously without undue degradation and reduction in useful life span.

Polyvinyl chloride (PVC)
At present, a conductor temperature of 70 °C is considered satisfactory for general-purpose insulation. It is emphasised that this temperature refers to continuous use and should not be confused with short circuit temperature, which is somewhat greater. PVC insulation compounds are available that can operate continuously at temperature in excess of 70 °C, but cables having current ratings calculated for operation at these higher temperatures must be specially designed to have suitable high-temperature resistant coverings, and it is important also to consider the reduced margin of safety between the higher operating temperature and the maximum allowable short circuit temperature, which is close to the softening temperature of the insulation.

Polythene
At present a value of 70 °C is considered to be the maximum safe continuous operating temperature for insulation-grade polythene. This operating temperature, while ensuring a long life of the dielectric, is much closer to the softening temperature of the insulation than in that of PVC. While various types of polythene are available which have increased softening temperatures (e.g. high density, irradiated), the maximum continuous operating temperature remains virtually the same, because with prolonged operation at temperature much in excess of about 70 °C, oxidation of the polythene is difficult to control with available anti-oxidants.

Ambient temperature
The ambient temperature i.e., the temperature of the surrounding if the cable were absent, can be determined by measurement over a suitable period and the highest value taken for rating purposes. However, it is not always practicable to make these measurements, and standard values, applicable to given location or country, are usually assumed. As per IEC standards the accepted ambient temperature for cables in soil is 20 °C and for cables in air outside buildings 30 °C. The higher value of temperature 30 °C is used for the air temperature inside buildings. Rating factors are available to modify ratings, calculated on this assumption, for conditions known to be different.

6.4.2 Electrical resistivity

1. Conductor resistance
The conductor resistance R_0 is usually obtained by taking the resistance at 20 °C and conductor size in question and multiplying it by up to date temperature correction factor to obtain the value for cable at its working temperature.

2. Steel wire armour

The resistance of steel armour is usually based on the calculated cross-sectional area of the armour and a resistivity of 14.6×10^6 ohm-cm at 20 °C and the temperature co-efficient is 0.0045.

6.4.3 Thermal resistivity

Thermoplastic insulation and sheathing

The cable makers association has determined the thermal resistivity of various types of cable insulation and sheathing by laboratory tests. In these experiments, a copper conductor covered with the appropriate material under test was installed in a draught-free assembly, and a steady current of known value was passed through the conductor. The temperature of the conductor was accurately measured by thermocouples. The temperature of the outer face of the covering was obtained by measuring the resistance of a spiral of fine wire lapped around the outside of the core insulation, the temperature coefficient of the fuse wire spiral being known. The insulation core was covered by a layer of sheathing material which also completely covered the fine-wire spiral. From knowledge of the temperature of the inner surface of the insulation (i.e., conductor temperature) the insulation outer surface temperature, the power input into the conductor and the thickness of insulation, it is possible to calculate the thermal resistivity of various types of materials.

6.4.4 Bedding and serving materials

The compounded fibrous materials and bitumen compounds employed for bedding and serving are usually credited with a thermal resistivity of 500°C cm/W.

6.4.5 Soil thermal resistivities

It is normal practice to quote buried cable rating of soil thermal resistivity of 2.5 km/W. However, it is emphasised that it is preferable to take actual measurement of the soil thermal resistivities along the cable route if at all practicable. A very convenient method of measurement is to use a needle probe at intervals along the proposed route. The rating should be based on the highest values found. Non-homogeneous soil and those in which it is difficult to insert the probe are better measured by the use of a heated sphere. This method is lengthier than the probe method but is more reliable. It must be remembered that seasonal variations of thermal resistivity may be quite important and that certain soils show a tendency to dry out, not only from the action of atmosphere conditions but also from the action of heat from the buried cable, thus increasing the effective thermal resistance of the soil.

Investigations show that, among other things, the nature of soil and local back fill around the cable, compaction of these media, level of the water table and the characteristic of the load cycle are important factors which should be carefully taken into account when deciding the rating of a buried cable. In general, any of the following features indicates the need for special caution; sandy well-drained soil, particularly where the grains tend to be of core size; badly compacted soil and a cable carrying a continuous load or a load with a low cycle factor for long period during the dry season. It should also be noted that the most critical part of the thermal circuit in the soil is the first few millimeters around the cable, so that if sand is used to embed the cables, it should contain a wide range of particle sizes.

6.5 SHORT-CIRCUIT RATINGS

It happens frequently that the conductor size necessary for an installation is dictated by its ability to carry short circuit rather than sustained current. During a short circuit there is a sudden inrush of current for a few cycles followed by a steadier flow for a short period until the protective devices operates, normally between 0.2 and 3 seconds. During this period the current falls off slightly due to the increase of conductor resistance with temperature, but for calculation purpose it is assumed to remain steady.

At the commencement of the short circuit the cable may be operating at its maximum permissible continuous temperature and the increase in temperature caused by the short circuit is the main factor in deriving acceptable ratings. However, the current may be 20 or more times greater than the sustained current and it produces thermo-mechanical and electromagnetic forces proportional to the square of the current. The stresses induced may themselves impose an operating limit unless they can be contained adequately by the whole installation.

Short circuit ratings are regarded as internal ratings. Their calculation is based on an adiabatic equation and is not affected by external considerations. The short circuit ratings are derived from the equation

$$I_{sc} = \frac{K \times A}{\sqrt{t}} \tag{6.1}$$

where,

K = a constant combining temperature limits and properties of conductor material
A = area of conductor cross section mm^2
I_{sc} = short circuit current in ampere
t = time duration of the fault in seconds.

The values of the constant are given in Table 6.2.

Table 6.2 Values of conductor/temperature constant K

Insulation material	Conductor material	Operating temperature °C	Short circuit temperature °C	K factor
PVC	Copper	70	160	116
PVC	Aluminium	70	160	76
XLPE	Copper	90	250	143
XLPE	Aluminium	90	250	92

6.6 VOLTAGE DROP CALCULATION

Calculation of voltage drop on a power distribution network is another important aspect of system design. Voltage drop occurs on all parts of the power system network when a load current flows through its elements having finite impedance. When designing distribution systems, the system design engineer must see that the voltage drop from the point of commencement of supply to the farthest apparatus shall not exceed 3% of the system voltage. This is the statutory requirement as per the IE Rules 1956. The voltage drop limits specified in the standards refer to the normal operating conditions and need not be applied to situations of motor starting, simultaneous switching, etc.

The supply authority is under statutory obligation to supply the system voltage within certain percentage of the normal system voltage (See chapter 1 for details). A voltage much lower than the normal voltage can cause damages due to overheating of the motors, unwanted tripping of the starters, dimming of the lamps etc. A voltage much higher than the normal voltage also can cause damages like insulation deterioration, saturation of transformers and so on. Therefore it is the responsibility of the system designers to see that the right kind of cables and wires are selected so that the problems arising out of undue voltage drops are minimised.

The selection of cables for a distribution system is not only dependent on the voltage drop criterion. The short circuit withstand capability must also be taken into consideration. However, this aspect is more relevant in the upper stream of the distribution networks where the short circuit levels are high.

Voltage drop calculation is normally carried out on the assumption that the load on the three phase systems is balanced. This implies that the magnitude of the current in all the phases is the same and no current flows through the neutral wire. However, when calculating the voltage drop in a single phase system, the magnitude of the current is the same in the phase conductor and the neutral conductor. Hence voltage drop in both the conductors must be accounted for.

The voltage drop is determined by multiplying the complex value of the line current by the complex impedance value.

$$V_D = I_L \times Z_C \tag{6.2}$$

Where,

V_D = Voltage drop in volts
I_L = Complex line current in ampere
Z_C = Complex impedance of the cable in ohm

$$\%\text{Voltage Drop} = \%V_D = \frac{|V_D|}{|V_{LN}|} \times 100 \tag{6.3}$$

Where,

$|V_D|$ = magnitude of the voltage drop
$|V_{LN}|$ = magnitude of the line to neutral voltage.

The data regarding the cable impedances can be obtained from tables. Note that the data shown are the impedance values in ohm to neutral in three phase systems. The impedance values are normally expressed in ohm per kilometer length of the cable. Hence the actual cable impedance values are calculated using the formula:

For three phase systems

$$R_C = \frac{\text{Resistance in ohm per kilometer}}{1000} \times \text{Length of cable (m)} \tag{6.4}$$

$$\text{and } X_C = \frac{\text{Reactance in ohm per kilometer}}{1000} \times \text{Length of cable (m)} \tag{6.5}$$

For single phase system

$$R_C = 2 \times \frac{\text{Resistance in ohm per kilometer}}{1000} \times \text{Length of cable (m)} \tag{6.6}$$

$$\text{and } X_C = 2 \times \frac{\text{Reactance in ohm per kilometer}}{1000} \times \text{Length of cable (m)} \tag{6.7}$$

***Note:** The resistance of the cable is negligible for cross sections above 500 mm^2 and the inductive reactance is negligible for cable cross sections below 50 mm^2. In the absence of information regarding the inductive reactance, the value can be assumed to be equal to 0.08 per kilometer.*

6.6.1 Simplified method

Calculation of the voltage drops can be carried out very easily with the help of Table 6.3, which gives *the phase to phase voltage drop* per metre length of cable in terms of:

- Type of utilisation: Motor application ($\cos\phi = 0.8$), motor starting ($\cos\phi = 0.35$), and lighting ($\cos\phi = 1$)
- Type of cable: Single phase or three phase.

The voltage drop is given by

$$V_D = k \times I_L \times l \text{ mV} \tag{6.8}$$

Where,

k = Value of constant as given in the table
I_L = Load current in ampere
l = Length of cable in meters

Table 6.3a Voltage drops of aluminium cables in milli-volts per ampere per meter

Aluminium cross section mm^2	Single-phase circuits: Motor loads: Normal running (PF)=0.8	Single-phase circuits: Motor loads: Starting (PF)=0.35	Single-phase circuits: Lighting (PF=1)	Three-phase circuits: Motor Loads: Normal running (PF)=0.8	Three-phase circuits: Motor Loads: Starting (PF)=0.35	Three-phase circuits: Lighting (PF=1)
6	9.1	4.1	11.2	8	3.6	9.5
10	6.1	2.9	7.5	5.3	2.5	6.2
16	3.7	1.7	4.5	3.2	1.5	3.6
25	2.36	1.15	2.8	2.05	1	2.4
35	1.5	0.75	1.8	1.3	0.65	1.5
50	1.15	0.6	1.29	1	0.52	1.1
70	0.86	0.47	0.95	0.75	0.41	0.77
120	0.64	0.37	0.64	0.56	0.32	0.55
150	0.48	0.3	0.47	0.42	0.26	0.4
185	0.39	0.26	0.37	0.34	0.23	0.31
240	0.33	0.24	0.3	0.29	0.21	0.27
300	0.29	0.22	0.24	0.25	0.19	0.2
400	0.24	0.2	0.19	0.21	0.17	0.16

Table 6.3b Voltage drops of copper cables in millivolts per ampere per meter

VOLTAGE DROP IN PVC INSULATED COPPER CABLES				
	Installation method (clipped direct)			
Conductor cross sectional area	Twin core		Three core	
	Current carrying capacity	Volt drop/A/m	Current carrying capacity	Volt drop/A/m
mm^2	A	mV	A	mV
1.5	20	29	18	25
2.5	29	18	24	16
4	37	12	31	9.5
6	48	7.4	41	6.3
10	66	4.3	56	3.8
16	86	2.7	73	2.3
25	115	1.8	97	1.6
35	142	1.3	119	1.1
50	168	0.92	147	0.81
70	209	0.65	180	0.57
95	257	0.48	219	0.42
120	295	0.4	257	0.34
150	337	0.32	295	0.29
185	390	0.29	333	0.24
240	461	0.25	399	0.18
300	523	0.23	451	0.18
400	589	0.22	523	0.17

EXAMPLE 6.1

A three-phase, three-core 70 sq.mm aluminium cable, 50 meters long, supplies a three phase 415 V induction motor with the following loading:

100 A at 0.85 power factor on normal loading and

560 A at 0.35 power factor at starting.

Determine the voltage drop at the motor terminals: a) in normal operation and b) during starting?

Solution:

From Table 6.3a the voltage drop in milli-volt per ampere per metre for normal operation is 0.75.

$$\text{Therefore, the voltage drop in the cable} = 0.75 \times 100 \times 50$$

$$= 3750 \text{ mV} = 3.75\text{V}$$

$$\text{The percentage voltage drop } \%V_D = \frac{3.75}{415} \times 100$$

$$= 0.9\%$$

This value is less than the permissible limit of 3%. From Table 6.3a the voltage drop in milli-volts per ampere per metre for starting condition is 0.41 mV.

Therefore, the voltage drop during starting $= 0.41 \times 560 \times 50$ mV

$= 11.48$ V

$$\text{and the \% voltage drop \%}\ V_D = \frac{11.48}{415} \times 100$$

$= 2.766\%$, a value which is satisfactory.

EXAMPLE 6.2

A three-phase 3.5 core 120 sq.mm aluminium cable is connected between the MSB and SSB of a commercial building with predominant lighting loads. The length of the cable is 50 m and is carrying a balanced load current of 150 A. From the SSB a single phase light distribution board is connected using a 20 m long 2 core 6 sq.mm PVC insulated Al cable. The LDB is supplying a load current of 20A. Calculate the voltage drop up to the LDB.

Solution:

Voltage drop in the aluminium cable

From table

$V_D = 0.55$ mV per ampere per metre

$= 0.55 \times 150 \times 50$ mV

$= 4.125$ V (phase to phase)

$= 2.38$ V (phase to neutral)

Voltage drop in the single phase 6 mm^2 Al cable (from Table 6.3a)

$V_D = 11.2$ mV per ampere per metre

$= 11.2 \times 20 \times 20$ mV

$= 4.48$ V

Therefore total voltage drop $= 2.38 + 4.48$

$= 6.86$ V

$$\text{Total percentage voltage drop from the MSB to the LDB} = \frac{6.86}{240} \times 100$$

$= 2.86\%$.

This is within the allowable limits.

6.6.2 Voltage drop due to motor starting

The starting of induction motors generally produces a voltage dip on the system. This voltage dip sometimes may be severe enough to cause nuisance to other customers and also may cause control relays and motor starter coils to drop out, sensitive electronic devices to malfunction etc. The excessive voltage drop caused during the starting of the motors is due to the large starting current drawn by the motor which can be as high as 6 to 7 times the full load current of the motor.

Two different methods are available for determining the voltage drop due to motor starting. In the first method, the motor may be represented as a constant impedance load during starting as shown in Figure 6.1 and the voltage drop can be calculated using the voltage divider principle. The second method which is a more common and simpler approach represents the motor during starting as a constant current load and is shown in Figure 6.1.

Here,

V_{LN} = line to neutral voltage of the supply
V_{MN} = line to neutral voltage across the motor
I_{BR} = blocked rotor current of the motor (starting current)
Z_S = equivalent impedance of the source and cables.

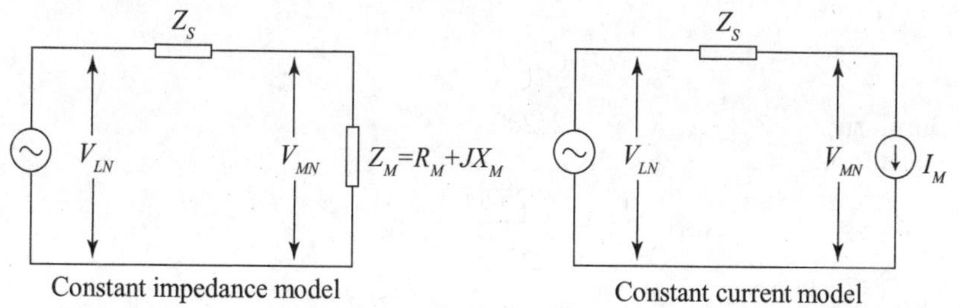

Figure 6.1

EXAMPLE 6.3
A 37 kW (50 HP), 415 V, three-phase squirrel cage induction motor is started against full load using a direct on line starter with rated voltage applied from a source whose impedance including the impedance of the cables is 0:01 + j 0:025 ohm per phase. Assuming a blocked rotor power factor of 0.4 lagging and a blocked rotor kVA per HP of 6, calculate the voltage drop during starting using (a) constant impedance model and (b) constant current model.

Solution:
a) Constant impedance model
At the instant of starting the motor is in blocked rotor state. The blocked rotor kVA rating of the motor at the instant of starting

$$kVA_{BR} = \text{kVA per HP} \times \text{HP}$$
$$= 6 \times 50$$
$$= 300 \text{ kVA}$$

The starting current (blocked rotor) $= \dfrac{300000}{\sqrt{3} \times 415}$

$$= 417.362 \text{ A}$$

$$\text{Active power during starting} = 300 \times 0.4$$
$$= 120 \text{ kW}$$
$$\text{Reactive power during starting} = 300(1 - 0.4^2)$$
$$= 252.00 \text{ kVAr}$$

The resistance and the reactance values of the motor can be calculated using the above results.

$$3 \times I_{BR^2} R = 120 \text{ kW}$$
$$R = \frac{120000}{3 \times 417.362^2}$$
$$= 0.23\ \Omega$$

Similarly, $3 \times I_{BR^2}\, X = 252000$

$$X = \frac{252000}{3 \times 417.362^2}$$
$$= 0.482\ \Omega$$

Applying the voltage divider rule,

$$\text{The voltage across the motor } V_{MN} = \frac{V_{LN}}{(Z_S + Z_M)} Z_M$$
$$= \frac{239.6(0.23 + j0.482)}{(0.01 + j0.025 + 0.23 + j0.482)}$$
$$= 239.6\left(\frac{0.23 + j0.482}{0.24 + j0.507}\right)$$
$$= 227.53 \text{ V}$$
$$\text{The percentage voltage drop } \%V_D = \frac{239.6 - 227.53}{239.6} \times 100$$
$$= 5.03\%$$

b) Constant current model

Motor current at starting = 417.362 A

$$\%V_D = I_{BR} \times Z_S$$
$$= 417.362(0.01 + j0.025)$$

$$= 417.362 \times 0.02693$$
$$= 11.24 \text{ V}$$
$$\text{Percentage voltage drop } \%V_D = \frac{11.24}{239.6} \times 100$$
$$= 4.69\%$$

It can be seen that the results of both methods are in close agreement and the second method need much lesser computations.

6.6.3 Voltage drop due to transformers

Another source of voltage drop in the power distribution is the presence of transformers. The voltage drop across the transformer can be calculated using the name plate details of the transformer. Normally the percentage impedance or the per-unit impedance of the transformer will be available from the name plate.

$$\text{The per unit impedance} = \frac{\text{Percentage impedance}}{100} \tag{6.9}$$
$$\text{Actual impedance} = \text{Per unit impedance} \times \text{Base impedance} \tag{6.10}$$
$$\text{and, base impedance} = \frac{\text{Base kV}^2}{\text{Base MVA}} \tag{6.11}$$

When computing the voltage drop, it is a common practice to refer the transformer impedance to the low voltage side of the transformer. If the percentage impedance and the $\frac{X}{R}$ ratio are available, the resistance and the inductive reactance can be calculated as follows.

$$\theta = \tan^{-1} \frac{X}{R} \tag{6.12}$$
$$R_T = Z_T \cos\theta \tag{6.13}$$
$$X_T = Z_T \sin\theta \tag{6.14}$$
$$\%R_T = \%Z_T \cos\theta \tag{6.15}$$
$$\%X_T = \%Z_T \sin\theta \tag{6.16}$$
$$Z_T = R_T + jX_T \tag{6.17}$$

And the voltage drop across the transformer $= I_L \times Z_T$ (6.18)

EXAMPLE 6.4

Determine the voltage drop across a 250 kVA, 11 kV/415 V, three phase, delta-star connected transformer having a percentage impedance of 4.5% and an $\frac{X}{R}$ ratio of 6. The transformer is delivering full load at 0.8 power factor lagging.

Solution:

Applying equation (6.9) to (6.18)

Per unit impedance of the transformer $= \frac{4.5}{100} = 0.045$

Base impedance of the transformer with reference to the LV side $= \frac{0.415^2}{0.250} = 0.6889$

The impedance of the transformer with reference to the LV side

$$Z_T = \text{Base impedance} \times \text{per unit impedance}$$

$$= 0.6889 \times 0.045$$

$$= 0.0310\ \Omega$$

Using equation (6.12)

$$\theta = \tan^1 6$$

$$= 80.537°$$

Full load current of the transformer on the LV side $= \frac{250 \times 1000}{\sqrt{3} \times 415}$

$$= 347.80 \text{ A}$$

Therefore, the voltage drop across the transformer $= 347.8 \times 0.031$

$$V_D = 10.782 \text{ V}$$

Percentage voltage drop $\%V_D = \frac{10.782}{239.6} \times 100$

$$= 4.5\%$$

Typical values of the resistance, reactance and the impedance of 11 kV/415 V distribution transformers are given in Table 6.4.

Table 6.4 Typical impedance values of 11 kV/415 V distribution transformers

Rated power (kVA)	Oil immersed transformers				Cast resin transformers			
	Impedance voltage (%)	R (mW)	X (mW)	Z (mW)	Impedance voltage (%)	R (mW)	X (mW)	Z (mW)
100	4	37.9	59.5	70.6	6	37.0	99.1	105.8
160	4	16.2	41.0	44.1	6	18.6	63.5	66.2
200	4	11.9	33.2	35.3	6	14.1	51.0	52.9
250	4	9.2	26.7	28.2	6	10.7	41.0	42.3
315	4	6.2	21.5	22.4	6	8.0	32.6	33.6
400	4	5.1	16.9	17.6	6	6.1	25.8	26.5
500	4	3.8	13.6	14.1	6	4.6	20.7	21.2
630	4	2.9	12.9	13.2	6	3.5	16.4	16.8
800	6	2.9	10.8	11.2	6	2.6	13.0	13.2
1000	6	2.3	10.3	10.6	6	1.9	10.4	10.6
1250	6	1.8	8.3	8.5	6	1.5	8.3	8.5
1600	6	1.4	6.5	6.6	6	1.1	6.5	6.6
2000	6	1.1	5.2	5.3	6	0.9	5.2	5.3

Table 6.5 650/1100V, 3.5 Core, Aluminium Conductor, PVC insulated and Sheathed, Un armoured (AYY) and Armoured (AYFY and AYWY)

Nominal area of conductor cross-section mm^2	Max DC conductor resistance at 20 °C W/km	A.C Current rating			
		In air at 40 °C		In Ground at 30 °C	
		PVC Insulation A	HRPVC Insulation A	PVC Insulation A	HRPVC Insulation A
25	1.200	70	84	76	87
35	0.868	86	103	92	105
50	0.641	105	126	110	125
70	0.443	130	156	135	154
95	0.320	155	186	165	188
120	0.253	180	216	185	211
150	0.206	205	246	210	239
185	0.164	240	288	235	268
240	0.125	280	336	275	314
300	0.100	315	378	305	348
400	0.0778	375	450	335	382

The general procedure for voltage drop calculation is summarised below.

Step 1: From the given data, obtain the system impedance values.
Step 2: Determine the load supplied at each segment of the system.
Step 3: Determine the load current magnitude and phase angle.
Step 4: Calculate the voltage drop along each segment of the circuit starting from the point of commencement of supply (Point of common coupling).

Step 5: Add the percentage voltage drops along each segment and check for conformity with standard values.

The details of aluminium cables are given in Tables 6.5 to 6.8 (Courtesy: Reliance Cables Ltd.)

Table 6.6 650/1100 V 3 Core Aluminium Conductor PVC Insulated and Sheathed Unarmoured (AYY) and Armoured (AYFY and AYWY)

Nominal Area of Conductor sq.mm	Max D.C Conductor Resistance at 20 °C Ω/km	A.C Current Rating							
		In air at 40 °C				In ground at 30 °C			
		PVC insulation Amps		HRPVC insulation Amps		PVC insulation Amps		HRPVC insulation Amps	
		2C	3C	2C	3C	2C	3C	2C	3C
10	3.080	56	47	67	56	59	51	67	58
16	1.910	72	64	86	77	75	66	86	75
25	1.200	99	84	119	101	97	86	111	98
35	0.868	120	105	144	126	120	100	137	114
50	0.641	150	130	180	156	145	120	165	137
70	0.443	185	155	222	186	170	140	194	160
95	0.320	215	190	258	228	205	175	234	200
120	0.253	240	220	288	264	230	195	262	222
150	0.206	270	250	324	300	265	220	302	251
185	0.164	305	290	366	348	300	240	342	274
240	0.125	350	335	420	402	335	270	382	308
300	0.100	395	380	474	456	370	295	422	336
400	0.0778	455	435	546	522	410	325	467	371
500	0.0605	490	480	588	576	435	345	496	393
630	0.0469	560	550	672	660	485	390	553	445
800	0.0367	640	630	768	756	530	440	604	502
1000	0.0291	740	720	888	864	580	490	661	559

Table 6.7 650/1100V, 4 Core, Aluminium Conductor, PVC insulated and Sheathed, Unarmoured (AYY) and Armoured (AYFY and AYWY)

Nominal area of conductor mm^2	Max DC Conductor Resistance at 20 °C Ω/km	A.C Current Rating			
		In air at 40 °C		In ground at 30 °C	
		PVC Insulation A	HRPVC Insulation A	PVC Insulation A	HRPVC Insulation A
4	7.410	23	28	28	32
6	4.610	30	36	35	40
10	3.080	40	48	46	52
16	1.910	51	61	60	68
25	1.200	70	84	76	87
35	0.868	86	103	92	105
50	0.641	105	126	110	125
70	0.443	130	156	135	154
95	0.320	155	186	165	188
120	0.253	180	216	185	211
150	0.206	205	246	210	239
185	0.164	240	288	235	268
240	0.125	280	336	275	314
300	0.100	315	378	305	348
400	0.0778	375	450	335	382

Table 6.8 11/11 kV, Three Core Aluminium conductor, XLPE insulated and PVC sheathed, Screened, Flat Strip armoured cables. [A2XFY (P)]

Conductor Nominal Area	Appx. Weight of Cable	Max DC Resistance at 20 °C	Max DC Resistance at 90 °C	Appx. Reactance at 50 Hz	Appx. Capacitance per phase	Continuous Current Rating		Short circuit rating for one second
						In ground at 20 °C	In air at 30 °C	
mm^2	kg/km	W/km	W/km	W/km	F/km	A	A	kA (rms)
35	3560	0.8680	1.110	0.136	0.159	120	146	3.29
50	4000	0.6410	0.8220	0.129	0.175	141	174	4.7
70	4510	0.4430	0.5680	0.119	0.196	172	216	6.58
95	5130	0.3200	0.4110	0.113	0.224	205	260	8.93
120	5670	0.2530	0.3250	0.110	0.241	233	282	11.3
150	6220	0.2060	0.2650	0.106	0.259	261	321	14.1
185	6970	0.1640	0.2110	0.103	0.281	294	364	17.4
240	7990	0.1250	0.1620	0.0985	0.314	340	429	22.6
300	9050	0.1000	0.1300	0.0952	0.340	384	492	28.2
400	10570	0.0778	0.1020	0.0929	0.375	439	575	37.6
500	12280	0.0605	0.0810	0.0890	0.417	497	661	47

EXAMPLE 6.5

For the power distribution network shown in Figure 6.2, determine the percentage voltage drops in

a) 300 sq.mm cable
b) 185 sq.mm cable
c) 50 sq.mm cable and
d) Total %voltage drop up to the 22 kW motor.

The load on the MSB is 400 kVA at 0.9 pf lagging and that on the SSB is 100 kVA at 0.85 pf lagging.

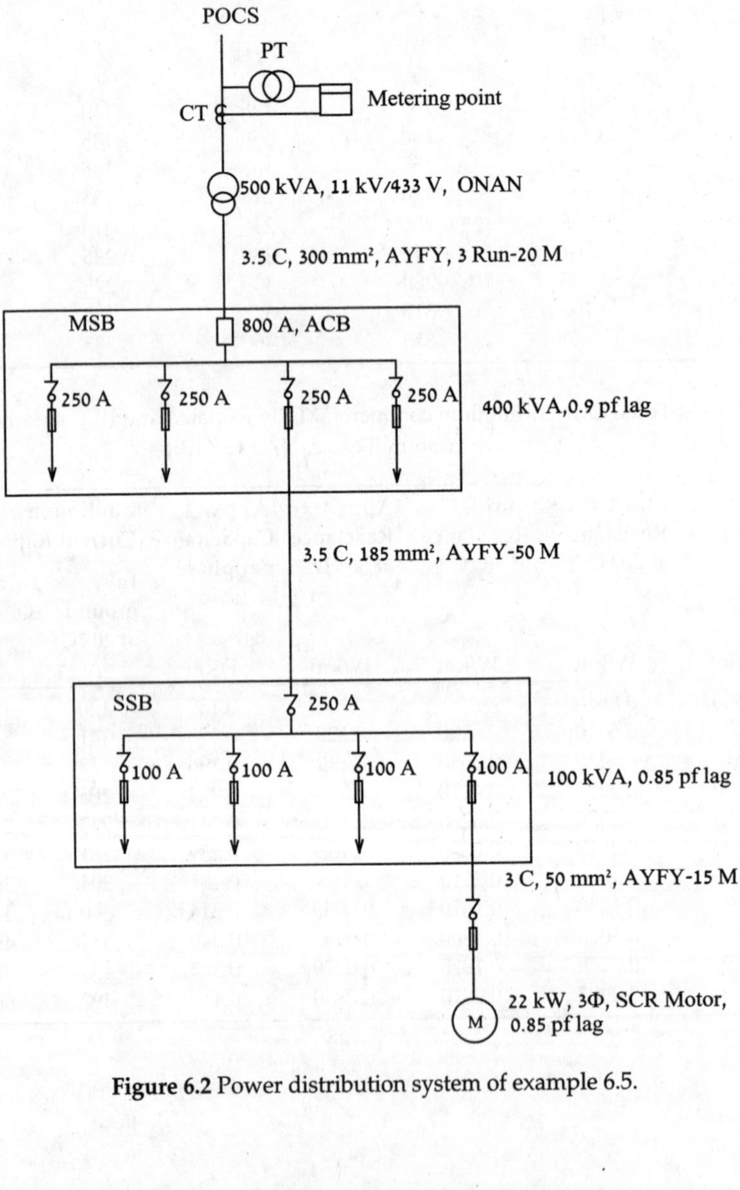

Figure 6.2 Power distribution system of example 6.5.

Step 1: Determination of system impedance values

From Table 6.4, the impedance of the 500 kVA oil filled transformer is

$$R_T = 0.0038\ \Omega$$
$$X_T = 0.0136\ \Omega$$

For the 300 sq.mm 3.5 core cable, from Table 6.7,

$$R_C = 0.1 \text{ ohm/km}$$
$$X_C = 0.08 \text{ ohm/km (assumed)}$$

Therefore the effective resistance of the three runs of 300 sq.mm cable

$$R_C = \frac{1}{3} \times 0.1 \times \frac{20}{1000}$$

$$= 0.00066\ \Omega$$

$$X_C = \frac{1}{3} \times 0.08 \times \frac{20}{1000}$$

$$= 0.000533\ \Omega$$

Similarly for the 185 sq.mm cable

$$R_C = 0.164 \times \frac{50}{1000}$$

$$= 0.0082\ \Omega$$

$$X_C = 0.08 \times \frac{50}{1000}$$

$$= 0.004\ \Omega$$

For the 50 sq.mm cable

$$R_C = 0.641 \times \frac{15}{1000}$$

$$= 0.0096\ \Omega$$

$$X_C = 0.08 \times \frac{15}{1000}$$

$$= 0.0012\ \Omega$$

Step 2: Determination of the power flow in each segment of the system

The loading on the MSB is given as 400 kVA at 0.9 pf lagging. This also will be the power supplied by the transformer. The load on the 185 sq.mm cable as well as the SSB is given as 100 kVA at 0.85 pf lagging.

The load in the 50 sq.mm cable is due to the 22 kW induction motor delivering normal load at 0.85 pf lagging. Assuming a full load efficiency of 0.85, the power drawn by the motor

$$P_M = \frac{22}{0.85} = 25.88 \text{ kW}$$

Solution:

$$\text{Therefore the kVA loading on the 50 sq.mm cable} = \frac{25.88}{0.85}$$

$$= 30.45 \text{ kVA}$$

Step 3: Determination of the line currents and their phase angles

$$\text{The line current on the secondary side of the transformer} = \frac{400000}{\sqrt{3} \times 415}$$

$$= 556.48 \text{ A}$$

Power factor angle ϕ $= \cos^{1}(0.9)$ lagging
$= 25.84°$lagging

$$\text{Current in the 185 sq.mm cable corresponds to 100 kVA} = \frac{100000}{\sqrt{3} \times 415}$$

$$= 139.12 \text{ A}$$

Power factor angle ϕ $= \cos^{1}(0.85)$ lagging
$= 31.79°$lagging

$$\text{Current in the 50 sq.mm cable} = \frac{30.45 \times 1000}{\sqrt{3} \times 415}$$

$$= 42.36 \text{ A}$$

Power factor angle ϕ $= \cos^{-1}(0.85)$ lagging
$= 31.79°$lagging

Step 4: Calculation of voltage drop and the percentage voltage drop and percentage voltage drop
The values of resistances obtained from the tables correspond to a temperature of 20 °C. These values will have to be modified to the practical working temperature of the cable, say 70 °C, using the formula

$$R_{70} = R_{20}(1 + \alpha\Delta_{\theta})$$

where, α = Temperature coefficient of aluminium
Δ_{θ} = Difference in temperature

For the 300 sq.mm cable $R_{70} = 0.00066(1 + 0.0036 \times 50)$
$= 0.00078\ \Omega$

For the 185 sq.mm cable $R_{70} = 0.0082(1 + 0.0036 \times 50)$
$= 0.0097\ \Omega$

For the 50 sq.mm cable $R_{70} = 0.0096(1 + 0.0036 \times 50)$
$= 0.0113\ \Omega$

Therefore

The impedance of the 300 sq.mm cable $= 0.00078 + j\,0.000533$
$= 0.000945\backslash 34.34\ \Omega$

The impedance of the 185 sq.mm cable $= 0.0097 + j\,0.004$
$= 0.0105\backslash 22.41\Omega$

The impedance of 50 sq.mm cable $= 0.0113 + j\,0.0012$
$= 0.0114\backslash 6.06\Omega$

Magnitude of the voltage drop in 300 sq.mm cable $= 556.48 \times 0.000945$
$= 0.526$ V

Percentage voltage drop $\%V_D = \frac{0.526}{240} \times 100$

$= 0.22\%$

Magnitude of the voltage drop in 185 sq.mm cable $= 139.12 \times 0.0105$
$= 1.461$ V

Percentage voltage drop $\%V_D = \frac{1.461}{240} \times 100$

$= 0.61\%$

Magnitude of the voltage drop in 50 sq.mm cable $= 42.36 \times 0.0114$
$= 0.482$ V

Percentage voltage drop $\%V_D = \frac{0.482}{240} \times 100$

$= 0.201\%$

Therefore total voltage drop from the transformer to the 22 kW motor terminal

Total $\%V_D = 0.22 + 0.61 + 0.201$
$= 1.031\%$

Simplified method
From Table 6.3
The voltage drop per ampere per meter length of 300 sq.mm cable $= 0.25$ mV

Therefore the total voltage drop in the 300 sq.mm cable $= (0.25 \times 556.48 \times 20) \times \frac{1}{3}$

$= 927.5$ mV

The factor $\frac{1}{3}$ takes care of the voltage drop in 3 runs of 300 sq.mm cable.

$$\%V_D = \frac{0.9275}{415} \times 100$$

$$= 0.223\%$$

The voltage drop per ampere per meter of 185 sq.mm cable $= 0.34$ mV

Therefore voltage drop in 185 sq.mm cable $= 0.34 \times 139.12 \times 50$

$= 2.365$ V

$$\%V_D = \frac{2.365}{415} \times 100$$

$$= 0.569\%$$

The voltage drop per ampere per meter of 50 sq.mm cable $= 1$ mV

Therefore total voltage drop in 50 sq.mm cable $= 1 \times 42.36 \times 15$

$= 0.635$ V

Therefore

$$\%V_D = \frac{0.635}{415} \times 100$$

$$= 0.153\%$$

Total % voltage drop up to the 22 kW motor $= 0.223 + 0.569 + 0.153$

$= 0.945\%$

The slight discrepancy between the results of the two methods is due to the fact that in the simplified approach the power factor of the currents are taken as 0.8, where as in actual situation the power factors are 0.9, 0.85 and 0.85 respectively.

EXAMPLE 6.6

If the three phase fault level on the 11 kV side of the transformer is 250 MVA, determine whether the cables specified in the example 6.5 will satisfy the short circuit withstand capacity or not.

Solution:

From Table 6.4, the percentage impedance of the transformer is 4%.

Assuming a base of 100 MVA, the per unit impedance of the source

$$Z_S = \frac{100}{250} = 0.4 \text{ pu}$$

And the per unit impedance of the transformer

$$Z_T = \frac{0.04 \times 100}{0.05} = 8.0 \text{ pu}$$

From example, the impedance of the 300 sq.mm cable $= 0.00945W$

Therefore, the per unit impedance of 300 sq.mm cable $= \dfrac{\text{Actual impedance}}{\text{Base impedance}}$

$$= \frac{0.00945}{(0.433)^2} \times 100$$

$$= 5.04 \text{ pu}$$

The fault level on the MSB $= \dfrac{\text{Base MVA}}{\text{Total pu.imp upto MSB}}$

$$= \frac{100}{0.4 + 8.0 + 5.04}$$

$$= 7.44 \text{ MVA}$$

Corresponding fault current $= \dfrac{7.44 \times 10^6}{\sqrt{3} \times 433}$

$$= 9920.29 \text{ A}$$

For PVC aluminium cables, the short circuit capacity for 3 seconds short circuit duration is given by

$$I_{SC} = \frac{K \times A}{\sqrt{t}}$$

$$= \frac{76 \times 300}{\sqrt{3}} = 13.163 \text{ kA}$$

This value is much above the expected fault current of 9.920 kA.

Similarly, the impedance of the 185 sq.mm cable from Table 6.5 is 0.0105 ohm.

The per unit impedance of the 185 sq.mm cable $= \dfrac{0.0105}{(0.433)^2} \times 100$

$$= 5.6 \text{ pu}$$

So the fault level at the SSB

$$MVA_{SC} = \frac{100}{0.4 + 8 + 5.04 + 5.6}$$

$$= 5.252 \text{ MVA}$$

Fault current at SSB $\quad I_{SC} = \dfrac{5.252 \times 10^6}{\sqrt{3} \times 433} = 7003.00 \text{ A}$

For the 185 sq.mm cable permissible short circuit rating $= \dfrac{76 \times 185}{\sqrt{3}}$

$= 8117.5$ A

This value is again higher than the expected fault current of 7003A.

For the 50 sq.mm cable, the impedance, from example 6.5 is 0.0114 ohm

Per unit impedance of 50 sq.mm cable $= \dfrac{0.0114 \times 100}{0.433^2}$

$= 6.08$ pu

Therefore the fault level on the motor terminal $= \dfrac{100}{0.4 + 8. + 5.04 + 5.6 + 6.08}$

$= 3.98$ MVA

Prospective fault current at the motor terminals $= \dfrac{3.98 \times 10^6}{\sqrt{3} \times 415}$

$= 5538$ A

For the short circuit capacity of the 50 sq.mm cable$= \dfrac{76 \times 50}{\sqrt{3}}$

$= 2193$ A

Here, the short circuit capacity of the cable for 3 second duration is below the actual fault current level and hence a faster protective device which will operate within 0.5 second of the fault will be required to provide the necessary protection for the cable. If the fault duration is limited to 0.5 seconds,

I_{SC} of the 50 sq.mm cable $= \dfrac{76 \times 50}{\sqrt{0.5}}$

$= 5374$ A

This value is very close to the calculated value of the prospective short circuit current 5460 A.

6.7 GENERAL METHODS OF CABLE LAYING

Different methods are adopted for laying of cables. The admissible methods of installation are listed in Table 6.9. There are many different ways of wiring systems and methods of installation of cables.

Table 6.9 Methods for laying the cables and conductors (IEC 60364-5-52)

Type of cables or conductors		Methods of cable installation						
		Clipped direct	**Conduit**	**Cable trunking**	**Cable ducting**	**Cable trays/ ladders**	**On insulators**	**On support wire**
Bare conductors		No	No	No	No	No	Yes	No
Insulated conductors		No	Yes	Yes	Yes	No	Yes	No
Sheathed including armoured cables	One core cables	Yes	Yes	Yes	Yes	Yes	-	Yes
	Multi core	Yes	Yes	Yes	Yes	Yes	-	Yes

Table 6.10 Different methods of installation (Ref: IEC 60364-5-52)

Item No.	Description	Reference method of installation to obtain current carrying capacity
1	Insulated conductors or single core cables in conduit in a thermally insulated wall	A1
2	Multi-core cables in conduit in a thermally insulated wall	A2
3	Insulated conductors or single core cables in conduit on a wooden or masonry wall or spaced less than 0.3 times diameter of cables from the wall	B1
4	Multi-core cables on a wooden or masonry wall or spaced less than 0.3 times the diameter of the cable from it	B2
5	Single core or multi-core cables fixed on, or spaced less than 0.3 times the cable diameter from the wall	C
6	Cables on non-perforated trays	C
7	Multi-core cables in conduit or in cable ducting in ground / Single core cables in conduit or in cable ducting in ground	D
8	Cables on perforated trays	E or F
9	Bare uninsulated conductors on insulators	G

IEC 60364-5-52 standard lists the definition of these methods and these methods are identified with reference letters. Table 6.10 lists these reference methods and these identifications are very useful in sizing the cable with respect to their current carrying capacity.

In determining the actual current carrying capabilities of the cables and conductors for the various types of installation methods, the cable insulation should not be subjected to more than the permissible limits of temperature for the respective types of insulation material. The maximum permissible operating temperature of different insulating materials is given in Table 6.11.

Table 6.11 Maximum operating temperatures of insulation materials (IEC: 60364-5-52)

Type of insulation	Temperature limit °C
Polyvinyl Chloride (PVC)	70 at the conductor
Cross-linked polyethylene (XLPE) and ethylene propylene rubber (EPR)	90 at the conductor
Mineral (PVC coated or exposed to touch)	70 at the sheath
Mineral (bare not exposed to touch and not in contact with combustible material)	105 at the sheath

In order to account for the environmental and other conditions of installation, correction factors are to be introduced to determine the correct cross sectional area of the cables. If I_L is the actual load current, the corrected value of the load current to be taken into account for determining the cable cross-section,

$$I'_L = \frac{I_L}{k_1 k_2 \ldots}$$

where, $k_1, k_2, \ldots$ etc., are the correction or de-rating factors of the cables.

Derating factor for the ambient temperature

The current carrying capacities of the cables when they are laid in air are based on the assumption that the average temperature is 30 °C.

Table 6.12 Derating factor k_1 for cables laid in air (IEC 60364-5-52)

Ambient temperature °C	Insulation material of the cable	
	PVC	XLPE and EPR
15	1.17	1.12
20	1.12	1.08
25	1.06	1.04
35	0.94	0.96
40	0.87	0.91
45	0.79	0.87
50	0.71	0.82
55	0.61	0.76
60	0.50	0.71
65	–	0.65
70	–	0.58

If the cables are located in an atmosphere where the ambient temperature is other than 30 °C, minor correction is to be applied to the current carrying capacity of the cables. The derating factor k_1 is given in Table 6.12.

On the other hand the current carrying capacities of the cables when they are laid in the ground are based on the average temperature equal to 20 °C. The correction factor k_2 required to be applied is given in Table 6.13.

Table 6.13 Derating factor k_2 for cables laid in ground (IEC 60364-5-52)

Ground temperature °C	Insulation material of the cable	
	PVC	XLPE and EPR
15	1.05	1.04
25	0.95	0.96
30	0.89	0.93
35	0.84	0.89
40	0.77	0.85
45	0.71	0.80
50	0.63	0.76
55	0.55	0.71
60	0.45	0.65
65	–	0.60
70	–	0.53

Derating factor for soil thermal resistivity

The current carrying capacities of the cables are based on a soil thermal resistivity of 2.5 k.m/W. When the soil conditions are different suitable correction is to be made in the current carrying capacity of the cables. Based on experience, an empirical relation has been established between the nature of the soil and its resistivity. The empirical value of the correction factor k_3 is given in Table 6.14.

Table 6.14 Correction factor k_3 for the nature of the soil (Source: Schneider electric)

Nature of the soil	Correction factor k_3
Very wet soil	1.21
Wet soil	1.05
Damp soil	1.0
Very dry soil	0.86

Derating factor for bunching or grouping of cables

When conductors or cables are laid in groups or bunched together, there will be a reduction in the overall current carrying capacity of the cables due to the reduction in heat dissipation from the cables. Bunching relates to number of circuits consisting of the following numbers of loaded conductors.

1. Two insulated conductors, two single core cables or one twin core cable.
2. Three insulated conductors or three single core cables or one three core cable.

When more than one circuit are laid in proximity, the current carrying capacity of the cables will have to be derated and the derating factor k_4 is given in Table 6.15.

Table 6.15 De-rating factor k_4 for the bunching of cables (IEC 60364-5-52)

Bunching arrangement	Number of circuits or multi core cables						Reference method
	1	**2**	**3**	**4**	**6**	**9**	
Bunched in air, on a surface, embedded or enclosed	1.0	0.80	0.70	0.65	0.57	0.50	A to F
Single layer on wall, on floor or on perforated tray	1.0	0.85	0.79	0.75	0.72	0.70	C
Single layer fixed directly under a wooden ceiling	0.95	0.81	0.72	0.68	0.64	0.61	C
Single layer on a perforated horizontal or vertical tray	1.0	0.88	0.82	0.77	0.73	0.72	E and F
Single layer on a ladder support or cleats etc.	1.0	0.87	0.82	0.80	0.79	0.78	E and F

When the cables are laid directly in the ground, the bunching factor or the correction factor k_4 shall be as given in Table 6.16.

Table 6.16 Derating factor k_4 for more than one circuit, single-core or multi-core cables laid direct in the ground (IEC 60364-5-52)

Number of circuits	Cable to cable clearance				
	Touching each other	**One cable diameter**	**0.125 m**	**0.25 m**	**0.5 m**
2	0.75	0.8	0.85	0.9	0.9
3	0.65	0.7	0.75	0.8	0.85
4	0.6	0.6	0.70	0.75	0.80
5	0.55	0.55	0.65	0.7	0.80
6	0.50	0.55	0.60	0.7	0.80

Derating for the presence of harmonic currents

Normally, the current carrying capacity of a three phase 4 core cable is on the assumption that only the phase conductors are fully loaded. However, when harmonic currents are present in the system, the neutral current can be significant and even higher than the phase current. This is because, all the third harmonic currents adds up and flow through the neutral. This will definitely affect the current carrying capacity of the cable and a derating is to be applied. This derating factor k_5 is dependent on the percentage of third harmonics present. Table 6.17 gives the derating required for the presence of harmonics.

Table 6.17 Derating factor k5 for accounting third harmonic current (IEC 60364-5-52)

Third harmonics%	Correction factor	
	Selection based on phase current	Selection based on neutral current
0–15	1.0	
15–33	0.86	
33–45		0.86
>45		1.0

EXAMPLE 6.7

An industrial substation with a 1000 kVA, 11 kV/0.433 kV, delta/Star connected ONAN type transformer is being provided with the 11 kV supply from a public distribution network 2.5 km away from the industry through underground cable. Choose the appropriate current capacity of the XLPE cable, if the cable is laid 1m below the ground level. The soil condition is very dry type. The cable is to be laid along with an existing cable with a spacing of 0.125 m. The average ground temperature is given as 30 °C. The short circuit level at the transformer location is 250 MVA.

Solution:

Selection of cable cross section based on current carrying capacity

Full load current of the transformer $= \dfrac{1000}{\sqrt{3} \times 11} = 52.49$ A

Derating factors for the environmental and laying conditions

Buried cable with ground temperature of 30 °C	k_2	= 0.93
Soil type-(very dry)	k_3	= 0.86
Cable configuration, 2 cables with a spacing of 0.125 m	k_4	= 0.85
Presence of third harmonics-nil for delta connected system	k_5	= 1
Effective de rating factor	k	$= k_2k_3k_4k_5$
		= 0.69

Therefore current carrying capacity of the cable $= \dfrac{52.49}{0.69}$

$= 76.07$ A

From Table 6.8, the size of the cable that will safely carry 76.07 A is three-core, 35 mm^2 XLPE cable with a current rating of 120 A.

The resistance per km length of this cable $= 1.10\ \Omega$

And the inductive reactance per km $= 0.136\ \Omega$

Impedance per kilometer $= 1.10 + j0.136$

$= 1.1084\ \Omega$

Total impedance of the cable Z $= 1.1084 \times 2.5 = 2.771\ \Omega$

The voltage drop in the cable at full load $= \sqrt{3}IZ$

$$= \sqrt{3} \times 52.49 \times 2.771$$

$$= 251.93 \text{ V}$$

Therefore % voltage drop $\%V_D$ $= \dfrac{251.93}{11000} \times 100$

$$= 2.29\%$$

This value is well within the permissible limit.

Check for short-circuit withstand capability

Prospective short circuit current at the HT terminal of the transformer $= \dfrac{250000}{\sqrt{3} \times 11}$

$$= 13.122 \text{ kA}$$

From Table 6.8 it is seen that only a cable size of 150 sq.mm will satisfy these short circuit withstand capability. Hence we have to choose a cable with a cross section of 150 sq.mm three core, XLPE cable with a short circuit capacity of 14.1 kA for the purpose.

6.8 SIMPLIFIED APPROACH FOR CABLE SELECTION

In order to make the selection easier, simplified tables are given in IEC 60362-5-42 for most commonly used cable configurations. Tables 6.18 summarises the most frequently used cable configurations for un buried cables and Table 6.19 gives the current ratings of cables for buried cables. Depending on the number of loaded conductors and type of insulation of the cables, the practical value of current rating for a particular cross section of the cables can be directly read from the tables. For example, for a three core 50 sq.mm XLPE Aluminium cable mounted on wall, the current rating will correspond to reference method C, and the current rating will be 136 A. The method is to search along the row corresponding to the reference method up to the type of cable and then descend to the row corresponding to the specified cross section of the cable. In this case move up to column 9 of Table 6.18, along the row corresponding to reference method C and then descend to the row corresponding to the row corresponding to 50 sq.mm Aluminium cable.

Table 6.18 Current carrying capacity of unburied cables (IEC 60364-5-52)

Reference method	Number of loaded conductors and type of insulation											
A1		2 PVC	3 PVC		3 XLPE	2 XLPE						
A2	3 PVC	2 PVC		3 XLPE	2 XLPE							
B1				3 PVC	2 PVC		3 XLPE					
B2			3 PVC	2 PVC		3 XLPE	2 XLPE		2 XLPE			
C					3 PVC		3 PVC	3 XLPE		2 XLPE		
E						3 PVC		2 PVC	3 XLPE		2 XLPE	
F							3 PVC		2 PVC	3 XLPE		2 XLPE
1	2	3	4	5	6	7	8	9	10	11	12	13
Copper (mm²)	**Current carrying capacity ampere**											
1.5	13	13.5	14.5	15.5	17	18.5	19.5	22	23	24	26	-
2.5	17.5	18	19.5	21	23	25	27	30	31	33	36	-
4	23	24	26	28	31	34	36	40	42	45	49	-
6	29	31	34	36	40	43	46	51	54	58	63	-
10	39	42	46	50	54	60	63	70	75	80	86	-
16	52	56	61	68	73	80	85	94	100	107	115	-
25	68	73	80	89	95	101	110	119	127	135	149	161
35	-	-	-	110	117	126	137	147	158	169	185	200
50	-	-	-	134	141	153	167	179	192	207	225	242
70	-	-	-	171	179	196	213	229	246	268	289	310
95	-	-	-	207	216	238	258	278	298	328	352	377
120	-	-	-	239	249	276	299	322	346	382	410	437
150	-	-	-	-	285	318	344	371	395	441	473	504
185	-	-	-	-	324	362	392	424	450	506	542	575
240	-	-	-	-	380	424	461	500	538	599	641	679
Aluminium (mm²)	**Current carrying capacity-ampere**											
2.5	13.5	14	15	16.5	18.5	19.5	21	23	24	26	28	-
4	17.5	18.5	20	22	25	26	28	31	32	35	38	-
6	23	24	26	28	32	33	36	39	42	45	49	-
10	31	32	36	39	44	46	49	54	58	62	67	-
16	41	43	48	53	58	61	66	73	77	84	91	-
25	53	57	63	70	73	78	83	90	97	101	108	121
35	-	-	-	86	90	96	103	112	120	126	135	150
50	-	-	-	104	110	117	125	136	146	154	164	184
70	-	-	-	133	140	150	160	174	187	198	211	237
95	-	-	-	161	170	183	195	211	227	241	257	289
120	-	-	-	186	197	212	226	245	263	280	300	337
150	-	-	-	-	226	245	261	283	304	324	346	389
185	-	-	-	-	256	280	298	323	347	371	397	447
240	-	-	-	-	300	330	352	382	407	439	470	530

Table 6.19 Current carrying capacity of buried cables (IEC 60364-5-52)

Installation method	Size mm^2	No. of loaded conductors and type of insulation			
		Two PVC	**Three PVC**	**Two XLPE**	**Three XLPE**
D			Copper		
	1.5	22	18	26	22
	2.5	29	24	34	29
	4	38	31	44	37
	6	47	39	56	46
	10	63	52	73	61
	16	81	67	95	79
	25	104	86	121	101
	35	125	103	146	122
	50	148	122	173	144
	70	183	151	213	178
	95	216	179	252	211
	120	246	203	287	240
	150	278	230	324	271
	185	312	258	363	304
	240	361	297	419	351
	300	408	336	474	396
D			Aluminium		
	2.5	22	18.5	26	22
	4	29	24	34	29
	6	36	30	42	36
	10	48	40	56	47
	16	62	52	73	61
	25	80	66	93	78
	35	96	80	112	94
	50	113	94	132	112
	70	140	117	163	138
	95	166	138	193	164
	120	189	157	220	186
	150	213	178	249	210
	185	240	200	279	236
	240	277	230	322	272
	300	313	260	364	308

6.9 QUESTIONS

1. What is the significance of cable sizing in electrical system design?
2. What are the different configurations in which cables are available for electrical power distribution?
3. Distinguish between 3 core, 3.5 core and 4 core cables.
4. What is the importance of voltage drop calculations in system design?
5. Determine the voltage drop in a 415 V, three phase feeder consisting of a 70 sq.mm 3.5 core AYFY cable, 150 m long and carrying a load of 75 kVA, 415 V at 0.85 power factor lagging.
6. A 22 kW/30 HP, 415 V, squirrel cage induction motor is started using a direct on line starter. The equivalent impedance per phase of the motor is 0:01 + j0:02 ohms. Assuming a blocked rotor kVA/HP of 5.6 and blocked rotor power factor of 0.35, calculate the percentage voltage drop at starting using a) constant impedance model and b) constant current model.
7. A 37 kW/50 HP, 415 V, three phase squirrel cage motor is having a blocked rotor kVA/HP of 6 and a blocked rotor power factor of 0.35. If the motor is started with a DOL starter, determine the percentage voltage drop at starting. Also determine the maximum number of starts permitted per hour based on the voltage flicker curve.
8. For the power distribution network shown in worked example 6.5, if the loading on the MSB is 300 kVA at 0.85 lagging power factor, on the SSB is 125 kVA with 0.9 power factor lagging and the motor capacity is 15 kW with a full load power factor of 0.88 lagging, calculate the percentage voltage drops at the MSB, SSB and the motor terminals.
9. What will be the percentage voltage drops at the MSB, SSB and the motor terminals of the example 6.5 at starting, if the motor is started using direct on line starting. Assume that the loads on the MSB and SSB are nil and a blocked rotor kVA/HP of 5.6 and a blocked rotor power factor of 0.35 lagging.
10. An industrial facility is supplied from the electricity utility distribution network at 11 kV. The short circuit data of the utility feeder is 150 MVa with an X/R ratio of 1. The service transformer at the industry is 500 kVa, 11 kV/433 V, %Z = 6, delta/star connected with an X/R ratio of 4. The secondary terminal of the transformer is connected to the Main switch board (MSB) with three runs of 300 sq.mm 3.5 AYFY cable of 15 m length. The demand on the MSB is 420 kVA at 0.9 pf lagging. Connection to a sub-switchboard (SSB) is taken from the MSB with a 185 sq.mm, 3.5 AYFY cable of 50 m length. The demand on the SSB is 100 kVA at 0.85 pf lagging. An outgoing feeder from the SSB supplies a three phase induction motor of 15 kW with a full load power factor of 0.85 lagging. The cable size of this feeder is 16 sq.mm, 3 core, AYFY with a length of 10 m. Determine the percentage voltage drops at the MSB, SSB and at the motor terminals when the motor is delivering full load.

QUESTIONS

CHAPTER 7

EXTERIOR LIGHTING

OBJECTIVES

At the end of this chapter you will

- Understand the importance of exterior lighting.
- Know the general requirements for exterior lighting.
- Know how to select the lamp sources for exterior lighting.
- Understand the importance of road lighting.
- Understand the basic design criterion for road and area lighting.
- Be able to solve simple problems in road and area lighting.

7.1 INTRODUCTION

Exterior lighting design is an important aspect of illumination engineering. Lighting design for exterior application will have to consider the following aspects of lighting.

1. Functional lighting and
2. Decorative lighting.

Road lighting, yard lighting (area lighting), flood lighting etc. come under the first category whereas decorative exterior lighting include monumental lighting, facade lighting, and special lighting for festivals. This chapter mainly deals with the design aspects of the first group.

7.2 GENERAL CONSIDERATIONS

The primary aims of exterior lighting are safety and security. Different types of luminaires are used for exterior lighting. These luminaires can be loosely classified into "Static" and "Adjustable" luminaires. Luminaires permanently fixed on top of a pole to produce light in a predetermined manner are called static luminaires. Luminaires mounted with adjustable brackets are referred to as adjustable luminaires. Adjustable luminaires are mostly used for flood lighting and area lighting, although use of static luminaires for area lighting is also common.

The mounting height of the luminaire is an important factor in external lighting. For average roads and public areas pole heights are chosen such that the luminaires can be serviced from the ground. For highway intersections or large parking areas, high mast lighting systems are preferred. Maintenance and servicing of luminaires on a high mast system is difficult. Therefore high mast systems generally use lamps with very high average life or incorporate a lowering mechanism for servicing of lamps. A lighting system with pole height of more than 16 metre is generally referred to as high mast system.

The basic design technique for area and roadways lighting system is the same, whether they use high mast system or regular poles. The design is based on the concept of having the luminaire installed on a pole with an arm to produce light on the surface beneath. Figure 7.1 shows the details of a typical luminaire mounted on a pole for roadway lighting.

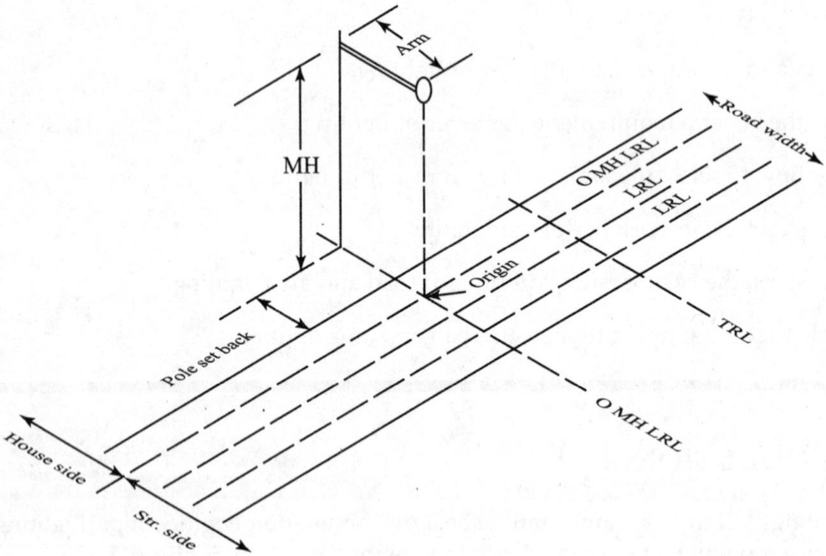

Figure 7.1

7.2.1 Definition of terms

Origin: The point on the road directly under the luminaire.

Longitudinal road line (LRL): Imaginary line running parallel to the curb, expressed as a multiple of mounting height (MH).

Transverse road line (TRL): Imaginary lines perpendicular to the curb or LRL, is expressed in terms of multiples of MH.

Reference line (0 MH LRL and 0 MH TRL): Reference line passing through the origin.

Street side (SS): The space located on the street side of 0 MH LRL.

House side (HS): Space located at the house side of 0 MH LRL.

Longitudinal distance (LD): A distance measured between two TRL in a direction parallel to the curb.

Transverse distance (TD): A distance measured between two LRL in a direction perpendicular to the curb.

7.2.2 Photometric report

Luminaires used for external lighting are always supplied with a photometric report by the manufacturers. These photometric reports contain valuable information regarding the characteristic features of the luminaire. The most important features are:

- Luminaire identification and description
- Luminaire classification (ANSI/IES)
- Utilisation curve
- Iso-lux curve
- Luminous flux distribution.

7.2.3 Luminaire identification and description

This part of the photometric report gives the manufacturer's name, product number, type of lamps to be used, type of reflector, socket position and shows a cross sectional view of the luminaire with dimensions.

7.2.4 Classification of the luminaires

All luminaires used for area lighting and roadway lighting are classified into three types according to their light distribution. They are:

- Vertical light distribution or the Spread
- Lateral light distribution and
- Control of light distribution at high angles.

7.2.5 Vertical light distribution or the spread

The vertical distribution of light from a luminaire is expressed as Short (S), Medium (M) or Long (L) depending on where the maximum candle power occurs. If the maximum candle power occurs between a transverse road line (TRL) of 1 MH and 2.25 MH, the luminaire is classified as Small spread. If the maximum candle power occurs between a TRL of 2.25 MH and 3.75 MH, the spread of the luminaire is Medium and if the maximum candle power occurs between TRL of 3.75 MH and 6 MH, the spread is called Long.

7.2.6 Lateral light distribution

The lateral light distribution of a luminaire is classified into five types and is indicated by roman numerals I, II, III, IV and V. These represent roughly the shape of the beam pattern occupying a space on the horizontal surface of the road. The type of distribution is determined by knowing the exact location of the half maximum candle power contour on the road. All points on the road receiving one half the maximum candle power are joined together to form the contour.

Type I Half maximum CP contour lies within 0 MH to 1 MH LRL
Type II Half maximum CP contour lies within 1 MH to 1.75 MH LRL
Type III Half maximum CP contour lies within 1.75 MH to 2.75 M LRL
Type IV Half maximum CP contour lies beyond 2.75 MH LRL
Type V Circular symmetrical CP distribution around the luminaire in all lateral angles.

Distribution characteristics of a luminaire are usually decided by the reflector design or simply by changing the position of the lamp holder. This can also be achieved by changing the angle of luminaire position.

7.2.7 Control of light distribution at high angles

A luminaire is classified as Cutoff, Semi cutoff and Non-cutoff depending on how much lumen is emitted at or about 80° and 90° angles, expressed as a percentage of rated lamp lumens. This is illustrated in Figure 7.2 shown below.

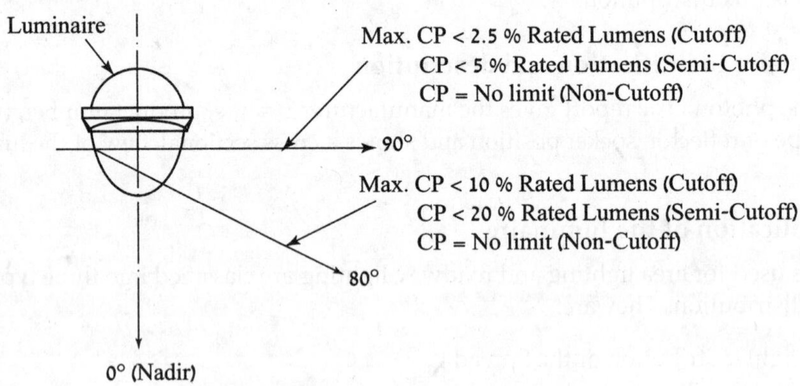

Figure 7.2 Luminaire classification.

The control is 'Cutoff', if a maximum of 2.5% of rated lamp lumens occurs at or above 90° and a maximum of 10% of rated lamp lumen occurs at or above 80°. For a 'Semi cutoff' type luminaire, a maximum of 5% of rated lumens occurs at or above 90° and a maximum of 20% of rated lamp lumens occurs at or above 80°. Luminaire with no such limitations are classified as 'Non-cutoff'.

7.2.8 Utilisation curve

The utilisation curve gives a graphical representation of coeffcient of utilisation data. It helps us to find out the actual amount of light falling on the work plane compared to the total bare lamp lumen. For a luminaire used for road lighting, there will be two utilisation curves, one for the street side and another for the house side or curb side. Most luminaires are designed to have a higher utilisation curve for the street side than for the house side. The x-axis of this curve is the ratio of Transverse distance to Mounting height (TD/MH) and the y-axis gives the percentage of bare lamp lumens. In order to determine the CU value, it is necessary to have the ratio of TD to MH for the street as well as house side. Corresponding to these values, there will be two values of CU from the curve. The net CU value is the sum of these two CU values. Figure 7.3 shows the arrangement of road lighting.

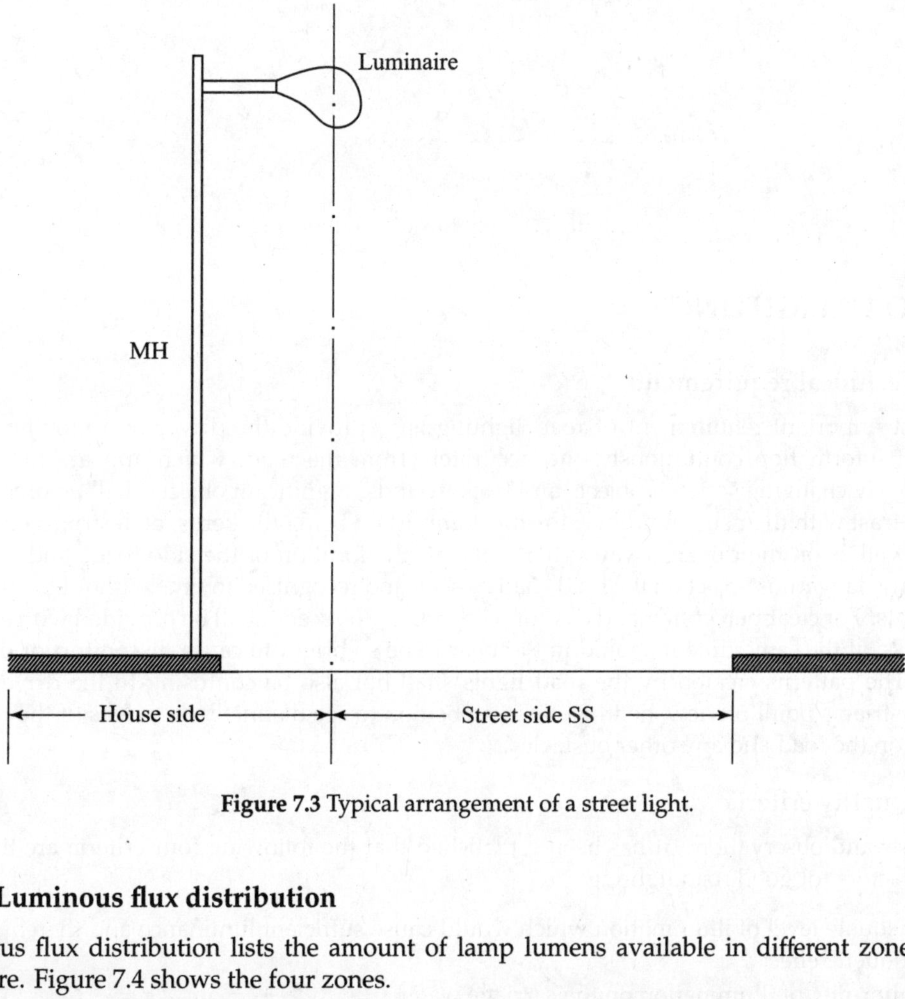

Figure 7.3 Typical arrangement of a street light.

7.2.9 Luminous flux distribution

Luminous flux distribution lists the amount of lamp lumens available in different zones of the luminaire. Figure 7.4 shows the four zones.

The two zones below 90° are the downward house side (DNHS) and downward street side (DNSS). Those above the 90° are the upward house side (UPHS) and upward street side (UPSS). The zonal lumens may also be expressed as a percentage of the total lumens. To minimise the wastage of light output, it is essential that the luminaires used for exterior lighting have very little or no light output above 90°. The amount of light produced between the angles 70° and 90° will have a marked effect on the glare.

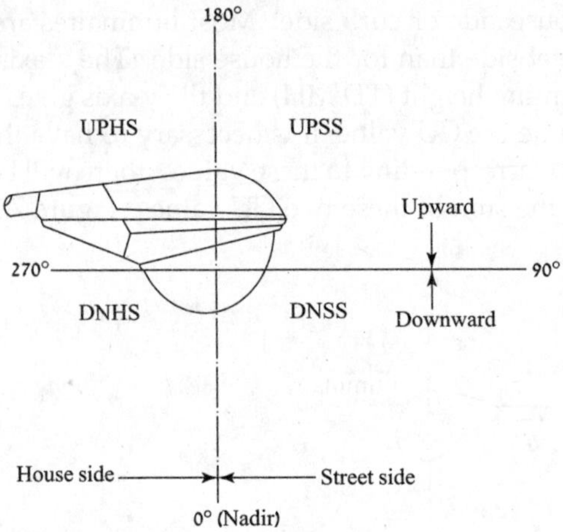

Figure 7.4 Light flux distribution.

7.3 ROAD LIGHTING

7.3.1 Technical requirements

The most important requirement of road lighting is to provide the driver of a moving vehicle, the exact information continuously and accurately from the road, which appears to him as a continuously changing scene of objects and backgrounds. Significant objects shall be present with clear contrast with their surroundings, the most important being the kerbs, pedestrians on the foot path as well as on the carriage way, other vehicles, the location of the side roads and the traffic signs. Any dangerous object on the road shall be seen and recognised in time without the use of the head lights. For clear perception and ease of recognition, the scene shall be provided with adequate brightness, at the same time it should not appear to be so bright to cause discomfort or disability to see. The patterns created by the road lights shall not also be confusing to the driver. From the pedestrian's point of view, he must be in a position to see distinctly the edges of the footpath, vehicles on the road and any other obstacles.

7.3.2 Quality criteria

From tests and observations, it has been established that the following four criteria are the determining factors for good road lighting:

- Adequate level of illumination which would cause sufficient luminance and thereby desired silhouette effect
- Uniformity of illumination on the carriage way
- Limiting the glare to the minimum possible level and
- Optical guidance.

The level of illumination is the density of luminous flux on the road surface on the assumption that the light flux is uniformly distributed. The average level of illumination to be provided is given in ISS 1944 (part I and II). The luminance is defined as the luminous flux per unit projected area in the direction of the observer. Luminance is the product of illumination and the road surface reflection factor and is measured in candela per square metre. It would be too expensive to provide such a lighting level on the road way such that all objects are seen and recognised. Therefore, road lighting is normally designed such that an object on the road is seen by the driver as a dark outline or silhouette against the bright road surface.

The uniformity of illumination is required to provide visual comfort to the driver and to make the objects on the roadway distinctly visible. Therefore longitudinal as well as transverse uniformity is very important. The level of uniformity can be expressed as the ratio of maximum or minimum illumination to the mean illumination on the road surface. As per ISS 1944, the ratio of minimum illumination to the mean illumination shall not be less than 0.4. This is made possible by choosing the right kind of light fixture and the optimum spacing to mounting height ratio.

The glare on the road originates from the light fixtures. There are two types of glare. The disability glare and the discomfort glare. Disability glare reduces the ability to see properly while discomfort glare produces a sense of discomfort without reducing significantly the ability to see. The amount of glare can be minimised by selecting the right type of cut off fixtures and choosing a higher level of mounting the light fixtures.

Optical guidance is required firstly to recognise the border line between the road surface and the foot path (especially on curves) and secondly to recognise from a distance, the road intersections, round about etc. This is also required to get a clear distinction between major roads and secondary roads. Proper location of the street light poles will provide excellent optical guidance.

7.3.3 Classification of roads

Roads are classified based on the volume of traffic, speed and composition of the traffic. The National Electric Code classifies the roads into seven groups. The listing is given in Table 7.1.

Typical arrangement of lamp posts for road lighting is shown in Figure 7.5. For roads of width ranging from 6 to 9 metres it is economical to arrange the lamp posts on side of the road. For roads of width ranging from 9 to 12 metres, zigzag spacing is recommended to get a more or less uniform distribution of light on the road surface. For the roads of width from 12 to 20 metres, face to face spacing of lamp posts on both sides of the road becomes necessary. For multi lane roads, spacing of the lamp posts on the median, or on the median plus the curbs on both sides may have to be chosen, depending on the number of lanes of traffic, planned for the road.

Table 7.1 Classification of roads as per NEC 2011 (as in IS -1944)

Group	Description
A_1	Very important roads with rapid and dense traffic, where safety, speed of traffic and comfort to drivers are the only consideration
A_2	Other main roads with considerable mixed traffic like main city streets, arterial roads etc.
B_1	Secondary roads of considerable traffic such as principal local traffic routes, shopping streets etc.
B_2	Secondary roads with comparatively light traffic
C	Residential and unclassified roads not included in group A and B
D	Grade separated junctions, bridges and elevated roads
E	Town and city centres and areas of civic importance
F	Roads with special requirements
G	Road tunnels

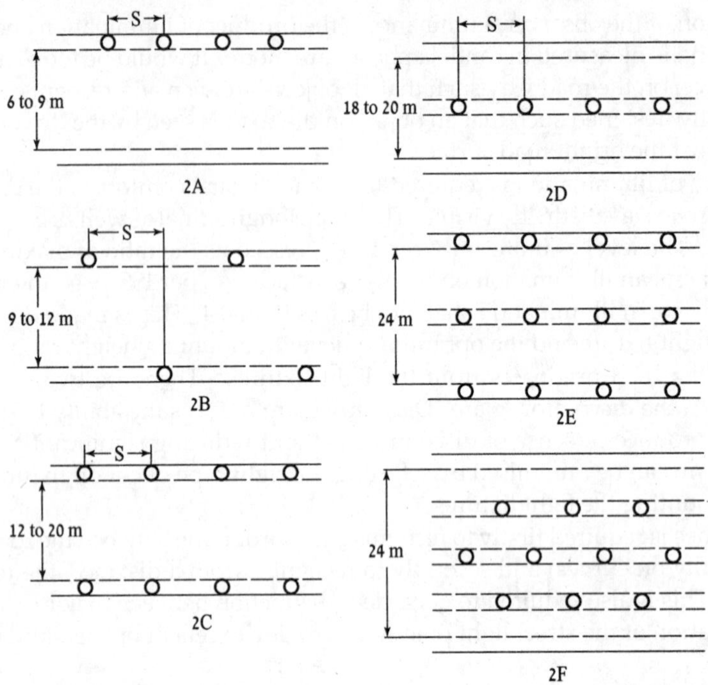

Figure 7.5 Details of lamp spacing for different classes of roads (NEC 2011).

7.3.4 Selection of light sources

Over the past century, road lighting systems have gone through the complete range of light sources commercially available, from the incandescent through the fluorescent to the high intensity discharge lamps. The growing demand for energy conservation has forced the designers to choose the HID sources as the only alternative.

Use of incandescent lamps for the road lighting is the most inefficient solution and should be avoided as far as possible. This is because the incandescent lamps have the lowest efficacy (lumens per watt output) and shortest life span. Even though their initial cost is very low, because of their short life span, maintenance cost can be high.

Fluorescent lamps have better efficacy and life span when compared to incandescent lamps. But their main drawbacks are difficulty in light control, relatively large size and sensitive to temperature. For external use, these light sources have to be enclosed in sealed housing.

Mercury vapour lamps have been in use for years primarily because of their extremely long life. But their efficacy and light quality are not so good when compared to other High Intensity Discharge lamps (HID). This light source has already been phased out because of the reasons mentioned above.

Metal Halide lamps offer good colour quality and efficacy. But they have relatively lower life span when compared with Mercury vapour lamps and High Pressure Sodium vapour lamps (HPS). These sources are however the best choice when colour rendering and energy conservation are of equal importance.

High Pressure Sodium vapour lamps offer the maximum efficacy among the 'white light' sources and have extremely long life. Their colour rendering is termed as golden white and is acceptable for most of the exterior applications.

Low Pressure Sodium (LPS) lamps have the highest efficacy, but their biggest disadvantage is its colour which is mono chromatic yellow, under which all colours other than yellow look gray, brown or black. Because of this, the use of LPS for interior application is ruled out. However, they find application where colour rendering is not of much importance.

All road way lighting is now bonded by MH, HPS or LPS sources. Where colour rendering and energy saving are important, MHis probably the only solution. If colour rendering is not important, but energy saving and long life are the main concern, the choice is limited to HPS or LPS. Power LED light sources are being considered for external lighting because of their inherent qualities like very long life and low power consumption.

7.3.5 Design considerations

The spacing between luminaires can be determined using the equations of average lumen method.

$$\textit{Total initial lumen } L_n = \frac{\text{lux} \times \text{area}}{CU \times LLF}$$

$$\textit{Area of the road} = W \times S$$

Where,

W = Width of the road in metre
S = Spacing between lamp poles.
LLF = Light loss factor
CU = Coefficient of utilisation

The light loss factor has two components, lamp lumen depreciation (LLD) and lamp dirt depreciation (LDD).

$$LLF = LLD \times LDD$$

Therefore,

$$S = \text{spacing between poles} = \frac{L_n \times LLF \times CU}{\text{lux} \times W}$$

If all parameters are known, it is evident that the spacing between poles is directly proportional to the lamp lumens. Theoretically speaking, use of the highest wattage lamp will space the lamps farthest apart. In reality this is not a feasible solution because, uniformity of light and glare will have a deciding role over the permissible spacing.

In order to get perfect uniformity, it is necessary that the illuminance directly under the luminaire be the same as that midway between the poles. Satisfying this condition necessitates a luminaire that emits higher amount of light sideways rather than directly downwards.

In reality it is almost impossible to provide perfect uniformity of light on the road. As a general rule, the maximum light occurs underneath the luminaire and the minimum is at the midpoint between the poles. A measure of acceptable degree of uniformity is expressed as a ratio of the average to minimum illuminance levels. According to IES the ratio of average to minimum illuminance should not be greater than 3 for most of the applications. For residential areas, this ratio can be as high as 6.

Three popular models of pole placement along a roadway are:

1. Spaced continuously on one side of the road with a spacing of S metre
2. Staggered spacing on both sides of the road with a spacing of S metre between consecutive poles and
3. Spacing on opposite sides of the road with a spacing of 2S metre between consecutive poles on the same side.

Of the three arrangements, item 1 is the least expensive since it involves less wiring and trenching. As a general rule for the purpose of energy saving, economy and uniformity of illumination, the ratio of spacing to mounting height should not be less than 3 or larger than 5. Closer spacing means higher cost and power consumption, whereas with larger spacing it is impossible to obtain the specified ratio of average to minimum illumination of 3:1.

EXAMPLE 7.1

Design a road way lighting scheme with the following data.

Width of the road way = 12 m
Illumination required = 15 lux
Mounting height of poles = 9 m
Arm length = 2 m
(Assume that the lamps are placed on one side of the road.)

Table 7.2 Luminaire details for example 7.1

Lamp type	CU	LLF	Wattage	Lumen output
HPS	0.6	0.7	100 150 250	9500 16000 26000
LPS	0.3	0.95	90 135 180	12500 21500 33000

Solution:

The spacing between poles is given by the equation

$$S = \frac{L \times LLF \times CU}{\text{lux} \times W}$$

Where,

L = Lumen output of the lamp
LLF = Light loss factor
CU = Coefficient of utilisation
lux = Illumination level required
W = Width of the roadway.

Here, in this problem, assuming HPS lamps are used for the purpose,

$L = 9500$, $LLF = 0.7$, $CU = 0.6$, $W = 12$ m and lux = 15

Substituting we get

$$S = \frac{9500 \times 0.7 \times 0.6}{15 \times 12}$$

$$= 22.166 \text{ m}$$

$$\text{Therefore spacing to mounting height ratio} = \frac{22.166}{9}$$

$$= 2.46$$

This is less than 3 and hence, the design is not satisfactory. Now consider 150 W HPS lamp source.

$L = 16000$, $LLF = 0.7$, $CU = 0.6$, $W = 12$ m and lux = 15

Therefore,

$$S = \frac{16000 \times 0.7 \times 0.6}{15 \times 12}$$

$$= 37.333 \text{ m}$$

$$\text{Spacing to mounting height ratio} = \frac{37.333}{9}$$

$$= 4.148$$

This ratio is greater than 3 and less than 5. Hence, the design is satisfactory.

7.3.6 Guidelines for specific locations

The NEC guidelines for group A and Group B roads are listed in Table 7.3.

Table 7.3 Guidelines for group A and group B roads (NEC 2011)

<table>
<tr><th rowspan="2">Description</th><th colspan="4">Group of road</th></tr>
<tr><th>A₁</th><th>A₂</th><th>B₁</th><th>B₂</th></tr>
<tr><td>Average level of illumination on road surface-lux</td><td>30</td><td>15</td><td>8</td><td>4</td></tr>
<tr><td>Ratio of minimum to average illumination</td><td>0.4</td><td>0.4</td><td>0.4</td><td>0.4</td></tr>
<tr><td>Type of luminaire</td><td></td><td></td><td></td><td></td></tr>
<tr><td>Preferred</td><td>Cutoff</td><td>Cutoff</td><td>Cutoff or semi-cutoff</td><td>Cutoff or semi-cutoff</td></tr>
<tr><td>Permitted</td><td>Semi-cutoff</td><td>Semi-cutoff</td><td>Non-cutoff</td><td>Non-cutoff</td></tr>
<tr><td>Mounting height, m</td><td>9-10</td><td>9-10</td><td>7.5-9</td><td>7.5-9</td></tr>
<tr><td>Spacing to mounting height</td><td colspan="2">Cutoff = 3
Semi-cutoff = 3.5</td><td colspan="2">Cutoff = 3
Semi-cutoff = 3.5 Non-cutoff = 4</td></tr>
</table>

7.4 AREA LIGHTING

Area lighting means the illumination of a large area of space with an average level of lighting. This is quite different from that of flood lighting, where the illumination required has to meet a very high level of quality and quantity. Typical examples of area lighting are

- Airport parking space
- Railway yards
- Vehicle parking space etc.

All luminaires designed for road lighting can also be used for area lighting. Type III and type IV luminaires are the most suitable for area lighting. Some of the limiting factors of area lighting are

- Restriction on the mounting height
- Colour rendering property of the lamp source
- Spacing limitations.

Each situation has its own characteristics that may be different from others. In order to make a good preliminary selection and to simplify the design procedure, the following factors shall be taken into consideration.

1. Spacing between poles shall not be more than 4.5 times the height of the poles.
2. Spacing between the edges of the area and the nearest pole shall not be greater than 2.25 times the mounting height.
3. A minimum of two lights per pole shall be employed in order to achieve even distribution of lighting.

EXAMPLE 7.2

Design a lighting scheme for an area measuring 180 m × 80 m. The design requirements are

1. Illumination level 15 lux
2. Mounting height of poles 10 m
3. Luminaires per pole 2 nos.

The details regarding the luminaires available are given in Table 7.4.

Table 7.4 Luminaire details for example 7.2

Type of lamp	CU	LLF	Watts	Lumen output
HPS	0.7	0.75	100	9500
			150	16000
			250	26000
LPS	0.65	0.95	90	12500
			135	21500
			180	33000

Solution:

Assume the ratio of spacing to mounting height to be 4.5,

The allowable maximum spacing is = 4.5 × 10 m = 45 m

Therefore, we can accommodate four numbers of poles along the longer side of the area. The poles are placed at 22.5, 45, 45, 45 and 22.5 m respectively from one of the shorter edges. This distribution satisfies the spacing to mounting height criterion as well as the spacing between the edges and the nearest poles. Since the width of the area is only 80 metres the pole distribution can be 17.5, 45 and 17.5 metres. In all there will be 4 × 2 poles in the area. The location of the poles is shown in the layout Figure 7.6.

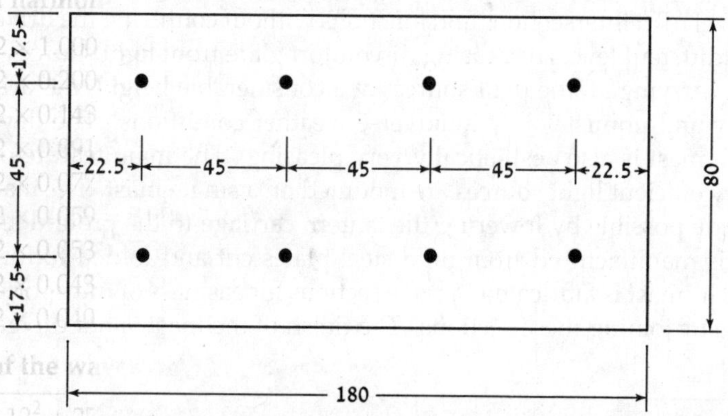

Figure 7.6

Since the number of luminaires per pole is two,

Total number of luminaires = 8 × 2 = 16 nos.

Case 1:

If the option is for High pressure Sodium vapour lamps,

$$\textit{Lumen output of each lamp} = \frac{A \times E}{N \times CU \times LLF}$$

$$= \frac{80 \times 180 \times 15}{16 \times 0.7 \times 0.75} = 22714 \text{ lumens.}$$

From the given data, 250 W HPS lamps with a lumen output of 26000 lumens will satisfy the design requirement.

Case 2:

If Low Power Sodium vapour lamps are considered,

$$\textit{Lumen output of each lamp} = \frac{A \times E}{N \times CU \times LLF}$$

$$= \frac{180 \times 80 \times 15}{16 \times 0.65 \times 0.95} = 21862 \text{ lumens.}$$

From the given data, 135 W LPS lamp with an initial lumen output of 21500 lumens will satisfy the design requirement.

7.5 HIGH MAST LIGHTING

High mast lighting is ideal for areas where a high standard of lighting is required from a relatively few number of units. A few examples are: Air ports, high way intersections, Railway marshalling yards, Car parks etc. The concept of high mast lighting was developed to overcome the difficulties in lighting multi level road intersections and junctions without confusing the driver with an array of light luminaires at different levels or causing discomfort glare from light sources at different levels. With a single mast carrying all the light sources at a considerable height, the driver is able to pick out his route safely and unmistakably in adverse weather conditions. The day light appearance created by the high mast is also aesthetically very pleasing. The main feature of the high mast is that a group of very efficient light sources are mounted on a single mast. The maintenance of these light sources is made possible by lowering the lantern carriage to the ground level by means of a winch. The mast is manufactured from mild steel plates cut and folded to form a continuously tapered polygon. The mast is fabricated in three sections for easiness of transportation and the mast is assembled at site by joining these sections. The finish of the mast is hot dip galvanised to make it weather proof.

7.6 QUESTIONS

1. What is the difference between external lighting and internal lighting?
2. Explain the different types of light sources suitable for external lighting.
3. What are the basic design considerations for road lighting.
4. How are roads classified according to National Electrical Code for illumination purpose?
5. How is area lighting different from road lighting?
6. Design a road way lighting scheme with the following data.

   Width of the road way = 18 m

   Illumination required = 15 lux

   Mounting height of poles = 9 m

   Arm length = 2 m

   (Assume that the lamps are placed on the median at the centre of the road and each lamp post carries two lamps on either side.)

Table 7.5 Luminaire details for example 7.6

Lamp type	CU	LLF	Wattage	Lumen output
			100	9500
HPS	0.6	0.7	150	16000
			250	26000
			90	12500
LPS	0.3	0.95	135	21500
			180	33000

7. Design a lighting scheme for an area measuring 225 m × 90 m. The design requirements are

    1. Illumination level 15 lux
    2. Mounting height of poles 10 m
    3. Luminaires per pole 2 nos

    The details regarding the luminaires available are given below.

Table 7.6 Luminaire details for example 7.7

Lamp type	CU	LLF	Wattage	Lumen output
HPS	0.7	0.75	120	8000
			250	24000
			400	46000
LPS	0.65	0.95	90	12500
			135	21500
			180	31500

Typical values of k are:

- Current with rectangular waveform $k = 0.86$
- Frequency converter current (THD ≈ 50%) $k = 0.80$

Figure 9.8 shows the typical de-rating required for a transformer supplying non-linear and electronic loads.

Figure 9.8 De-rating required for a transformer supplying electronic loads.

Asynchronous machines

International Electrotechnical Commission (IEC) specifies a weighted harmonic factor—Harmonic Voltage Factor (HVF) to be satisfied for all asynchronous machines. The expression for HVF is given by equation (9.7)

$$HVF = \sqrt{\sum_{n=2}^{13} \frac{V_n^2}{n}} \leq 0.02 \qquad (Source: IEC - 60892) \tag{9.7}$$

EXAMPLE 9.3

A supply voltage has a fundamental frequency component of V_1 and harmonic voltages of $V_3 = 2\%$, $V_5 = 3\%$ and $V_7 = 1\%$ of V_1. Find the value of THD_V and HVF.

$$THD_V = \frac{\sqrt{\sum_{h=2}^{n} V_h^2}}{V_1}$$

$$= \frac{\sqrt{2^2 + 3^2 + 1^2}}{100}$$

$$= \frac{\sqrt{14}}{100} = 0.03742 = 3.742\%$$

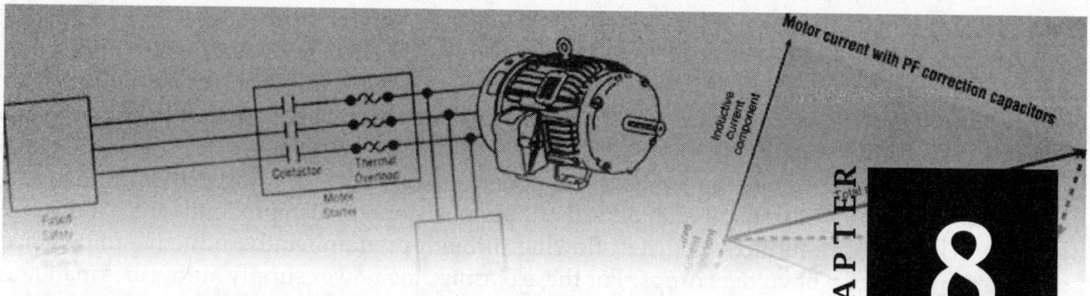

CHAPTER 8

POWER FACTOR IMPROVEMENT

OBJECTIVES

At the end of this chapter you will

- Understand the principle behind the power factor compensation.
- Understand the ill effects of poor power factor.
- Realise the different methods of power factor improvement.
- Understand the techno-economic benefits of improved power factor.
- Know where to locate the power factor improving capacitors.
- Know how to choose the control and protective devices for capacitors.
- Be able to properly size the cables for the capacitor banks.
- Be able to calculate the voltage rise due to capacitor installation.
- Be able to calculate the inrush current into a capacitor.

8.1 NATURE OF REACTIVE ENERGY

All inductive machines and devices that operate on alternating current system convert electrical energy into mechanical work and heat. This energy which is recorded by the energy meter (kWh meter) is referred to as the active energy. In order to perform energy conversion, magnetic fields are to be set up in the machines and devices. These magnetic fields are associated with another form of energy to be supplied by the source, known as reactive (wattless) energy. Inductive circuits cyclically absorb energy from the system (during magnetic build-up) and re-injects this energy back to the system (during collapse of the magnetic fields), twice in every cycle of the supply frequency. A similar phenomenon takes place in the capacitive elements in a power system, such as power cables, capacitance of the cables etc. However, in this case the energy is stored electro-statically. The cyclic charging and discharging of the capacitive elements are similar to what happens in inductive elements. But the current flow to and from the capacitive elements are in exact phase opposition to the inductive elements. This wattless (reactive) current or the wattless component of the load current does not consume power from the system, but it causes power loss in transmission and distribution systems.

In practical power systems, the wattless component of load current is invariably inductive in nature. The impedance of transmission and distribution systems is also predominantly inductive in nature. Therefore, an inductive current flowing through predominantly inductive impedance produces the worst case of voltage drops. For these reasons, the power supply authorities mandate that measures are taken to reduce the ill effects of wattless component of load as much as possible.

8.2 POWER FACTOR

The power factor is the ratio of active power to apparent power.

$$\text{Power factor} = \frac{\text{Active power}}{\text{Apparent power}} = \frac{P}{S} \tag{8.1}$$

If the voltage and current are perfect sinusoidal signals, the power factor also is equal to the cosine of the angle between the voltage and current wave forms.

$$\text{Power factor} = \cos\phi$$

where, ϕ = Phase angle between voltage and current wave forms.
P = Active power
S = Apparent power
Q = Reactive power.

The active power, apparent power and the reactive power are related by the equation

$$\text{Apparentpower}^2 = \text{Activepower}^2 + \text{Reactivepower}^2 \tag{8.2}$$

$$S^2 = P^2 + Q^2$$

$$\tan\phi = \frac{Q}{P} \tag{8.3}$$

$$Q = P\tan\phi$$

Improvement of power factor of an installation has several advantages. There are technical as well as economic advantages. The economic advantages are mainly due to the reduction in the demand kVA due to the improved power factor. For the same loading condition, if the power factor is high, the kVA demand will be less than a situation where the power factor is low. The economic benefit can be realised by considering an example.

EXAMPLE 8.1

The monthly demand charge on an industry is Rs. 300/kVA maximum demand. The industry has a maximum power consumption of 100 kW at an average power factor of 0.75 lagging. If the power factor of this industry is improved to 0.93 lagging by installing power factor improving capacitors, find the monthly saving in demand charges.

Maximum demand of the industry before the PF improvement $= \dfrac{100}{0.75}$

$= 133.33$ kVA

Maximum demand after PF improvement $= \dfrac{100}{0.93}$

$= 107.53$ kVA

Reduction in demand kVA $= 133.33 - 107.53$

$= 25.8$ kVA

Saving in monthly demand charges $=$ Rs. 25.8×300

$=$ Rs. 7740

This analysis has not taken into consideration the additional cost factor for installing and operating the power factor improving capacitors.

The technical advantages are:

- Reduction in the size of the cables
- Reduction in the I^2R losses in the cables
- Reduction in the voltage drop in the system
- Increase in the available power.

Since the magnitude of the load current is inversely proportional to the power factor, for the same amount of delivered power, it can be seen that if the power factor is improved, the magnitude of the load current will be decreased. A lower load current will allow the designer to opt for a smaller cross-section of the cable. As the load current carried by the cable is reduced, the power loss in the cable will also be reduced. Similar is the case with the voltage drop in the cables. A better power factor also guarantees increased availability of power. This is especially true in the case of transformers. If an additional load is anticipated on the existing transformer, it may be better to improve the power factor of the existing load rather than going for replacing the existing transformer with a larger capacity transformer. This can be explained with the help of a numerical example.

EXAMPLE 8.2

An industry is receiving electric supply through a 630 kVA 11 kV/433 V transformer and the present load is 450 kW at an average power factor of 0.8 lagging. If an additional load of 100 kW at a power factor of 0.7 lagging is anticipated, examine the possibility of handling the situation without upgrading the capacity of the transformer.

Apparent power of the existing load $= \dfrac{P_1}{0.8} = \dfrac{450}{0.8}$

$S_1 = 562$ kVA

Reactive power of the existing load $= Q_1 = \sqrt{S_1^2 - P_1^2}$

$= 337$ kVAr

Apparent power due to the additional load $= \dfrac{P_2}{0.7} = \dfrac{100}{0.7}$

$S_2 = 143$ kVA

Reactive power of the additional load $= Q_2 = \sqrt{S_2^2 - P_2^2}$

$= 102$ kVAr

Total apparent power required $S = 562 + 143$

$= 705$ kVA

If the additional load is shared by the existing transformer, the transformer will be overloaded since the total kVA demand exceeds the limit of the transformer capacity. If power improving capacitors are installed, the overloading situation can be over come.

Total load to be supplied P $= 450 + 100$

$= 550$ kW

Capacity of the transformer S $= 630$ kVA

Maximum reactive power the transformer can supply $= Q$

$= \sqrt{S^2 - P^2}$

$= \sqrt{630^2 - 550^2}$

$= 307$ kVAr

Total reactive power required by the load $= Q_1 + Q_2$

$= 337 + 102$

$= 439$ kVAr

Therefore, capacity of the power factor compensator $= 439 - 307$

$= 132$ kVAr

Thus, if a power factor correcting capacitor bank of 132 kVAr is installed, the existing transformer can handle the new situation without overloading the transformer. The analysis has not taken momentary overloading due to load variations.

8.3 HOW TO IMPROVE THE POWER FACTOR

Different methods are available for effecting power factor improvement. They are:

- Using fixed capacitors under favourable conditions
- By means of automatically controlled stepped capacitors.

8.3.1 Fixed Capacitors

In this case one or more fixed capacitors are connected to the power system to provide a fixed level of compensation. The control may be:

- Manual using a load break switch or circuit-breaker
- Semi-automatic using a contactor
- Directly connected to the device and switched with the device.

These types of compensation are generally employed at the terminals of inductive loads or at the bus bars supplying numerous small motors or inductive appliances for which individual compensation may be costly. They can also be employed where the load is reasonably constant.

8.3.2 Automatically Switched Capacitor Banks

This type of compensation provides automatic control of power factor correction and maintains the power factor close to the optimum level. This method is suitable on installations where active and/or reactive power variations are relatively large. A bank of capacitors is divided into a number of sections, each of which is controlled by a contactor. The size of the capacitor bank can therefore be increased or decreased by suitably switching the contactors. The switching operation is done by a control relay which monitors the power factor of the installation.

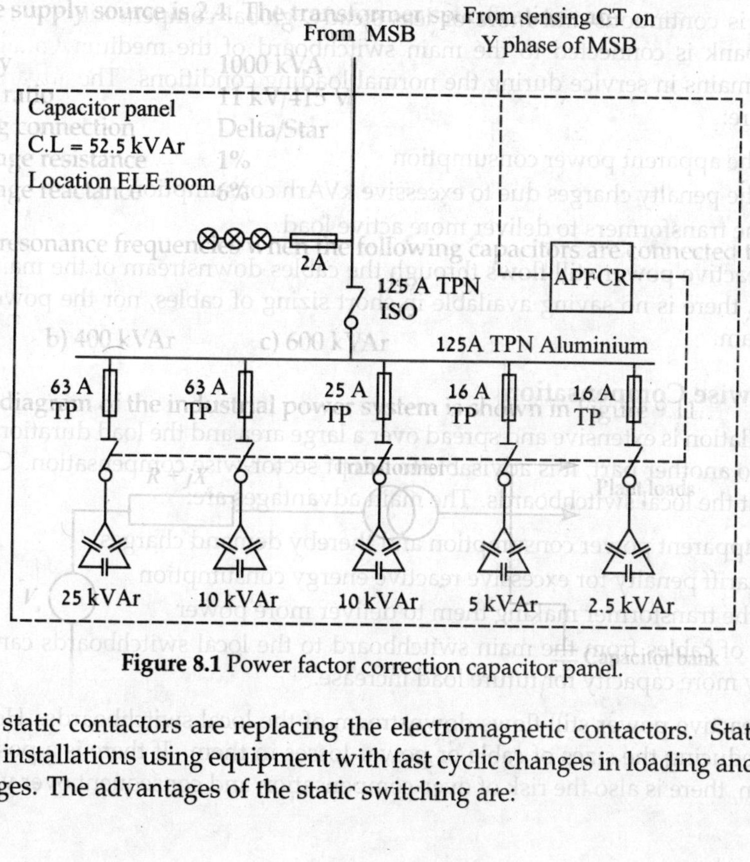

Figure 8.1 Power factor correction capacitor panel.

Nowadays, static contactors are replacing the electromagnetic contactors. Static switching is preferable with installations using equipment with fast cyclic changes in loading and/or is sensitive to transient surges. The advantages of the static switching are:

- Immediate response to power factor change
- Unlimited operations
- Elimination of transients
- Silent in operation.

A typical automatic power factor correction switchboard is shown in Figure 8.1.

In general, if the kVAr rating of the capacitor bank is less than or equal to 15% of the transformer rating, it will be preferable to have a fixed value of compensation. Otherwise, it is advisable to install an automatically controlled power factor compensator.

8.4 LOCATION OF CAPACITORS

The location of the compensating capacitors will generally decide the mode of compensation. The locations can be:

- Global: one location for the entire installation
- Partial: section by section compensation
- Local: at each inductive load device
- A combination of the latter two.

8.4.1 Global Compensation

When the load is continuous and more or less steady, global compensation is the ideal choice. The capacitor bank is connected to the main switchboard of the medium voltage distribution network and remains in service during the normal loading conditions. The advantages of global compensation are:

1. Reduces the apparent power consumption
2. Reduces the penalty charges due to excessive kVArh consumption
3. Enables the transformers to deliver more active load.

However, reactive power still flows through the cables downstream of the main switchboard. Because of this, there is no saving available in short sizing of cables, nor the power losses in the cable downstream.

8.4.2 Sector-wise Compensation

When the installation is extensive and spread over a large area and the load duration pattern differs from one part to another part, it is advisable to adopt sector-wise compensation. Capacitor banks are connected at the local switchboards. The main advantages are:

- Reduces apparent power consumption and thereby demand charges
- Reduces tariff penalty for excessive reactive energy consumption
- Relieves the transformer making them to deliver more power
- The sizes of cables from the main switchboard to the local switchboards can be reduced or will allow more capacity for future load increase.

However, reactive power still flows downstream of the local switchboards. Hence there is no possibility of reducing the sizes of cable or power losses in them. If there is a possibility of large power variation, there is also the risk of over-compensation and consequent over voltage.

8.4.3 Individual Compensation

Individual compensation is always the best choice if the rating of the motors is significant with respect to the total load of the installation. In this case, the capacitors are directly connected to the terminals of the individual loads (significantly induction motors). The kVAr rating of the capacitors required for different rating of motors is given in Table 8.1. Compensation at the main switchboard also will be beneficial as a complement to the individual compensation. The main advantages are:

- Reduces apparent power consumption and thereby demand charges
- Reduces tariff penalty for excessive reactive energy consumption
- Relieves the transformer making them to deliver more power
- Reduces the size of all cables and the I^2R loss in them
- The entire installation is relieved of excessive reactive current.

Table 8.1 Capacitor rating for direct connection to induction motors in kVAr

Motor HP/kW	Capacitor rating (kVAr) for motor speed in r.p.m						Cable size AYFY
	3000	1500	1000	750	600	500	(for capacitor) (mm²)
3/2.5	1	1	1.5	2	2.5	2.5	4
5/3.7	2	2	2.5	3.5	4	4	4
7.5/5.7	2.5	3	3.5	4.5	5	5.5	6
10/7.5	3	4	4.5	5.5	6	6.5	6
15/11.2	4	5	6.7	7.5	8.5	9	10
20/15	5	6	7	9	11	12	16
25/18.7	6	7	9	10.5	13	14.5	25
30/22.8	7	8	10	12	15	17	25
50/37	11	12.5	16	18	23	25	50
75/57	16	17	21	23	29	32	95
100/75	21	17	21	23	29	32	185
130/102	31	33	36	38	48	55	185
200/150	40	42	45	47	60	67	185
250/187	46	50	53	55	68	76	240

8.5 OPTIMAL COMPENSATION

Different methods are available for determining the optimal compensation required for an installation. An elaborate method is to determine the active and reactive power loading of the installation at the design stage itself. This requires a detailed study of the actual loading conditions of the various equipment in the installation. For an existing installation, the optimum rating of the compensating capacitor can be decided based on the following criteria.

- Energy charges prior to the installation of compensating capacitors
- Reduction in future energy charges after installing the capacitors
- Cost of installing the capacitors and accessories, its maintenance charges, reduction in the I^2R losses in the cables etc.

8.5.1 Determination of the Rating of the Capacitor for PF Improvement

The rating of the capacitor for improving the power factor of a system can be obtained by analysing a simple power system. Let

S_L = Apparent power consumed by the load before installing the capacitor
S_D = Apparent power supplied by the source after installing the capacitor
Q_L = Reactive power consumed by the load before adding the capacitor
Q_D = Reactive power supplied by the source after adding the capacitor
P_L = Active power consumed by the load
Q_C = Reactive power rating of the capacitor bank

It is assumed that the active power of the system remains unchanged after the addition of the capacitor bank. The reactive power supplied by the capacitor

$$Q_C = Q_L - Q_D \tag{8.4}$$

The apparent power of the load

$$S_L = \frac{P_L}{PF_L} \tag{8.5}$$

The desired apparent power after compensation

$$S_D = \frac{P_L}{PF_D} \tag{8.6}$$

Where PF_L and PF_D are the actual load power factor and the desired load power factor after compensation.

The reactive power can also be expressed as

$$Q_L = \sqrt{S_L^2 - P_L^2} \tag{8.7}$$

$$Q_D = \sqrt{S_D^2 - P_L^2} \tag{8.8}$$

Substituting equation 8.5 into 8.7 and 8.6 into 8.8

$$Q_L = \sqrt{\left(\frac{P_L}{PF_L}\right)^2 - P_L^2} \tag{8.9}$$

$$Q_D = \sqrt{\left(\frac{P_L}{PF_D}\right)^2 - P_L^2} \tag{8.10}$$

Substituting equation 8.9 and 8.10 into equation 8.4

$$Q_C = \sqrt{\left(\frac{P_L}{PF_L}\right)^2 - P_L^2} - \sqrt{\left(\frac{P_L}{PF_D}\right)^2 - P_L^2} \tag{8.11}$$

$$= P_L\left[\sqrt{\left(\frac{1}{PF_L^2}\right) - 1} - \sqrt{\left(\frac{1}{PF_D^2}\right) - 1}\right] \tag{8.12}$$

EXAMPLE 8.3

An industrial power system has a demand of 250 kW at a power factor of 0.7 lagging. Determine the rating of the capacitor bank required to improve the power factor to 0.9 lagging.

Solution :

Applying equation 8.12 directly

$$Q_C = 250\left[\sqrt{\left(\frac{1}{0.7^2}\right)-1}-\sqrt{\left(\frac{1}{0.9^2}\right)-1}\right]$$

$$= 133.97 \approx 135 \text{ kVAr}$$

8.5.2 Simplified Method

An approximate method is generally adequate for most of the practical installations. If the average power factor of the installation is known and if the power factor is to be improved to a level to avoid penalty tariff (for example pf = 0.93), Table 8.2 can be used to determine the value of kVAr rating of the capacitor bank per kW of the load.

Table 8.2 kVAr rating of the capacitor bank per kW rating of the load required for improving the power factor

Initial power factor	Corrected value of power factor										
	0.90	**0.91**	**0.92**	**0.93**	**0.94**	**0.95**	**0.96**	**0.97**	**0.98**	**0.99**	**1.0**
0.70	0.536	0.564	0.591	0.625	0.657	0.691	0.729	0.769	0.811	0.878	1.02
0.71	0.508	0.536	0.563	0.597	0.629	0.663	0.701	0.741	0.783	0.850	0.992
0.72	0.479	0.507	0.534	0.568	0.600	0.634	0.672	0.712	0.754	0.821	0.963
0.73	0.452	0.480	0.507	0.541	0.573	0.607	0.645	0.685	0.727	0.794	0.936
0.74	0.425	0.453	0.480	0.514	0.546	0.580	0.618	0.658	0.700	0.767	0.909
0.75	0.398	0.426	0.453	0.487	0.519	0.553	0.591	0.631	0.673	0.740	0.882
0.76	0.371	0.399	0.426	0.460	0.492	0.526	0.564	0.604	0.652	0.713	0.855
0.77	0.345	0.373	0.400	0.434	0.466	0.500	0.538	0.578	0.620	0.687	0.829
0.78	0.319	0.347	0.374	0.408	0.440	0.474	0.512	0.552	0.594	0.661	0.803
0.79	0.292	0.320	0.347	0.381	0.413	0.447	0.485	0.525	0.567	0.634	0.776
0.80	0.266	0.294	0.321	0.355	0.387	0.421	0.459	0.499	0.541	0.608	0.750
0.81	0.240	0.268	0.295	0.329	0.361	0.395	0.433	0.473	0.515	0.582	0.724
0.82	0.214	0.242	0.269	0.303	0.335	0.369	0.407	0.447	0.489	0.556	0.698
0.83	0.188	0.216	0.243	0.277	0.309	0.343	0.381	0.421	0.463	0.530	0.672
0.84	0.162	0.190	0.217	0.251	0.283	0.317	0.355	0.395	0.437	0.504	0.645
0.85	0.136	0.164	0.191	0.225	0.257	0.291	0.329	0.369	0.417	0.478	0.620
0.86	0.109	0.140	0.167	0.198	0.230	0.264	0.301	0.343	0.390	0.450	0.593
0.87	0.083	0.114	0.141	0.172	0.204	0.238	0.275	0.317	0.364	0.424	0.567
0.88	0.054	0.085	0.112	0.143	0.175	0.209	0.246	0.288	0.335	0.395	0.538
0.89	0.028	0.059	0.086	0.117	0.149	0.183	0.230	0.262	0.309	0.369	0.512
0.90	-	0.031	0.058	0.089	0.121	0.155	0.192	0.234	0.281	0.341	0.484

EXAMPLE 8.4

It is required to improve the power factor of a 500 kVA installation from the existing power factor of 0.8 lagging to a level of 0.93 lagging. Determine the rating of the capacitor bank required for this improvement.

Solution :

Active power of the installation before compensation = 500 × 0.8
= 400 kW

From Table 8.2, the value of the capacitor in kVAr/kW can be obtained from the meeting point of the row corresponding to 0.8 and the column corresponding to 0.93. The value is 0.355 kVAr/kW.

Therefore, total value of the capacitor bank = 400 × 0.355
= 142 kVAr
≈ 150 kVAr

8.5.3 Voltage Improvement

In addition to improving the power factor, addition of capacitors will also produce a voltage rise on the bus to which it is connected. The rise is due to the leading current of the capacitor supplied through the inductive reactance of the system. Consider a circuit as shown in Figure 8.2. This represents an industrial power system with a capacitor bank connected to the MV side of the transformer.

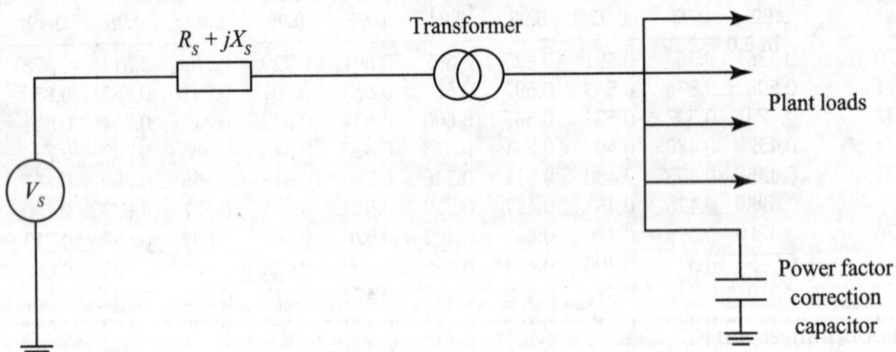

Figure 8.2 Single line diagram of a power system with PF improving capacitor.

The rated current of the capacitor

$$I_C = \frac{Q_c}{\sqrt{3} \times V_{LL}} \quad (8.13)$$

Where

Q_c = The three-phase VAr rating of the capacitor bank
V_{LL} = Rated line to line voltage of the capacitor bank

Assuming that the system resistance is very small compared to the inductive reactance, the voltage rise is given by

$$\text{Voltage rise} = V_\Delta = X_L \times I_C$$

Substituting equation (8.13), we get

$$V_\Delta = X_L \frac{Q_c}{\sqrt{3} \times V_{LL}} \tag{8.14}$$

The percentage voltage rise

$$\%V_\Delta = \frac{V_\Delta}{V_{LN}} \times 100 \tag{8.15}$$

Substituting equation (8.14) in equation (8.15) we get

$$\%V_\Delta = \frac{X_L \times Q_C}{\sqrt{3} \times V_{LL} \times V_{LN}} 100$$

For a balanced three-phase system, $V_{LN} = \frac{V_{LL}}{3}$

Therefore percentage voltage rise

$$\%V_\Delta = X_L \times \frac{Q_C}{V_{LL}^2} \times 100 \tag{8.16}$$

EXAMPLE 8.5

A 450 kVAr, 415 V, a three-phase capacitor bank is to be installed on the 415 volt side of a 630 kVA transformer rated at 11 kV/415 V delta/star connected and having an impedance of 0.0023 + *j* 0.0125 ohm per phase when referred to the 415 V side. Calculate the percentage voltage rise due to the installation of this capacitor.

Solution :

Substituting the given data directly in equation (8.16) yields

$$\text{Percentage voltage rise} = 0.0125 \left[\frac{4{,}50{,}000}{415^2} \right] \times 100$$

$$= 3.267\%$$

8.6 POWER FACTOR CORRECTION OF INDUCTION MOTORS

Individual compensation is always recommended where the capacity of the induction motor is comparable to the installation capacity. The power factor of an induction motor is very low at no load or at light load conditions and if the installation contains a large number of induction motors, care should be taken to see that these motors are not allowed to be kept on no load or under light load conditions. Otherwise, these loads will be drawing a large amount of reactive kVA. The general rules to be followed are:

- Capacitors should be directly connected to the motors
- Unloaded motors should be switched off as far as possible
- Motors should not be oversized so that they are loaded to near rated conditions.
- Motors for special applications like plugging, stepping, inching and reversing shall not be compensated for power factor.
- As the magnitude of the current drawn from the supply will be lower for a motor with compensation, the overload setting shall be adjusted by the ratio = $\frac{\cos\varphi \text{ before compensation}}{\cos\varphi \text{ after compensation}}$.

Another issue to be addressed in connection with the power factor improvement of induction motor is the possibility of self-excitation of the motor. If the motor is driving a high inertia load, the motor will continue to rotate some time after the power is switched off. The "magnetic inertia" of the rotor current will cause an emf to be induced in the stator windings which causes a current to flow through the capacitor connected across the motor. This capacitor current will produce a rotating magnetic field which acts exactly along the same axis and in the same direction as the decaying magnetic field of the motor. Consequently, the magnetic flux increases, naturally stator current also increases and the voltage at the motor terminals also increases, sometimes to a dangerous level. This phenomenon is called self-excitation.

In order to avoid self-excitation, the kVAr rating of the capacitor shall be limited to a value given by

$$Q_c \leq 0.9\sqrt{3}VI_0 \tag{8.17}$$

where,

Q_c = kVAr capacity of the capacitor
V = Line to line voltage of the supply in kV
I_0 = No load current of the motor

Additional precautions are also required for high inertia loads driven by induction motors. To avoid the possible build-up of self-excitation, high speed circuit-breakers/contactors shall be employed to trip the circuit in the event of power loss of power supply. The protection scheme for such application should also include an over voltage tripping relay together with a reverse power relay (as the motor will feed power to the rest of the system until the inertial energy is dissipated). If for any other reasons, the capacitor used for compensation is higher than the recommended value as in Table 8.1, it is advisable to connect the capacitor through a circuit-breaker/contactor which trips simultaneously with the motor circuit-breaker/contactor. But the wiring shall be so arranged that the capacitor contactor/CB are closed before the motor contactor/CB is closed for starting operation.

8.7 PROTECTION AND CONTROL

Special care is required in the choice of upstream cables and control devices of capacitor banks. For the capacitor, the current is a function of the following:

1. The applied voltage and its harmonics
2. The capacitance value.

From equation 8.13,
The normal current through a capacitor

$$I_C = \frac{Q_C}{\sqrt{3} \times V_{LL}}$$

However, in practice the supply voltage has a permissible range of tolerances at fundamental frequency. The voltage also may contain higher order harmonics. The capacitance will also have a manufacturing tolerance from its declared value. All these factors can contribute to an increase in the value of the current to about 50% of the normal value. Out of this approximately 30% of the increase is due to the increase in voltage and 15% is due to the manufacturing tolerance of the capacitor value. Therefore, all components carrying capacitor current must be rated to carry the worst case conditions at normal ambient temperature. If the ambient temperature is higher than the normal, further de-rating will be required.

The size and setting of the circuit-breakers can be chosen in order to allow a long time (overload) setting of

$1.36\ I_n$ to $1.5\ I_n$	for normal capacitors
$1.12\ I_n$	for detuned capacitors (tuned at $2.7\ f$)
$1.19\ I_n$	for detuned capacitors (tuned at $3.8\ f$)
$1.31\ I_n$	for detuned capacitors (tuned at $4.3\ f$)

Short time delay (short circuit) setting must be chosen so that it is not sensitive to inrush currents. The standard short time setting will be $10\ I_n$ for all types of capacitors. Table 8.1 also gives the details of cables to be used for the capacitors.

EXAMPLE 8.6

Determine the thermal and magnetic setting (long time delay setting and short time delay setting) for an ordinary capacitor bank with the following specification.
50 kVAr, 415 volts, 50 Hz, three-phase.

Solution :

$$\text{Full load current of the capacitor } I_n = \frac{50{,}000}{\sqrt{3} \times 415} = 69.56 \text{ A}$$

$$\text{Therefore, the long time delay setting } = 1.36\ I_n = 1.36 \times 69.56$$
$$= 94.6 \text{ A or say } 100 \text{ A}$$

$$\text{And short time delay setting } = 10\ I_n = 10 \times 69.56$$
$$= 695.6 \text{ A or say } 700 \text{ A}$$

EXAMPLE 8.7

Determine the thermal and magnetic setting of the circuit-breaker that is used to control a 100 kVAr, three-phase, 415 V detuned capacitor bank tuned at $2.7\ f$.

Solution :

$$\text{Full load current of the capacitor } I_n = \frac{1{,}00{,}000}{\sqrt{3} \times 415}$$

$$= 139.12 \text{ A}$$

For a detuned capacitor tuned at 2.7 f, the thermal setting will be 1.12 I_n

$$= 1.12 \times 139.12$$

$$= 155.81 \text{ A or } 160 \text{ A}$$

$$\text{And the magnetic setting} = 10\, I_n$$

$$= 1391.2 \text{ A say } 1400 \text{ A}$$

8.8 VOLTAGE TRANSIENTS

When capacitor banks are switched ON, high frequency voltage and current transients can occur. If harmonics are not present in the system, the maximum peak voltage that may be present during switching will not exceed twice the peak value of the supply voltage if the capacitors were uncharged before switching. If the capacitors were in a charged condition at the time of switching, the magnitude of the voltage transient can reach a value up to three times the peak supply voltage. This maximum will occur only if:

- The existing voltage across the capacitor before switching is equal to the peak value of the supply voltage
- Switch is closed at the instant of the peak supply voltage, and
- Polarity of the power supply voltage is opposite to that of the charged capacitor.

If such a situation arises, the current transient will be at its maximum possible value or will be twice the maximum value when closing the switch on to an uncharged capacitor. For any other situation, the voltage and current transient will be less than the extreme limits mentioned above. On the other hand, if the peak voltage across the capacitor and the supply voltage are of the same polarity and the capacitor is switched at the instant of supply voltage peak, then there will not be any voltage or current transients. When capacitors are switched automatically, care must be taken to ensure that the section of capacitor about to be switched is fully discharged through the discharging resistors. To achieve this, the discharge time of the capacitors can be modified by reducing the value of the discharge resistor.

8.9 SWITCHING CONSIDERATIONS

In practice, it is always required to install capacitor banks consisting of several steps of switched capacitors, rather than a single fixed capacitor. This is particularly true in situations where the reactive power requirements vary over a wide range during the day. When a totally de-energised capacitor is switched ON, the capacitor behaves like a short circuit resulting in a large inrush current. The inductive reactance of the system will try to limit the magnitude of this large inrush current. This type of switching a single bank is known as isolated bank switching.

When switching one step of a multi-step capacitor bank, a larger magnitude of current will result since the already energised capacitors also will contribute to the charging of the capacitor being switched. This type of switching is known as back to back switching. Back-to-back switching will definitely result in larger value of inrush current when compared to isolated bank switching.

Limiting the inrush current is normally done by installing current limiting reactors. The reactor used for this purpose is essentially air core in design. Iron core reactors are not recommended as they tend to saturate at high current conditions. The continuous current rating of these reactors shall be the same as the continuous current rating of the capacitor bank.

8.9.1 Isolated Bank Switching

Calculation of the exact values of capacitor switching currents is an extremely difficult process. A simplified approach is possible with the following assumptions:

- The power system is perfectly balanced and is represented by a single-phase equivalent circuit.
- The source of power supply is modelled as a DC source
- The magnitude of the voltage source remains truly constant
- The magnitude of the DC voltage source is equal to the peak line to neutral voltage of the system and
- Resistance of the system is neglected.

The above assumptions are valid on the fact that the duration of the transient period is much lower than one cycle of the system frequency. In practice, the magnitude of the system voltage is very unlikely to change during such short time period.

The peak value of the source voltage

$$V_0 = \sqrt{2}\left[\frac{V_{LL}}{\sqrt{3}}\right] \tag{8.18}$$

Since the resistance of the system is neglected, the source inductance can be calculated by using the equation:

$$L = \frac{1}{2\pi f} \times \frac{V_{LL}^2}{VA_{SC}} \tag{8.19}$$

where f = system frequency Hz
V_{LL} = line to line voltage of the system
VA_{SC} = three-phase short circuit volt ampere at location of the capacitor.

The capacitance value of the capacitor bank

$$C = \frac{1}{2\pi f} \times \frac{Q_C}{V_{LL}^2} \tag{8.20}$$

The maximum instantaneous peak value of the inrush current

$$I_{MAX} = V_0\sqrt{\frac{C}{L}} \tag{8.21}$$

(A detailed derivation of the equation is out side the scope of this book)

EXAMPLE 8.8

The three-phase short circuit level on the 11 kV side of a transformer in an industrial premise is 250 MVA. The transformer is rated at 500 kVA, 11 kV/415 V, 50 Hz, delta/star connected and with a percentage reactance of 6%. If a capacitor bank of 200 kVAr is connected on the MV side of the transformer, calculate the peak instantaneous switching current.

Solution:

Assume a base voltage of 11 kV and base MVA of 100 MVA on the primary side of the transformer.

The per unit impedance of the source $Z_{pu} = \dfrac{MVA_B}{MVA_S}$

$$= \frac{100}{250} = 0.4 \text{ pu} = X_S$$

The source inductance will be the same as source impedance as the resistance is being neglected.

The per unit inductance of the transformer on the selected base

$$X_T = 0.06 \times \left(\frac{11}{11}\right)^2 \left(\frac{100}{0.5}\right)$$

$$= 12 \text{ pu}$$

Total inductive reactance of the system $X_{pu} = X_T + X_S = 12.4$ pu

The base impedance on the 415 side of the transformer $= \dfrac{kV_B^2}{MVA_B}$

$$= \frac{0.415^2}{100}$$

$$= 0.0017223\ \Omega$$

Total inductive reactance $X_L = X_{pu} \times \text{base impedance}$

$$= 12.4 \times 0.0017223$$

$$= 0.0214\ \Omega$$

Therefore the total inductance of the system $L = \dfrac{X_L}{2\pi f}$

$$= \frac{0.0214}{2\pi \times 50}$$

$$= 68.15 \times 10^{-6} \text{ H}$$

The capacitance value of the capacitor bank $C = \dfrac{2,00,000}{2\pi \times 50 \times (415)^2}$

$= 3.69 \times 10^{-3}$ F

From equation (8.18)

$$V_0 = \sqrt{2}\left[\frac{V_{LL}}{\sqrt{3}}\right]$$

$$= \frac{\sqrt{2}\times 415}{\sqrt{3}}$$

$$= 338.80 \text{ V}$$

The peak instantaneous switching current $I_{MAX} = V_0\sqrt{\dfrac{C}{L}}$

$$= 338.8\sqrt{\frac{3.69\times10^{-3}}{68.15\times10^{-6}}}$$

$$= 2493 \text{ A}$$

8.9.2 Back-to-Back Switching

The analysis of the system becomes more complicated when capacitors are switched in steps. This is also the situation in most of the practical systems. The most severe situation is when the last step of a capacitor bank is being switched with all other steps already energised and if the switching instant is at the peak of the system voltage. Under these conditions all the energised capacitors will try to discharge into the capacitor step that is being energised. Two additional simplifying assumptions are made in the analysis. The contribution of current from the supply source is neglected and all the steps of the capacitors are equal in size. Let C be the capacitance of each step of the capacitor bank and L be the inductance of the cables connecting the capacitors. Let N be the number of steps of the capacitors in the bank.

The maximum peak instantaneous current flowing into the last step, I_{MAX} is given by

$$I_{MAX} = \left[\frac{N-1}{N}\right]V_0\sqrt{\frac{C}{L}} \tag{8.22}$$

EXAMPLE 8.9

If the capacitor bank of the example in 8.8 consist of four steps of 50 kVAr capacitors and the inductance of the lead cable is 5 μH, determine the peak inrush current into the last step of the capacitor.

Solution:

Using equation 8.20

Capacitance of each step of the capacitor bank $C = \dfrac{1}{2\pi f} \times \dfrac{Q_C}{V_{LL}^2}$

$$= \frac{50{,}000}{2\pi \times 50 \times 415^2} = 924.11 \times 10^{-6} \text{ F}$$

Applying equation 8.22,

The peak inrush current into the last step
$$= \left[\frac{N-1}{N}\right] V_0 \sqrt{\frac{C}{L}}$$
$$= \left[\frac{4-1}{4}\right] 338.80 \times \sqrt{\frac{924.11 \times 10^{-6}}{5 \times 10^{-6}}}$$
$$= 3454.47 \text{ A}$$

It is interesting to note that the normal rated current of the capacitor step is only 69.56 A, where-as the peak inrush current is of the order of 50 times, i.e. $\left(\frac{3454.47}{69.56}\right)$ of the normal current.

8.10 QUESTIONS

1. Explain the importance of reactive power in power distribution systems.
2. Power factor need not always be equal to cos φ. Explain why?
3. What are the economic benefits of power factor compensation?
4. What are the different methods of power factor improvement?
5. What are the main considerations in improving the power factor of induction motors?
6. Explain the ill effects of switching capacitors in banks.
7. An industrial plant has an active power demand of 800 kW at 0.75 power factor lagging. The supply voltage is 415 volts, three-phase 50 Hz. Determine:
   a) Apparent power demand in kVA
   b) Reactive power demand
   c) Rating of the capacitor bank to improve the power factor to 0.93 lagging
   d) Apparent power demand after installing the capacitor bank.
8. A 630 kVA, 11k V/414 V delta-star connected transformer supplies an industrial plant at 415 volts, three-phase 50 Hz. A 150 kVAr capacitor is connected to the 415 volt bus of the plant. The percentage impedance of the transformer is 4%. Determine the instantaneous value of the transient inrush current when the capacitor is switched ON. Neglect the source impedance and the transformer resistance.
9. A five-step 415 volt capacitor bank consists of five steps of 50 kVAr each. The lead inductance of the connecting leads between the capacitor steps is 5 micro henry. Determine the maximum peak instantaneous inrush current when the last step is switched ON. Neglect the current contribution from the source.
10. A 250 kVAr capacitor bank is to be installed at the main switchboard (MSB) of an industrial plant. The plant is supplied through a 11 kV/415 V, 750 kVA, delta-star connected transformer with a percentage reactance of 5%. Calculate the percentage rise in voltage due to the installation of the capacitor.

CHAPTER 9

Power Quality Issues and Resonance Problems in System Design

OBJECTIVES

At the end of this chapter you will

- Understand the power quality issues.
- Understand the voltage and current harmonic distortion.
- Understand the sources causing harmonic distortion.
- Be able to calculate total harmonic distortion of voltage and current waveforms.
- Understand the problem of resonance when installing capacitors in a harmonic environment.
- Be able to determine the required filter inductance for a detuned capacitor bank.
- Understand the voltage rise on capacitors used in a filter bank.
- Be able to check whether standard specifications are satisfied or not.

9.1 INTRODUCTION

The term *power quality* is an important aspect of electrical power delivery system. It has gained its importance from the second half of 1990's. That does not mean that it was not important in the past. Utilities all over the world have been working for decades on the improvement of what is now accepted as power quality. The main reasons for the abrupt increase in a problem that was there for quite a long time can be summarised as follows:

- **Modern equipment have become more sensitive to voltage disturbances**
  Last few decades have seen a definite change in the dynamics of consumer loads. The loading pattern has moved from simple linear loads such as heaters and lighting devices to more complicated non-linear loads like solid state converters, switched mode power supplies which are very widely used in industrial, commercial and domestic equipment.

- **Proliferation of power electronic equipment causing voltage disturbances**
  Modern electronic devices provide lots of flexibility and reduced electricity consumption. It is anticipated that the entire electric power consumption is going to be processed by some form of electronic equipment at one stage or other. These electronic devices are, however bad news for the utility because they draw highly distorted currents and cause distortion of the system voltage.

- **A growing need for standardisation and performance criteria**
  The drive towards privatisation and deregulation have forced the utilities to treat the consumers as customers and electricity as a product with certain characteristics, which needs to be improved, guaranteed, measured and predicted. Nowadays, the customer can buy electrical energy somewhere, then transport the energy to some other place and use the energy at that point. It is not very clear who is responsible for the reliability and power quality. No matter what the legal construction is, the reliability and quality will need to be well defined and addressed properly.

- **Utilities want to deliver a good product**
  Designing and maintaining a power system with a high reliability for a reasonable cost have been a technical challenge accepted by the utilities and they continue to do so with a commitment. Many a power quality development has been driven by the power utilities.

- **Power supply has become too good**
  Long interruptions have become things of the past in many advanced countries and consumers have the wrong concept that electricity is something that is always available with high quality. For a country like India, where long power interruptions are frequent, power quality may not appear to be a big issue.

- **Power quality can now be easily measured**
  The availability of electronic devices to measure and show the waveform distortions have certainly contributed to the interest in power quality studies. In the past there were no means to measure these parameters.

9.2 OVERVIEW OF POWER QUALITY ISSUES

Power quality event	Cause	Impact on power converters
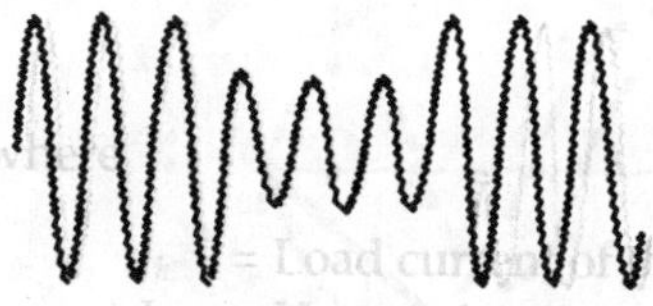 Voltage sag	A short term voltage drop in the range of 0.1 - 0.9 pu can be caused by a fault in adjacent voltage line or large current drawn across-the-line or the branch line in an internal distribution system.	This can cause a voltage drop or fluctuation on the DC link resulting in a shut down or disturbance at the output of the downstream converter depending upon the level of sag as well as capability of the downstream converter.
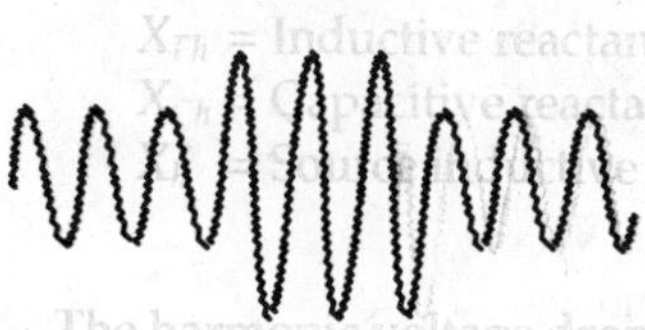 Voltage swell	A short-term increase in voltage from 1.1 to 1.8 pu may be caused by line to ground fault or dropping of large load across the line or in the branch line in an internal distribution.	Can impact the load in a more or less similar way as sags.
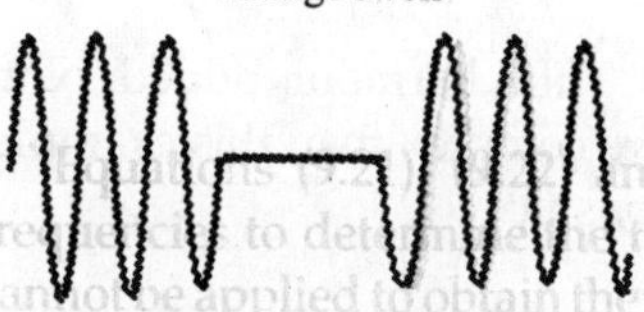 Voltage interruption	Natural disturbances such as lighting, storms, etc. can cause protection mechanism such as reclosers to trip. This results in temporary disruption of power lasting from 3 sec up to 1 min.	Generally, the hold-up time of the bulk capacitor on the DC link decides its impact on the downstream system.
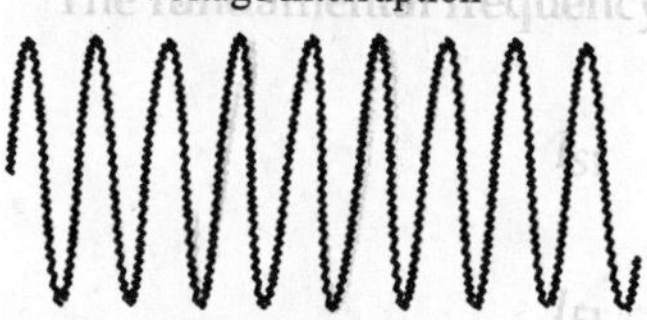 Voltage harmonics	Harmonic voltages result from harmonic current flowing in the system, that has a finite impedance.	The problems caused by harmonic are far more serious and are discussed in detail in subsequent section.

Figure 9.1 Power quality related events.

9.3 HARMONICS

In order to understand the problems created by harmonic currents and voltages, it is necessary to understand what is meant by the term harmonic. Any periodic, non-sinusoidal, time varying waveform may be expressed as a sum of several sinusoidal waveforms, referred to as harmonic components. Each of these harmonic components will have certain amplitude, frequency and phase shift. The amplitude and phase shift of the individual harmonic components can be determined by means of Fourier analysis applied to the original waveform.

The fundamental component of the waveform will have the same frequency as the waveform itself. Harmonic frequencies are integral multiples of the fundamental frequency, i.e., for a fundamental of 50 Hz, the third harmonic would be 150 Hz and the fifth harmonic would be 250 Hz. Figure 9.2 shows a fundamental sine wave with third and fifth harmonics.

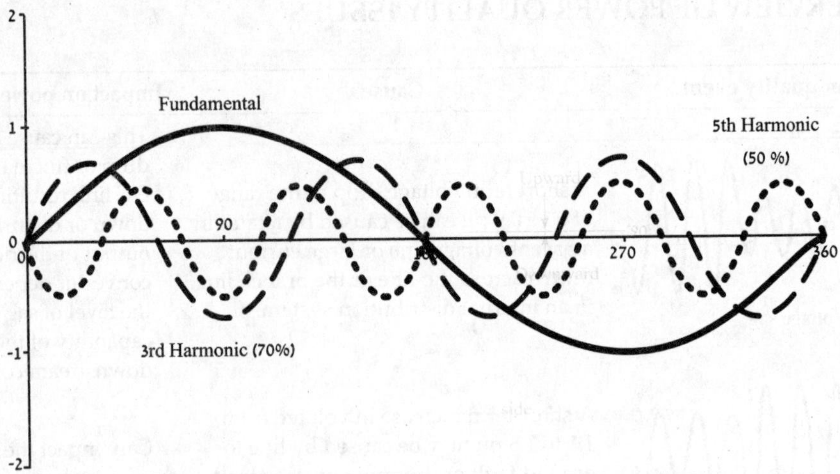

Figure 9.2 Fundamental with third and fifth harmonics.

Figure 9.3 shows a fundamental with 70% third harmonic and 50% fifth harmonic added. Note that in practice most distorted current waveforms will be much more complex than this example, containing many more harmonics with a more complex phase relationship.

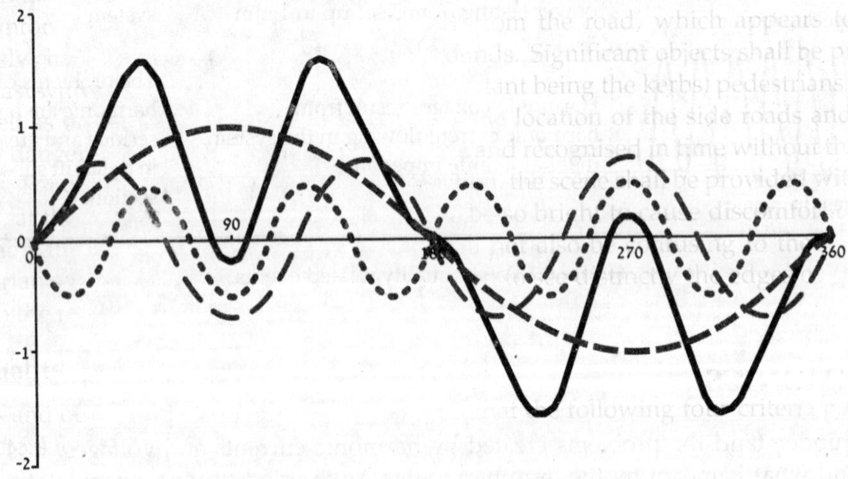

Figure 9.3 Distorted current waveform.

This waveform is clearly not a sine wave and that means that normal measurement equipment, such as averaging reading rms-calibrated multimetres, will give inaccurate readings. Note also that there are six zero crossing points per cycle instead of two, so any equipment that uses zero crossing as a reference will malfunction. The waveform contains non-fundamental frequencies and has to be treated accordingly. When talking about harmonics in power installations it is the *current* harmonics that are of most concern because the harmonics originate as currents and most of the ill effects are due to these currents.

A listing of the frequency, amplitude and phase angle of the individual harmonic components is referred to as the *harmonic spectrum* of the waveform. This information is critical in the design specification of harmonic filtering equipment. Table 9.1 represents the harmonic spectrum of a typical square wave current waveform up to 13th harmonic.

Table 9.1 Harmonic content of square wave

Frequency	Harmonic order	Magnitude	Phase shift
50	Fundamental	1.000	0°
150	3	0.333	0°
250	5	0.200	0°
350	7	0.143	0°
550	11	0.091	0°
650	13	0.077	0°

In addition to the harmonic spectrum of the waveform, several terms are used to quantify the nature of the harmonic waveforms, namely:

- Power factor
- Crest factor
- Distortion power
- Harmonic distortion values.

These indicators are very essential in determining any necessary corrective measures.

9.3.1 Power Factor

Definition

The power factor PF is the ratio between the active power and the apparent power

$$PF = \frac{P}{S}$$

There is always confusion among the practising engineers about the value of $\cos\varphi$ and the power factor PF. The value of $\cos\varphi$ is given by

$$\cos\varphi = \frac{P_1}{S_1}$$

Where

P = Active power
S = Apparent power
P_1 = Active power of the fundamental component of the frequency
S_1 = Apparent power of the fundamental frequency

Hence it can be seen that $\cos\varphi$ concerns exclusively with fundamental frequency and therefore differs from the power factor PF when there are harmonics in the system.

9.3.2 Crest Factor

Definition

The crest factor is the ratio between the peak value of the current or voltage to its rms values.

For a purely sinusoidal signal the crest factor is equal to $\sqrt{2}$. For a non-sinusoidal signal the crest factor can be greater than or less than $\sqrt{2}$. Typically all non-linear loads will have a crest factor much higher than $\sqrt{2}$. It is generally between 1.5 and 2 and can even reach up to 5 in critical cases. A high value of crest factor indicates high transient over currents which when detected by protective devices can lead to nuisance tripping.

9.3.3 Distortion Power

The active power of a signal comprising harmonics is the sum of the active powers resulting from the currents and voltages of the same harmonic order. The reactive power is defined exclusively in terms of the fundamental frequency.

The distortion power is defined as

$$D = \left(S^2 - P^2 - Q^2\right)^{1/2}$$

9.3.4 Total Harmonic Distortion (THD)

The *root of sum of squares or rss* value of the waveform represents the effective rms value of the waveform. The rss value is given by

$$V_{rss} = \sqrt{V_1{}^2 + V_2{}^2 + V_3{}^2 + \dots\dots + V_n{}^2} \quad (9.1)$$

$$= \sqrt{\sum_{h=1}^{n} V_h^2}$$

where

V_1 = rms magnitude of the fundamental component
V_2 = rms value of the second harmonic component and so on and
V_n = rms value of the nth harmonic component.

The percentage total harmonic distortion, or THD, of the waveform is a measure of how badly a waveform is distorted with respect to a pure sine wave. The THD is given by

$$THD_V = \frac{\sqrt{V_2{}^2 + V_3{}^2 + \dots\dots + V_n{}^2}}{V_1} \quad (9.2)$$

$$= \frac{\sqrt{\sum_{h=2}^{n} V_h^2}}{V_1}$$

If a waveform is a pure sine wave, the rss value of the waveform is equal to the rms value of the fundamental component. Also the THD of a pure sine wave is zero. Similarly THD of current waveforms can also be given as

$$THD_I = \frac{\sqrt{\sum_{h=2}^{n} I_h^2}}{I_1} \tag{9.3}$$

The above equation can be rewritten as

$$THD_I = \sqrt{\left(\frac{I_{rss}}{I_1}\right)^2 - 1} \tag{9.4}$$

9.3.5 Relation Between Power Factor and THD

When the supply voltage is sinusoidal or virtually sinusoidal, it may be stated that

$$P = P_1 = \sqrt{3} \times V_1 \times I_1 \times \cos\varphi_1$$

$$\text{Therefore, power factor } PF = \frac{P}{S} = \frac{\sqrt{3}V_1 I_1 Cos\varphi_1}{\sqrt{3}V_1 I_{rss}}$$

$$\text{From equation (9.4), } \frac{I_1}{I_{rss}} = \frac{1}{\sqrt{1+THD_I^2}}$$

$$\text{Hence } PF = \frac{\cos\varphi_1}{\sqrt{1+THD_I^2}} \tag{9.5}$$

Table 9.2 gives the relations between total harmonic distortion of the current waveform and the power factor.

Table 9.2 Relation between THD_I and PF

THD_I	0%	20%	40%	60%	80%	100%
$\frac{PF}{\cos\varphi}$	1.00	0.981	0.9285	0.8572	0.781	0.707

No useful conclusions can be drawn without knowledge of the spectrum of the current harmonics but it is still common to find only the total harmonic distortion (THD) figures quoted. When harmonics propagate around a distribution system, that is, to branch circuits not concerned with carrying the harmonic current, they do so as voltages. It is very important that both voltage and current values are measured and that quoted values are explicitly specified as voltage and current values.

Harmonic currents have been present in the electricity supply system for many years. Initially they were produced by the mercury arc rectifiers used to convert AC to DC current for railway electrification and for DC variable speed drives in industry. More recently the range of types and the number of units of equipment causing harmonics have risen sharply, and will continue to rise, so system designers must now consider harmonics and their side effects very seriously.

9.4 SOURCES OF HARMONICS

It is the characteristic of most power electronic control equipment to draw line currents that are non-sinusoidal. Several sources of harmonic currents that may be found on electrical power distribution network are listed below.

- Variable speed motor drives
- Rectifiers
- Arc furnaces
- Welding equipment
- Uninterruptible power supplies
- Switched mode power supplies
- Compact fluorescent lamps
- Electronic ballasts etc.

The majority of modern electronic units use switched mode power supplies (SMPS). These differ from older units in that the traditional step-down transformer and rectifier is replaced by direct controlled rectification of the supply to charge a reservoir capacitor from which the direct current for the load is derived by a method appropriate to the output voltage and current required. The advantage—to the equipment manufacturer—is that the size, cost and weight is significantly reduced and the power unit can be made in almost any required form factor. The disadvantage—to everyone else—is that, rather than drawing continuous current from the supply, the power supply unit draws pulses of current which contain large amounts of third and higher harmonics and significant high frequency components (see Figure 9.4). A simple filter is fitted at the supply input to bypass the high frequency components from line and neutral to ground but it has no effect on the harmonic currents that flow back to the supply. Single-phase UPS units exhibit very similar characteristics to SMPS.

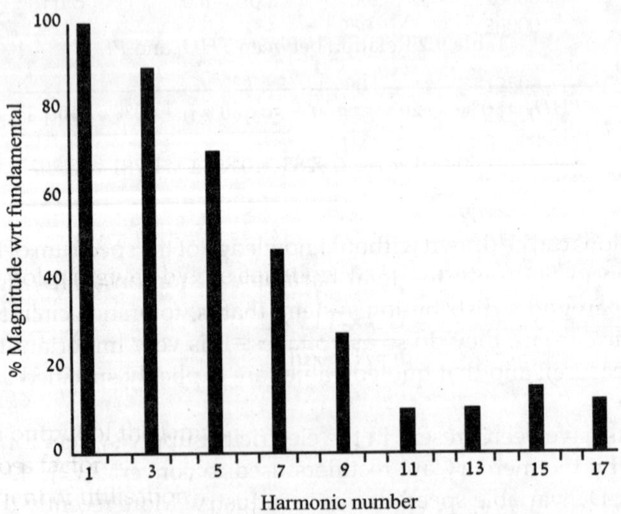

Figure 9.4 Harmonic spectrum of typical PC.

For high power units there has been a recent trend towards the power factor corrected inputs. The aim is to make the load look like a resistive load so that the input current appears sinusoidal and in phase with the applied voltage. It is achieved by drawing input current as a high frequency triangular waveform that is averaged by the input filter to a sinusoid. This extra level of sophistication is not yet readily applicable to the low-cost units that make up most of the load in commercial and industrial installations.

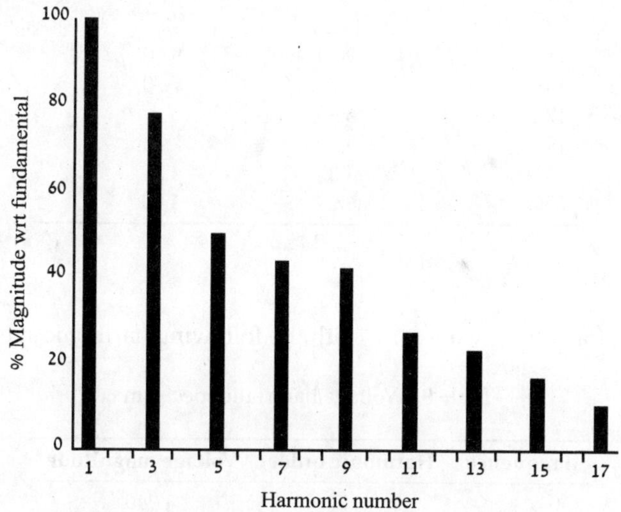

Figure 9.5 Harmonic spectrum of a typical CFL.

Electronic lighting ballasts have become popular in recent years following claims for improved efficiency. Overall they are only a little more efficient than the best magnetic ballasts and in fact, most of the gain is attributable to the lamp being more efficient when driven at high frequency rather than to the electronic ballast itself. Their chief advantage is that the light level can be maintained over an extended lifetime by feedback control of the running current—a practice that reduces the overall lifetime efficiency. Their main disadvantage is that they generate harmonics in the supply current. The power-factor corrected types are available at higher ratings that reduce the harmonic problems, but at a cost penalty. Smaller units are usually uncorrected. Compact fluorescent lamps (CFL) are now being sold as replacements for tungsten filament lamps. Miniature electronic ballast, housed in the connector casing, controls a folded 8 mm diameter fluorescent tube. CFL rated at 11 watt are sold as replacements for a 60 watt filament lamp and have a life expectancy of about 8000 hours. The typical harmonic current spectrum is shown in Figure 9.5. These lamps are being widely used to replace filament lamps in domestic buildings. Even though CF lamps are energy efficient retrofits for the incandescent lamps, the increased use of CF lamps introduces lots of harmonics in the supply system.

Table 9.3 lists typical harmonic spectrum of line currents drawn from the supply by some three-phase, variable speed motor drives and three-phase full wave rectifiers.

Table 9.3 Harmonic current spectrum of three-phase variable speed drive and three-phase full wave rectifiers

Harmonic order	Amplitude (% of fundamental)	
	6 pulse	12 pulse
1	100.00	100.00
5	20.00	
7	14.30	
11	9.10	9.10
13	7.70	7.70
17	5.90	
19	5.30	
23	4.30	4.30
25	4.00	4.00

EXAMPLE 9.1

Determine the THD for a voltage waveform with the following harmonic spectrum.

Table 9.4 Voltage harmonic spectrum

Frequency	Harmonic order	Voltage magnitude
50	1	240
250	5	20
350	7	11
550	11	5
650	13	2

All other harmonic components are zero.

Solution :

The rss value is given by equation (9.1)

$$V_{rss} = \sqrt{240^2 + 20^2 + 11^2 + 5^2 + 2^2}$$
$$= 241.143 \text{ V}$$

The THD is given by equation (9.2)

$$THD_V = \frac{\sqrt{20^2 + 11^2 + 5^2 + 2^2}}{240} \times 100\%$$

$$= 9.772\%$$

EXAMPLE 9.2

Using the values of harmonic current spectrum in Table 9.3, determine the following quantities for a 100 kVA, 415 V three-phase full wave (six pulses) rectifier.

a) Harmonic current spectrum
b) rss of the current waveform I_{rss}
c) THD of the current waveform THD_I.

Solution :

The fundamental component of line current $= \dfrac{1{,}00{,}000}{\sqrt{3}\times 415} = 139.12$ A

a) The individual harmonic currents are calculated and tabulated as follows:

$$I_1 = 139.12 \times 1.000 = 139.12 \text{ A}$$
$$I_5 = 139.12 \times 0.200 = 27.824 \text{ A}$$
$$I_7 = 139.12 \times 0.143 = 19.894 \text{ A}$$
$$I_{11} = 139.12 \times 0.091 = 12.659 \text{ A}$$
$$I_{13} = 139.12 \times 0.077 = 10.712 \text{ A}$$
$$I_{17} = 139.12 \times 0.059 = 8.208 \text{ A}$$
$$I_{19} = 139.12 \times 0.053 = 7.373 \text{ A}$$
$$I_{23} = 139.12 \times 0.043 = 5.982 \text{ A}$$
$$I_{25} = 139.12 \times 0.040 = 5.564 \text{ A}$$

b) The rss value of the waveform is

$$I_{rss} = \sqrt{139.12^2 + 27.824^2 + 19.894^2 + 12.659^2 + 10.712^2 + 8.208^2 + 7.373^2 + 5.982^2 + 5.564^2}$$
$$= 144.872 \text{ A}$$

c) The THD of the waveform is

$$THD_I = \frac{\sqrt{27.824^2 + 19.894^2 + 12.659^2 + 10.712^2 + 8.208^2 + 7.373^2 + 5.982^2 + 5.564^2}}{139.12} \times 100\%$$

$$= 29.05\%$$

9.5 DISTURBANCES CAUSED BY HARMONICS

The harmonics flowing in the distribution networks downgrade the quality of electrical power. This can have several negative effects on the operation of the power system.

- Increased losses on the distribution network due to increase in effective rms current.
- Overloads in neutral conductor due to cumulative increase in third harmonics created by single-phase loads.
- Overloads, vibration and premature ageing of generators, transformers and motors as well as increased noise levels (hum).
- Overloads and premature ageing of power factor correction capacitors.
- Distortion of the supply voltage that can disturb sensitive loads.
- Disturbances in communication networks and telephone lines.
- Resonance between the system inductance and the capacitance of the power factor correction capacitors.

A typical power distribution system with harmonic generating loads are shown in Figure 9.6.

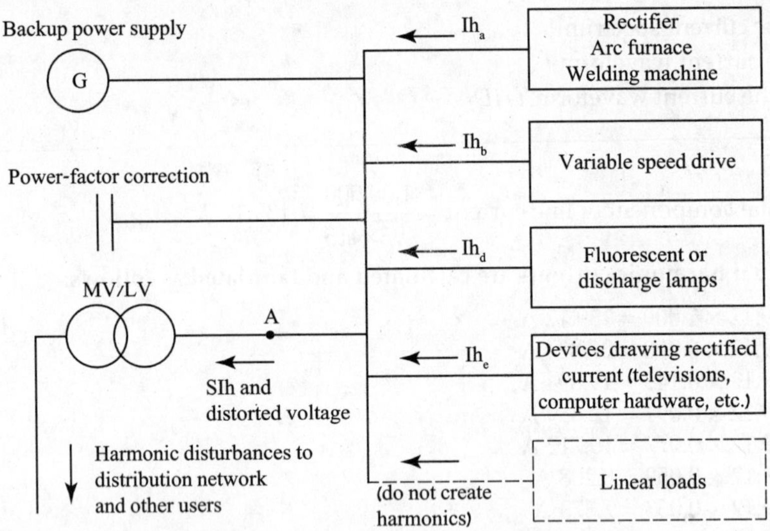

Figure 9.6 Sources of harmonics in a power system.

9.5.1 Increased Losses

Losses in conductors

The active power delivered to a load is a function of the fundamental component of current I_1.

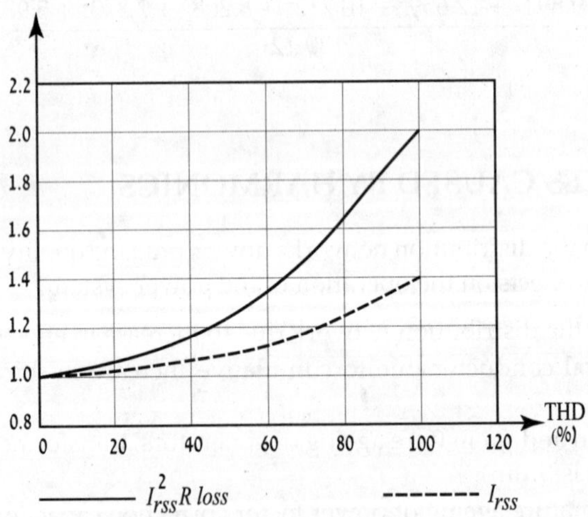

Figure 9.7 Impact of THD on effective rms current and I^2R losses.

When the current drawn by the load contains harmonics, the effective rms value of the current I_{rss} is greater than the fundamental current I_1. From equation (9.4) we can deduce that

$$I_{rss} = I_1\sqrt{1 + THD_I^2}$$

Figure 9.7 shows the impact of THD on the effective rms current and the I^2R losses in the conductors.

Losses in asynchronous machines

The harmonic voltages supplied to the asynchronous machines will generate harmonic currents with frequencies higher than 50 Hz in the rotor and cause additional losses. The presence of a THD_V of 10% can increase the losses by 6%.

Losses in transformers

Harmonic currents flowing through the transformer windings results in an increase in I^2R losses (copper loss) and also an increase in the iron loss due to eddy currents. The presence of harmonic voltage is responsible for an increase in the iron loss due to hysteresis. In general, losses in windings increase in proportion to the square of the THD_I and that in the core increases in proportion to the THD_V.

Losses in capacitors

The presence of harmonic voltages across the capacitors introduces currents in proportion to the frequency of the harmonics. These currents will generate additional losses.

9.5.2 Overloads on Equipment

Generators

Generators supplying non-linear loads must be derated, due to the additional losses caused by the harmonic currents. The level of de-rating needed is approximately 10% when the overall load is made of 30% of non-linear loads. In other words, it is necessary to oversize the generators when the load contains a major portion of non-linear loads.

Uninterrupted power supplies (UPS)

The current drawn by computer systems has a very high crest factor. A UPS sized to take into account only the rms current may not be capable of supplying the necessary peak current and may be overloaded.

Transformers

Transformers supplying comparable amount of non-linear loads needs de-rating and the de-rating factor is given by

$$k = \frac{1}{\sqrt{1 + 0.1\left(\sum_{h=2}^{40} h^{1.6} {T_h}^2\right)}} \tag{9.6}$$

where, $T_h = \dfrac{I_h}{I_1}$ *(Source: UTE Standard C 15-112)*

Typical values of 'k' are:

- Current with rectangular waveform $k = 0.86$
- Frequency converter current (THD ≈ 50%) $k = 0.80$

Figure 9.8 shows the typical de-rating required for a transformer supplying non-linear and electronic loads.

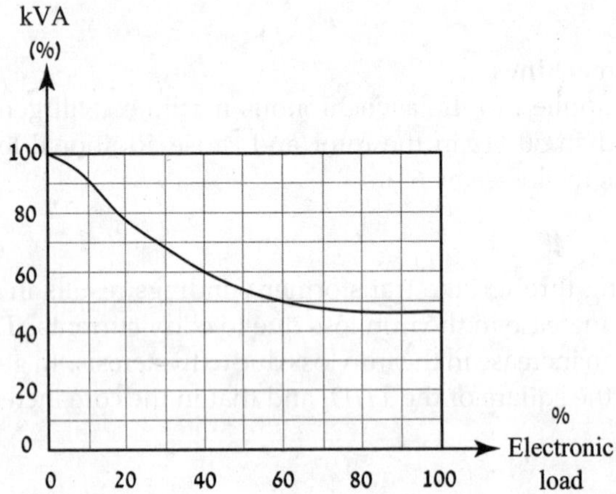

Figure 9.8 De-rating required for a transformer supplying electronic loads.

Asynchronous machines

International Electrotechnical Commission (IEC) specifies a weighted harmonic factor—Harmonic Voltage Factor-(HVF) to be satisfied for all asynchronous machines. The expression for HVF is given by equation (9.7)

$$HVF = \sqrt{\sum_{h=2}^{13} \frac{V_h^2}{h}} \leq 0.02 \quad (Source\ :\ IEC - 60892) \tag{9.7}$$

EXAMPLE 9.3

A supply voltage has a fundamental frequency component of V_1 and harmonic voltages of $V_3 = 2\%$, $V_5 = 3\%$ and $V_7 = 1\%$ of V_1. Find the value of THD_V and HVF.

$$THD_V = \frac{\sqrt{\sum_{h=2}^{n} V_h^2}}{V_1}$$

$$= \frac{\sqrt{2^2 + 3^2 + 1^2}}{100}$$

$$= \frac{\sqrt{14}}{100} = 0.03742 = 3.742\%$$

$$\text{Harmonic Voltage Factor, HVF} = \sqrt{\sum_{h=2}^{13} \frac{V_h^2}{h}}$$

$$= \sqrt{\frac{0.02^2}{3} + \frac{0.03^2}{5} + \frac{0.01^2}{7}} = 0.018$$

Capacitors

The effective rms current flowing through the capacitors needs to be limited to 1.3 times the rated current. (Ref: IEC standard IEC 60831-1).

EXAMPLE 9.4

A supply voltage having the following harmonic spectrum is applied to a capacitor bank. Calculate the effective rms voltage across the capacitor, THD_V and the effective rms current.

Harmonic spectrum V_5 = 8% of V_1, V_7 = 5% of V_1, V_{11} = 3% of V_1 and V_{13} = 1% of V_1.

$$\text{Effective rms voltage } V_{rss} = \sqrt{\sum_{h=1}^{n} V_h^2} = 1.0995 V_1$$

$$THD_V = \sqrt{\frac{\sum_{h=2}^{13} V_h^2}{V_1^2}}$$

$$= \frac{\sqrt{\left[(0.08)^2 + (0.05)^2 + (0.03)^2 + (0.01)^2\right] V_1^2}}{V_1}$$

$$= 0.09945 = 9.945\%$$

The capacitive reactance is inversely proportional to the frequency and hence the reactance of the capacitor at higher order frequencies will be as follows:

For the fifth harmonic voltage, capacitive reactance = $Z_5 = \dfrac{Z_1}{5}$

Similarly $Z_7 = \dfrac{Z_1}{7}$ $\quad Z_{11} = \dfrac{Z_1}{11}$

$$Z_{13} = \frac{Z_1}{13}$$

Therefore the currents at different frequencies will be

For the fundamental frequency $I_1 = \dfrac{V_1}{Z_1}$

For the fifth harmonic $I_5 = \dfrac{V_5}{Z_5} = \dfrac{0.08V_1}{\dfrac{Z_1}{5}} = 0.4I_1$

For the seventh harmonic $I_7 = \dfrac{V_7}{Z_7} = \dfrac{0.05V_1}{\dfrac{Z_1}{7}} = 0.35I_1$

For the eleventh harmonic $I_{11} = \dfrac{V_{11}}{Z_{11}} = \dfrac{0.03V_1}{\dfrac{Z_1}{11}} = 0.33I_1$

And for the thirteenth harmonic $I_{13} = \dfrac{V_{13}}{Z_{13}} = \dfrac{0.01V_1}{\dfrac{Z_1}{13}} = 0.13I_1$

The total harmonic distortion $THD_I = \dfrac{\sqrt{\sum_{h=2}^{n} I_h^2}}{I_1}$

$$= \frac{1}{I_1}\sqrt{(0.4^2 + 0.35^2 + 0.33^2 + 0.13^2)I_1^2}$$

$$= 0.6380$$

$$\frac{I_{rss}}{I_1} = \sqrt{1 + THD_I^{\,2}}$$

$$= \sqrt{1 + 0.638^2} = 1.1862$$

The effective rms current $I_{rss} = 1.1862I_1$

This means that the effective rms current will be 1.1862 times the fundamental component of the current at normal voltage. It can easily be seen that for this situation, if the supply voltage is 10% more than the normal voltage, the limit of 1.3 will be reached and will require resizing of the capacitors.

9.5.3 Overloading on Neutral Conductors

The presence of third harmonic components in the load current of a three-phase system creates a situation where all the third harmonic components add up in the neutral and the neutral current will be much more than the phase currents. This makes it necessary to provide a larger cross-section of conductor for the neutral wire. In extreme cases the neutral current can go up to $\sqrt{3}$ times the phase current.

9.5.4 Disturbances to Sensitive Loads

A distorted supply voltage can affect the normal operation of sensitive loads such as

- Temperature regulating devices
- Computer hardware
- Control and monitoring devices like protective relays, etc.

Harmonics can cause disturbances in low current control circuits. The level of disturbance depends on the distance between the power and control cables running in parallel, the distance between the cables and the frequency of the harmonics.

9.5.5 Resonance Problem with Power Factor Improving Capacitors

When power factor correction capacitors are employed in power supply systems, there is a possibility of resonance taking place between the inductive reactance of the source and the capacitive reactance of the power factor correction capacitors. Since the power factor correction capacitors are connected in parallel with the source, an infinite parallel impedance will occur when the inductive reactance of the source is equal to the capacitive reactance of the capacitor bank. If this parallel resonance occurs at or near a harmonic current produced by the non-linear loads, severe voltage distortion and harmonic current amplification may result. The increase in the harmonic currents may be large enough to cause nuisance fuse blowing and false tripping of circuit-breakers. It may be noted that the installation of capacitors to a power system will not create harmonics but the presence of capacitors on the system may cause amplification of the harmonic currents.

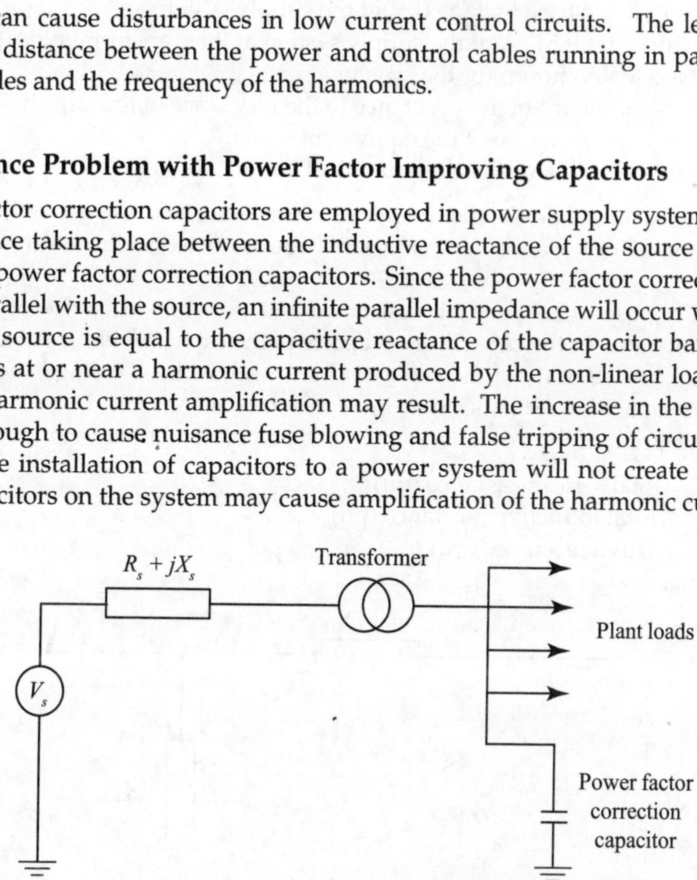

Figure 9.9 Single line diagram of a power system.

Consider the single line diagram of an industrial power system as shown in Figure 9.9. This diagram is representative of many of the practical industrial substations we come across. The substation is supplied from the electricity supply authority through a step down transformer. The total industrial load consists of both non-harmonic and harmonic producing loads. A capacitor bank is also connected to the MV bus for improving the power factor of the industrial load.

The supply system can be modelled using Thevenin's theorem with the help of the knowledge of the short circuit MVA of the supply system at the transformer location, the $\frac{X}{R}$ ratio of the supply system and the line to line voltage of the supply system.

$$R_S = \left[\frac{MVA_B}{MVA_{SC}}\right] \cos\left\{\tan^{-1}(X/R \text{ ratio}\right\} \quad (9.8)$$

$$X_S = \left[\frac{MVA_B}{MVA_{SC}}\right] \sin\left\{\tan^{-1}(X/R \text{ ratio}\right\} \quad (9.9)$$

where

kV_B = line to line base voltage (kV)of the supply system
MVA_{SC} = short circuit MVA of the supply system at the transformer location.
MVA_B = base MVA chosen for the system
X/R ratio = ratio of the inductive reactance to the resistance of the supply system.
R_S = per unit resistance of the equivalent system
X_S = per unit reactance of the equivalent supply system

From the transformer name plate details the per unit resistance and the per unit reactance of the transformer can be noted down and these values may be corrected to the assumed base values of MVA and kV.

Let, R_T and X_T represent the per unit resistance and per unit reactance of the transformer. Similarly, the per unit value of the capacitive reactance of the capacitor bank can also be calculated with the help of the kVAr capacity of the capacitor bank and the base voltage and base MVA. The per unit value of the capacitive reactance may be represented as X_C.

The single line diagram of the power system in per unit system is shown in Figure 9.10. Here

$R = R_S + R_T$ = Total series resistance (pu)
$X = X_S + X_T$ = Total inductive reactance (pu)
X_C = Total capacitive reactance (pu)

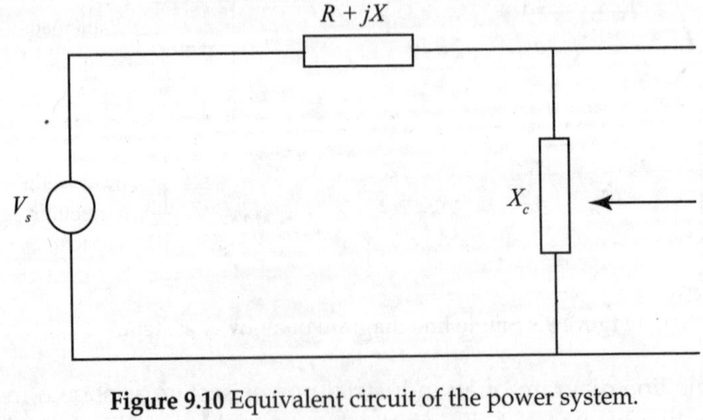

Figure 9.10 Equivalent circuit of the power system.

It can be visualised from the Figure 9.10 that resonance will occur when the inductive reactance of the system and capacitive reactance of the capacitor bank are equal. The frequency at which the resonance occurs is designated as f_o (Hz) or ω_o (rad/sec).

$$\omega_o L = \frac{1}{\omega_o C}$$

$$\omega_o = \frac{1}{\sqrt{LC}}$$

and the resonance frequency in Hz is given by

$$f_o = \frac{1}{2\pi\sqrt{LC}} \tag{9.10}$$

From Figure 9.10, the effective impedance looking into the system from the load bus consists of the source impedance in parallel with the capacitor impedance. The effective impedance is given by:

$$Z_{IN} = \frac{(R + jX)(-jX_C)}{R + j(X - X_C)} \tag{9.11}$$

This equation can be used to determine the system impedance at any other frequencies.

EXAMPLE 9.5

An industry is provided with the electric supply from a power supply authority through a 11 kV overhead line. The three-phase short circuit level at the end of the 11 kV feeder is 200 MVA. The X/R ratio of the supply source is 2.4. The transformer specifications are

Capacity	1000 kVA
Voltage ratio	11 kV/415 V
Winding connection	Delta/Star
Percentage resistance	1%
Percentage reactance	6%

Determine the resonance frequencies when the following capacitors are connected to the 415 V bus of the industry.

a) 200 kVAr b) 400 kVAr c) 600 kVAr

The single line diagram of the industrial power system is shown in Figure 9.11.

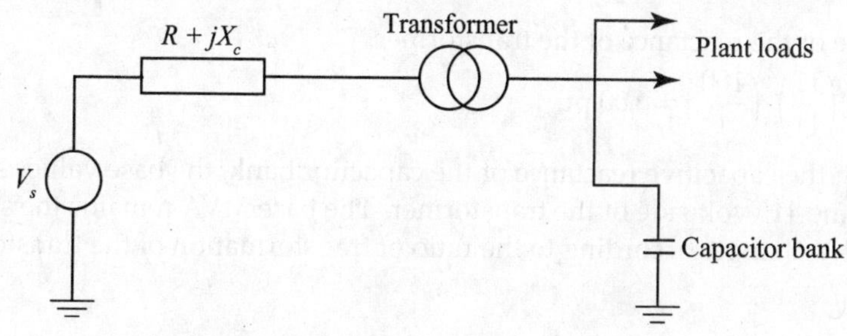

Figure 9.11 Single line diagram of example 9.5.

Solution :

Assume a base voltage of 11 kV and a base MVA of 100 MVA on the 11 kV side of the industrial power system.

The per unit impedance of the supply system $= \dfrac{MVA_B}{MVA_{SC}}$

$$= \frac{100}{200} = 0.5\ \Omega$$

Per unit resistance of the system R_S $= 0.5 \cos\left[\tan^{-1}\dfrac{X}{R}\right]$

$$= 0.19231\ \Omega$$

Per unit reactance of the system X_S $= 0.5 \cos\left[\tan^{-1}\dfrac{X}{R}\right]$

$$= 0.46154\ \Omega$$

The per unit resistance and reactance can be calculated on the newly selected base using the equation, where suffix '*o*' represents the old and suffix '*n*' represents the new quantities.

$$Z_{(n)} = Z_{(o)}\left(\frac{kV_{B(o)}}{kV_{B(n)}}\right)^2\left(\frac{MVA_{B(n)}}{MVA_{B(o)}}\right) \tag{9.12}$$

Here,
$R_{TO} = 0.01$ and $X_{TO} = 0.06$ (the pu resistance and reactance of the transformer on its own ratings)
$kV_{Bo} = kV_{Bn} = 11$ kV (old and new base voltages kV on the HT side of the transformer and)
$MVA_{Bo} = 1$ and $MVA_{Bn} = 100$ (old and new base MVA).

Therefore, the per unit value of the resistance of the transformer

$$R_T = 0.01\left(\frac{11}{11}\right)^2\left(\frac{100}{1}\right) = 1.00 \text{ pu}$$

and per unit value of the reactance of the transformer

$$X_T = 0.06\left(\frac{11}{11}\right)^2\left(\frac{100}{1}\right) = 6.00 \text{ pu}$$

For calculating the capacitive reactance of the capacitor bank, the base values selected needs to be transferred to the 415 volt side of the transformer. The base MVA remains the same through the system and base kV changes according to the ratio of transformation of the transformer.
Hence,

base MVA on the 415 V side = 100

and base kV on the 415 V side $= 11 \times \dfrac{0.415}{11} = 0.415$ kV

The base impedance on the 415 V side $Z_B = \dfrac{\text{Base kV}^2}{\text{Base MVA}} = \dfrac{(0.415)^2}{100}$

$$= 0.001722\ \Omega$$

The capacitive reactance of a three-phase capacitor bank is given by the equation

$$X_C = \frac{(V_{LL})^2}{Q_C} \tag{9.13}$$

where V_{LL} = line to line voltage across the capacitor
Q_C = Reactive VA capacity of the capacitor bank

The single line diagram of the power system can be redrawn as shown on Figure 9.12.

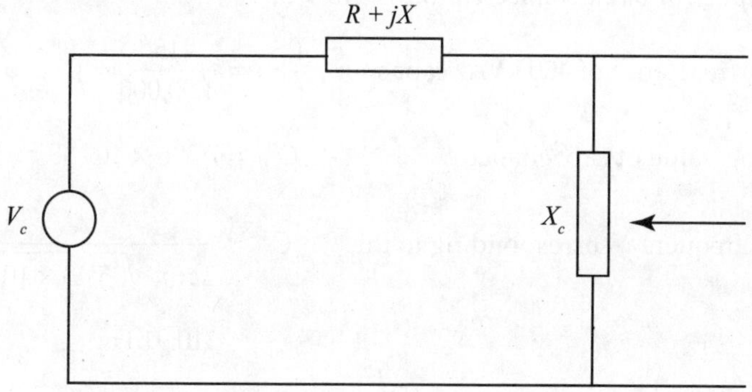

Figure 9.12 Single line diagram of the equivalent system.

The total inductive reactance of the system (pu) = $X_{pu} = X_S + X_T = 6.46154$ pu

Actual inductive reactance X (referred to 415 V side) $= X_{pu} \times Z_B$
$= 6.46154 \times 0.001722$
$= 0.011127\ \Omega$

Corresponding value of the inductance of the system $L = \dfrac{(0.011127)}{2 \times \pi \times 50}$

$$= 35.43 \times 10^{-6}\ \text{H}$$

Case a : 200 kVAr capacitor bank connected to the 415 V bus

The capacitive reactance of 200 kVAr capacitor bank $= \dfrac{(415)^2}{2{,}00{,}000} = 0.86113\ \Omega$

Corresponding value of capacitance C $= \dfrac{1}{2 \times \pi \times 50 \times 0.86113}$

$$= 3.698 \times 10^{-3}\ \text{F}$$

Therefore the resonant frequency corresponding to this value of capacitor connected to the system (from equation 9.10)

$$f_o = \frac{1}{2 \times \pi \sqrt{LC}}$$

$$= \frac{1}{2 \times \pi \sqrt{35.43 \times 10^{-6} \times 3.698 \times 10^{-3}}}$$

$$= 421.31 \text{ Hz}$$

The harmonic order at which resonance occurs $h = \frac{f_0}{f} = \frac{421.31}{50} = 8.426$

Case b : 400 kVAr capacitor bank connected to the 415 V bus

The capacitive reactance of 400 kVAr capacitor $= \frac{415^2}{4,00,000} = 0.4306\ \Omega$

Corresponding value of capacitance $C = 7.396 \times 10^{-3}$ F

The resonant frequency corresponding to this $C = \frac{1}{2 \times \pi \sqrt{35.43 \times 10^{-6} \times 7.396 \times 10^{-3}}}$

$$= 310.91 \text{ Hz}$$

Harmonic order of this resonant frequency $h = \frac{310.91}{50} = 6.218$

Case c : When 600 kVAr capacitor bank is connected to the 415 V bus

Capacitive reactance of the capacitor bank $= \frac{415^2}{6,00,000} = 0.287\ \Omega$

Corresponding value of capacitance $C = \frac{1}{2 \times \pi \times 50 \times 0.287}$

$$= 11.09 \times 10^{-3} \text{ F}$$

The resonant frequency corresponding to this $C = \frac{1}{2 \times \pi \sqrt{35.43 \times 10^{-6} \times 11.09 \times 10^{-3}}}$

$$= 253.91 \text{ Hz}$$

Harmonic order of this resonant frequency $h = \frac{253.91}{50} = 5.078$

From the above calculations, we see that there exists a resonant frequency very close to the fifth harmonic order when a 600 kVAr capacitor is connected to the system for power factor improvement.

If the industry is having a non-linear load which produces significant amount of fifth harmonic current, severe voltage distortions would result on the system. The next example illustrates the effect of a harmonic producing load on the system.

9.5.6 Short-Cut Method

In section 9.5.5 we have seen that the introduction of power factor correction capacitors for industries with non-linear loads can lead to total or partial resonant conditions at one or more harmonic frequencies. The harmonic order h or the natural resonant frequency between the system inductance and the power factor correction capacitor is also given by

$$h = \sqrt{\frac{kVA_{SC}}{Q_C}} \tag{9.14}$$

where, kVA_{SC} = short circuit kVA at the capacitor location
Q_C = kVAr rating of the capacitor bank and
h = harmonic order
$= \frac{f_0}{50}$ for a 50 Hz system

EXAMPLE 9.6

Solve the example 9.5 using equation (9.14) and check the validity of the equation. Assume that the system resistance can be neglected.

Solution :

From example 9.5, the total system reactance

$$X = X_S + X_T = 6.5 \text{ pu}$$

Therefore the short circuit level on the 415 V side

$$kVA_{SC} = \frac{\text{Base MVA}}{\text{reactance}}$$

$$= \frac{100}{6.5} = 15.385 \text{ MVA}$$

Using equation (9.14), the harmonic orders for different capacitors are calculated

For 200 kVAr capacitor $\quad h = \sqrt{\frac{15{,}385}{200}} = 8.7$

For 400 kVAr capacitor $\quad h = \sqrt{\frac{15{,}385}{400}} = 6.2$

and for 600 kVAr capacitor $\quad h = \sqrt{\frac{15{,}385}{600}} = 5.03$

These results are in close proximity to the results obtained in example 9.5.

EXAMPLE 9.7

If the industry considered in example 9.5 is supplying a 200 kVA, 415 V 50 Hz harmonic producing load with the following harmonic spectrum, calculate

a. Effective rms voltage and current, THD_V and THD_I without the 600 kVAr capacitor
b. Effective rms voltage and current, THD_V and THD_I with the 600 kVAr capacitor connected
c. Effective rms current in the capacitor and its THD_I

Table 9.5 Harmonic spectrum of 200 kVA load

S. No.	Harmonic order (h)	Line current (A)
1	5	50
2	7	30
3	11	15
4	13	7
5	17	3

Case a : When the 600 kVAr capacitor is not connected to the system

The system can now be represented as in Figure 9.13

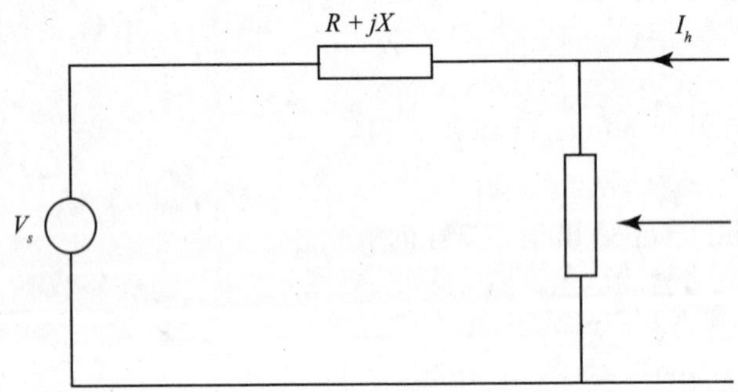

Figure 9.13 Representation of the power system without the capacitor.

From example 9.5,

The per unit value of the resistance of the total system $= 1.19231$ pu
$= 1.19231 \times 0.001722$
$= 0.00205\ \Omega$

The total inductive reactance of the system can be calculated for each harmonic frequency and is used to calculate the impedance of the network at each harmonic frequency. The voltage drop due to the harmonic currents can be computed and the details are summarised in Table 9.7.

From example 9.5, the total inductance of the system $L = 35.43 \times 10^{-6}$ H
Therefore the inductive reactance at fifth harmonic order ($f = 250$)

$$X_5 = 2 \times \pi \times 250 \times 35.43 \times 10^{-6}$$
$$= 0.05565\ \Omega$$

Therefore $$|Z_5| = |R + jX_5|$$

$$= |0.00205 + j0.05565|$$

$$= 0.05568\ \Omega$$

and the corresponding harmonic voltage drop

$$V_5 = I_5 \times |Z_5|$$
$$= 50 \times 0.05568$$
$$= 2.784\ \text{V}$$

To the harmonic currents are calculated and the details are tabulated in Table 9.6.

Table 9.6 Harmonic voltages produced due to harmonic currents

Harmonic order h	**Resistance** $R\,\Omega$	**Inductive reactance** $X\,\Omega$	**Impedance** $Z\,\Omega$	**Current** I_h A	**Voltage** V_h V
5	0.002065	0.05565	0.05568	50	2.784
7	0.002065	007791	0.07794	30	2.3382
11	0.002065	0.12243	0.12245	15	1.8368
13	0.002065	0.14469	0.14470	7	1.0130
17	0.002065	0.18921	0.18922	3	0.5670

For the 200 kVA harmonic producing load, the magnitude of the fundamental frequency load current

$$I_1 = \frac{2,00,000}{\sqrt{3} \times 415}$$

$$= 278.24\ \text{A}$$

Therefore, the effective rms value of the current

$$I_{rss} = \sqrt{I_1^2 + I_5^2 + I_7^2 + I_{11}^2 + I_{13}^2 + I_{17}^2}$$

$$= \sqrt{278.24^2 + 50^2 + 30^2 + 15^2 + 7^2 + 3^2}$$

$$= 284.78\ \text{A}$$

Total harmonic distortion

$$THD_I = \frac{\sqrt{50^2 + 30^2 + 15^2 + 7^2 + 3^2}}{278.24} \times 100$$

$$= 21.81\%$$

Similarly, the effective rms voltage

$$V_{rss} = \sqrt{V_1^2 + V_5^2 + V_7^2 + V_{11}^2 + V_{13}^2 + V_{17}^2}$$

$$= \sqrt{240^2 + 2.784^2 + 2.3382^2 + 1.8368^2 + 1.013^2 + 0.567^2}$$

$$= 240.037 \text{ V}$$

Total harmonic distortion

$$THD_V = \frac{\sqrt{2.784^2 + 2.3382^2 + 1.8368^2 + 1.013^2 + 0.567^2}}{240} \times 100$$

$$= 1.767\%$$

Case b : When 600 kVAr capacitor bank is connected to the system.

The equivalent circuit of the system with the capacitor bank connected is shown in Figure 9.14.

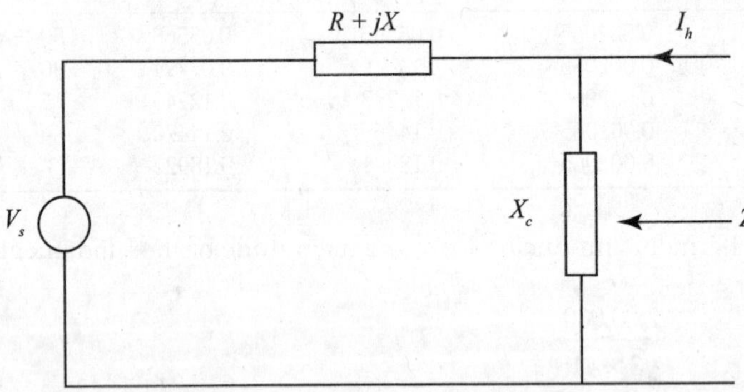

Figure 9.14 Equivalent single line diagram with the capacitor.

The effective impedance of the system when looking from the load end is given by

$$|Z_h| = \frac{(R + jX_h)(-jX_{ch})}{R + j(X_h - X_{ch})} \tag{9.15}$$

where

X_h represents the inductive reactance at respective harmonic frequencies,
X_{ch} represents the capacitive reactance at the respective harmonic frequencies and
h represents the harmonic order.

Using equation (9.15) the magnitude of the effective impedance of the system at various frequencies are calculated and the results are tabulated in Table 9.7.

Table 9.7 Effective impedance of the system at harmonic frequencies

Harmonic order	$\|Z_h\|$	I_h	V_h
5	1.1859	50	59.295
7	0.0864	30	2.592
11	0.03315	15	0.4973
13	0.02615	7	0.1831
17	0.01853	3	0.0556

The effective rms value of the bus voltage

$$V_{rss} = \sqrt{240^2 + 59.295^2 + 2.592^2 + 0.9973^2 + 0.1831^2 + 0.0556^2}$$
$$= 247.23 \text{ V}$$

and

$$THD_V = \frac{\sqrt{59.295^2 + 2.592^2 + 0.9973^2 + 0.1831^2 + 0.0556^2}}{240} \times 100$$
$$= 24.73\%$$

It is seen that when the capacitor is connected, the harmonic distortion of the voltage has risen to a value 24.73% as compared to a value of 1.767% when the capacitance was not connected. This is because of the fact that there exists a resonance condition very near the fifth harmonic order and the fifth harmonic current gets amplified by this phenomena leading to a large value of THD_V which is unacceptable.

Case c :

The current through the capacitor is calculated using the harmonic voltages that have been calculated in section b of this problem.

The capacitive reactance of the 600 kVAr capacitor

$$X_c = \frac{415^2}{6{,}00{,}000} = 0.2870\ \Omega$$

The capacitance value

$$C = \frac{1}{2 \times \pi \times f \times X_c}$$

$$= 11.0966 \times 10^{-3} \text{ F}$$

The capacitive reactance at other frequencies is calculated using the expression

$$X_{ch} = \frac{1}{2 \times \pi \times f_h \times C}$$

The results are tabulated as in Table 9.8.

Table 9.8 Harmonic currents in the capacitor bank

Harmonic order h	Harmonic frequency f_h	Capacitive reactance X_{ch}	Harmonic voltage V_h	Harmonic current I_h
5	250	0.0574	59.295	1033.014
7	350	0.0410	2.592	63.22
11	550	0.0261	0.4973	19.054
13	650	0.0221	0.1831	8.285
17	850	0.0169	0.0556	3.290

The fundamental frequency component of capacitor current I_1 $= \dfrac{6{,}00{,}000}{\sqrt{3} \times 415}$

$= 834.75$ A

The effective rms value of the current passing through the capacitor

$$I_{rss} = \sqrt{834.75^2 + 1033.014^2 + 63.22^2 + 19.054^2 + 8.285^2 + 3.29^2}$$

$$= 1329.798 \text{ A}$$

Note that the current actually existing in the capacitor is about 159.3% of the rated capacitor current. The presence of this current will definitely cause the capacitor fuse to blow or the capacitor circuit-breaker to trip. If the protective elements fail to act, the capacitor will be damaged due to the excessive over current.

From the above examples it is evident that a close monitoring is required to assess the possibility of resonance occurring in systems where power capacitors are employed for improving the system power factor. All system designers must take into account such aspects while designing new systems.

9.6 METHODS TO REDUCE THE IMPACT OF HARMONICS

Three different types of solution are available to mitigate the ill effects of harmonics. These are:

- Modifications in the installations
- Special devices in the supply system and
- Filtering the harmonics.

It is recommended that these solutions are adopted at the design stage itself. Some of the possible methods that can reduce the impact of harmonics are listed in the following sections.

9.6.1 Locate the Non-Linear Loads Upstream on the System

The effect of harmonic disturbances will be higher where the short circuit-power is lower. Hence it is preferable to connect all non-linear or harmonic producing loads as far upstream as possible economically on the system. Figure 9.15 shows a typical arrangement of connecting all non-linear loads at the upstream level.

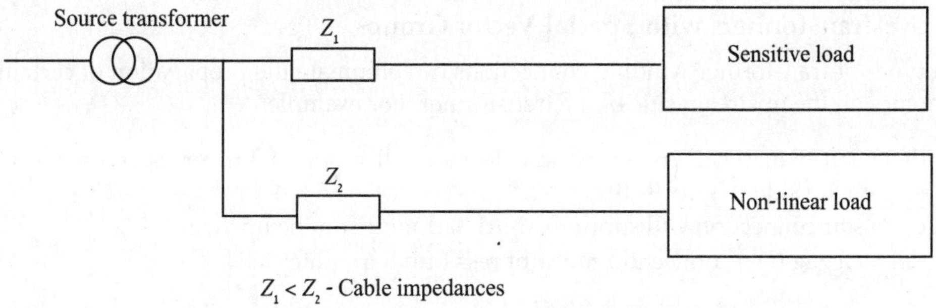

Figure 9.15 Non-linear loads at the upstream side of the system.

9.6.2 Group all Non-Linear Loads

While preparing the single line diagram of the industrial power system, all non-linear loads are separated from the linear loads and the two groups of loads are supplied from two different switch-boards. Figure 9.16 shows such an arrangement.

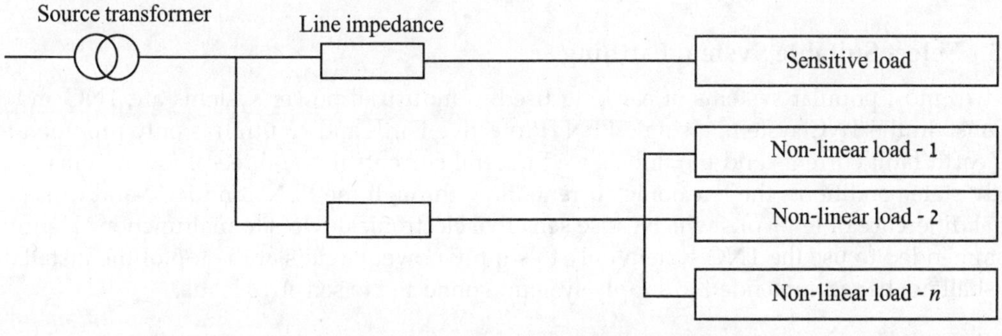

Figure 9.16 Non-linear loads grouped together.

9.6.3 Provide Separate Sources

A more effective but costly means of mitigating the ill effects of harmonics is to provide separate sources to cater to the linear and non-linear loads. Figure 9.17 shows such an arrangement.

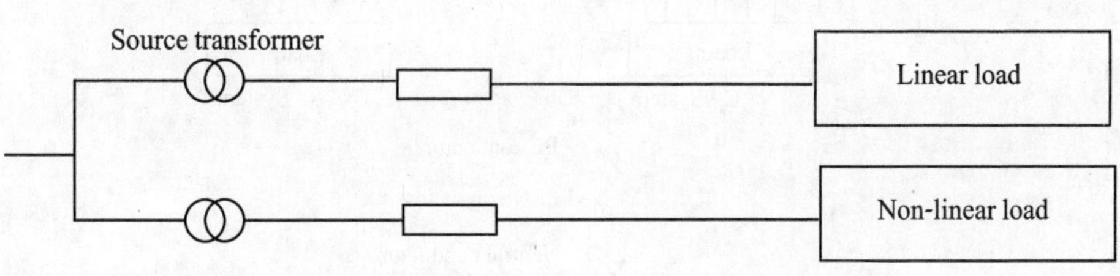

Figure 9.17 Separate source for non-linear loads.

9.6.4 Use Transformers with Special Vector Groups

Certain types of transformer winding connections can eliminate the propagation of certain orders of harmonics on the upstream side of the transformer. For example:

- A delta-star-delta (Dyd) connected transformer will suppress fifth and seventh harmonics on the upstream side (Figure 9.18)
- A delta star connection will suppress third harmonics on the upstream side.
- A delta zigzag (Dz) connection can suppress fifth harmonic etc.

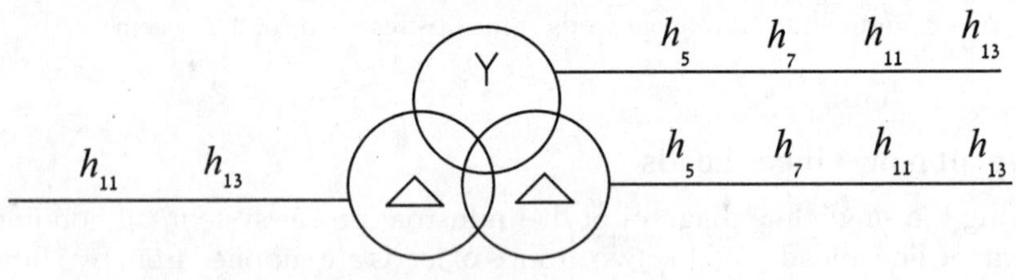

Figure 9.18 Use of transformers with special vector groups.

9.6.5 Select Suitable System Earthing

The two most popular systems of earthing used in industrial power systems are TNC and TNS systems. In the TNC system, a single PEN (Protective Earth and Neutral) is only provided to by pass earth fault currents and the unbalanced neutral currents in the three-phase systems. Under steady state conditions, the harmonic currents flow through the PEN conductor and cause a potential difference of few volts which cause sensitive electronic devices to malfunction. Hence, it is recommended to use the TNC system only to supply power circuits at the top of the installations and shall not be recommended to supply systems connected to sensitive loads.

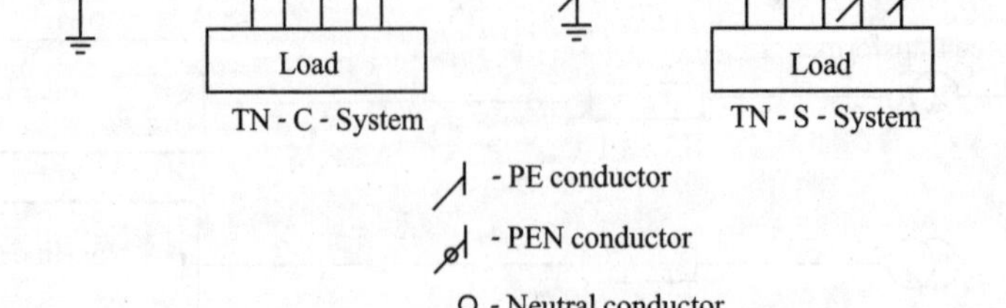

Figure 9.19

In TNS system, two separate conductors are provided for neutral and the protective earth function. The neutral conductor and protective conductor are completely separate and so the potential throughout the system is more uniform and this is the system of earthing highly recommended if the power system contains harmonic producing loads. Figure 9.19 shows the two types of system earthing.

9.6.6 Install Inductive Reactors

When variable voltage-variable frequency drives are used, it is possible to smoothen the supply current by installing line reactors. These reactors increase the line impedance and thereby limit the harmonic current. However, increase in the line impedance will increase the line voltage drop and may necessitate oversizing of cables to limit the voltage drop within permissible limits.

The application of power factor improving capacitors also is likely to cause resonance problems (as discussed in section 9.5.5). In order to reduce the ill effects of harmonics, harmonic suppression reactors are connected in series with the capacitor bank. The influence of reactors in series with the capacitor is to increase the impedance of the combination for higher order harmonics. This arrangement is known as *Detuned capacitors and is also known as passive filters*.

Other forms of harmonic filters are active filters and hybrid filters. Active filters are based on the application of power electronics and are generally installed in parallel with a non-linear load to reduce the ill effects of harmonics. They analyse the load current for the presence of harmonics and inject similar harmonics to the load with appropriate phase shift. As a result the harmonic currents are totally neutralised at the point of origin itself. The main advantages are:

- That they guarantee efficient harmonic compensation even when changes are made to the system.
- They are compatible with electrical generator sets.
- They can be connected to any point on the electrical network.
- Several filters can be used in the same installation.

The disadvantages are:

- They are costly and uses complex circuits.
- They are costly when compared to passive filters.
- They cannot be used for power factor correction.

Hybrid filters are a combination of active and passive filter elements and they combine the advantages of both the passive and active filters. This option can be chosen to filter out one specific frequency by passive filters and other frequencies by active filters. The detailed analysis of this type of filters is beyond the scope of this book.

9.7 DESIGN OF DETUNED CAPACITOR BANK (HARMONIC FILTER)

The occurrence of resonance due to the presence of harmonic currents can be avoided by placing a reactor in series with the capacitor. The idea is to shift the resonance frequency to a value lower than the lowest harmonic order of the load current. This series combination of reactor and capacitor is known as detuned capacitor or harmonic filter. Figure 9.20 shows the equivalent single line diagram of such an arrangement. The series resonant frequency of the detuned capacitor bank is generally selected as 3-10% lower than the lowest harmonic order produced by the load current. For a six-pulse power converter, the lowest harmonic order is the fifth harmonic which corresponds to 250 Hz on a 50 Hz supply system. A detuned capacitor applied to such a network may be tuned to resonate between 225 Hz and 242.5 Hz. Typically, the tuning frequency for a filter of this type will be 235 Hz corresponding to a harmonic order of 4.7.

An additional benefit of the detuned capacitor is that it supplies a portion of the harmonic current required by the load. The most significant part of the harmonic load current supplied will be corresponding to the tuning frequency. For example, a detuned capacitor tuned for operation at $4.7f$ will supply a major portion of the fifth harmonic current of the load. The filter will also supply a significantly lower amount of higher order harmonic load currents. This means that less harmonic currents need to be supplied by the source resulting in a reduction of total harmonic distortion THD_I.

The resonant frequency of the detuned filter circuit is given by:

$$f_0 = \frac{1}{2\pi\sqrt{L_f C}} \tag{9.16}$$

where,

L_f = Inductance of the tuning reactor (filter)
C = Capacitance of the PF improving capacitor

Solving for L_f using equation (9.16)

$L_f = \dfrac{1}{C(2\pi f_0)^2}$ Substituting for C

$$L_f = \frac{2\pi f X_C}{(2\pi f_0)^2} \tag{9.17}$$

$$= \frac{f X_C}{2\pi (f_0)^2} \tag{9.18}$$

$$= \frac{f}{(f_0)^2}\frac{V_{LL}^2}{2\pi Q_C}$$

where

V_{LL} = line to line voltage of the capacitor bank
f = system frequency
f_0 = tuning frequency of the detuned capacitor
Q_C = three-phase reactive volt ampere capacity of the capacitor

Equation (9.18) can be used to find out the value of the filter inductance to tune the capacitor bank.

EXAMPLE 9.8

Determine the values of filter inductances required to tune the following capacitor banks to a tuning frequency of 135 Hz.(2.7th harmonic)

a) 100 kVAr, 415 V, 50 Hz, three-phase capacitor bank
b) 1200 kVAr, 11 kV, 50 Hz, three-phase capacitor bank

a) Substituting equation (9.18) directly yields the value of the filter inductance.

$$L_f = \frac{50}{(135)^2}\frac{415^2}{2\pi \times 1{,}00{,}000} = 0.752 \text{ mH}$$

b) $$L_f = \frac{50}{(135)^2}\frac{(11{,}000)^2}{2\pi \times 12{,}00{,}000} = 44.02 \text{ mH}$$

The analysis of the system with detuned capacitor filter can be carried out using the equivalent diagram shown in Figure 9.20.

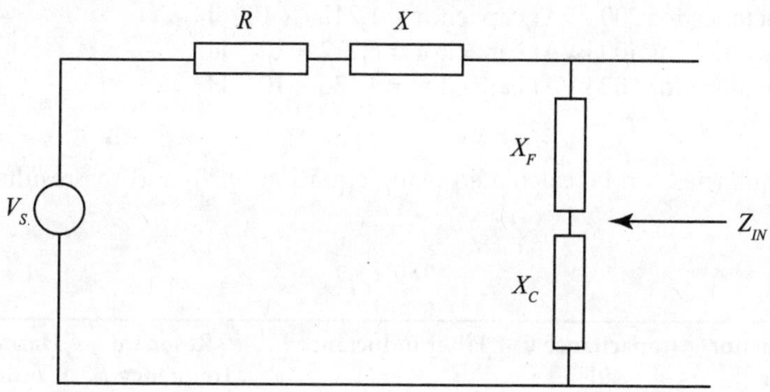

Figure 9.20 Equivalent circuit of the system with detuned capacitor filter.

The input impedance of the network as seen from the load end

$$Z_{IN} = \frac{(R + jX) \times j(X_F - X_C)}{R + j(X + X_F - X_C)} \quad (9.19)$$

The system will be at resonance when

$$X + X_F = X_C$$

$$\omega_0(L + L_f) = \frac{1}{\omega_0 C}$$

$$\omega_0 = \frac{1}{\sqrt{(L+L_f)C}}$$

and the resonant frequency

$$f_0 = \frac{1}{2\pi\sqrt{(L+L_f)C}} \quad (9.20)$$

EXAMPLE 9.9

If the power factor correction capacitors used in example 9.5 were provided with tuning reactors in series with the capacitors, determine;

a) Tuning inductance required to tune the filter to a harmonic order of 4.7.
b) Resonance frequencies with the tuning inductors installed and
c) Resonance frequencies with and without detuned capacitor filter.

Solution :

Case a :
The required values of inductors L_f can be calculated using the equation (9.18).

Filter inductance for 200 kVAr capacitor = 1.2426×10^{-6} H
Filter inductance for 400 kVAr capacitor = 6.212×10^{-5} H
Filter inductance for 600 kVAr capacitor = 4.136×10^{-5} H

Case b :
The resonant frequencies can be calculated using equation (9.20) and the results are tabulated in Table 9.9.

Table 9.9

Capacitor kVAr	Capacitance C μF	Filter inductance L_f mH	Resonant frequency f_0	Harmonic order h
200	3.698×10^{-3}	1.2426×10^{-6}H	207	4.14
400	7.396×10^{-3}	6.212×10^{-6}H	187.4	3.748
600	11.09×10^{-3}	4.136×10^{-6}H	173.6	3.472

Case c :
The resonant frequencies without tuning reactors as calculated in example 9.5 and the resonant frequencies with the tuning inductance as tabulated in Table 9.9 are consolidated in Table 9.10.

Table 9.10

Capacitor kVAr	Resonant frequency without filter Hz	Resonant frequency with filter inductance Hz
200	421.31	207
400	310.91	187.4
600	253.91	173.6

It can be seen that the resonant frequencies have been shifted to a value lower than the tuning frequency of the filter.

In order to calculate the bus voltage distortion, it will be necessary to compute the harmonic load currents shared by the source and the harmonic filter. The division of the total harmonic current is obtained by applying the current division rule in a parallel circuit. Referring to Figure 9.20,

$$I_{Fh} = I_h \frac{(R + jX_h)}{R + j(X_h + X_{Fh} - X_{Ch})} \tag{9.21}$$

$$I_{Sh} = I_h \frac{j(X_{Fh} - X_{Ch})}{R + j(X_h + X_{Fh} - X_{Ch})} \tag{9.22}$$

where

I_h = Load current of the harmonic order h
I_{Fh} = Harmonic current through the filter element of order h
I_{Sh} = Harmonic current supplied by the system of order h
X_{Fh} = Inductive reactance of the filter inductance at the harmonic frequency of order h
X_{Ch} = Capacitive reactance of the capacitor at the harmonic frequency of order h
X_h = Source inductive reactance corresponding to harmonic frequency of order h

The harmonic voltage drop in the system is given by

$$V_h = I_{Sh}(R + jX_h) \tag{9.23}$$

Equations (9.21), (9.22) and (9.23) are to be evaluated corresponding to different harmonic frequencies to determine the total harmonic distortion of the voltage at the bus. These equations cannot be applied to obtain the current division of the fundamental frequency component of current.

The fundamental frequency components of source and filter currents are

$$I_{S1} = \frac{\text{Three-phase volt ampere of the load}}{\sqrt{3} \times V_{LL}} \tag{9.24}$$

$$I_{F1} = \frac{\text{Line to neutral bus voltage}}{(X_F - X_C)} \tag{9.25}$$

EXAMPLE 9.10

For the industrial power system of the example 9.5 if a detuned capacitor filter of capacity 600 kVAr is installed and if the load current has a harmonic spectrum as given in example 9.7, determine

a) Harmonic current spectrum of the filter current and its effective rms current
b) Total harmonic distortion of the bus voltage and
c) Effective rms bus voltage.

Solution :

From examples 9.5 and 9.7

The total system resistance $R = 0.00205\ \Omega$
Total system inductance $L = 35.43 \times 10^{-6}$ H and
The capacitance of the 600 kVAr capacitor $C = 11.09 \times 10^{-3}$ F

From example 9.9
The inductance of tuning reactor for 600 kVAr capacitor bank

$$L_f = 4.136 \times 10^{-5} \text{ H}$$

The fundamental component of the 200 kVA harmonic producing load

$$I_1 = 278.24 \text{ A}$$

The harmonic spectrum of the load current is reproduced from example 9.7.

Table 9.11 Harmonic component of the 200 kVA load

S. No.	Harmonic order h	Line current A
1	5	50
2	7	30
3	11	15
4	13	7
5	17	3

Case a :

$$\text{Fundamental component of the filter current } I_{F1} = \frac{\frac{415}{\sqrt{3}}}{(X_F \sim X_C)}$$

$$X_F = 2\pi \times 50 \times 35.43 \times 10^{-6}\ \Omega$$
$$= 0.01113\ \Omega$$

$$X_C = \frac{1}{2\pi \times 50 \times 11.09 \times 10^{-3}}$$

$$= 0.2872\ \Omega$$

Therefore $\quad I_{F1} = 868.02$ A

The harmonic current spectrum of the filter circuit is obtained using equation (9.21) and is tabulated in Table 9.12.

Table 9.12 Harmonic spectrum of the filter current

Harmonic order h	Harmonic current of the load I_h (A)	Harmonic current in the filter circuit I_{Fh} (A)
5	50	44.02
7	30	18.29
11	15	7.676
13	7	3.474
17	3	1.444

The effective rms current of the filter circuit

$$I_{Frss} = \sqrt{I_{F1}^2 + I_{F5}^2 + I_{F7}^2 + I_{F11}^2 + I_{F13}^2 + I_{F17}^2}$$

$$= 869.37 \text{ A}$$

Rated current of the 600 kVAr capacitor bank

$$I_C = \frac{6,00,000}{\sqrt{3} \times 415}$$
$$= 834.75 \text{ A}$$

The percentage increase in the capacitor current due to the introduction

of the tuning inductor $= \dfrac{(869.37 - 834.75)}{834.75} \times 100$

$$= 4.15\%$$

Case b :

To find the total harmonic distortion of the voltage, we have to first calculate the harmonic currents supplied by the supply source. Using equation (9.22) the component of the source current at different harmonic frequencies are calculated and the results are tabulated in Table 9.13.

Table 9.13 Harmonic spectrum of the source current

Harmonic order *h*	**Harmonic current of the load** I_h (A)	**Harmonic current in the supply source** I_{Sh} (A)	**Harmonic voltage in the source** V_h(V)
5	50	5.98	0.33
7	30	11.72	0.91
11	15	7.32	0.89
13	7	3.53	0.51
17	3	1.56	0.29

The fundamental component of the bus voltage to neutral $V_1 = \dfrac{415}{\sqrt{3}} = 239.61$ V

Total harmonic distortion of the bus voltage $= \dfrac{\sqrt{V_5^2 + V_7^2 + V_{11}^2 + V_{13}^2 + V_{17}^2}}{V_1} \times 100$

$$= \frac{\sqrt{0.33^2 + 0.91^2 + 0.89^2 + 0.51^2 + 0.29^2}}{239.61}$$

$$= 0.6\%$$

Case c :

$$\text{The effective rms voltage of the bus} = \frac{\sqrt{239.61^2 + 0.33^2 + 0.91^2 + 0.89^2 + 0.51^2 + 0.29^2}}{239.61}$$

$$= 239.6143 \text{ V}$$

The performance of the industrial power system has been analysed for three different situations:

a) Without the power factor improving capacitor
b) With 600 kVAr power factor improving capacitor and
c) With 600 kVAr capacitor with a tuning reactor (harmonic filter)

The resulting values of the total harmonic distortion in the bus voltages are tabulated in Table 9.14.

Table 9.14 Comparison of total harmonic distortion voltage

S.No	System state	THD_V
1	Without power factor improving capacitor	1.767%
2	With 600 kVAr power factor improving capacitor	24.73%
3	With 600 kVAr capacitor harmonic filter	0.6%

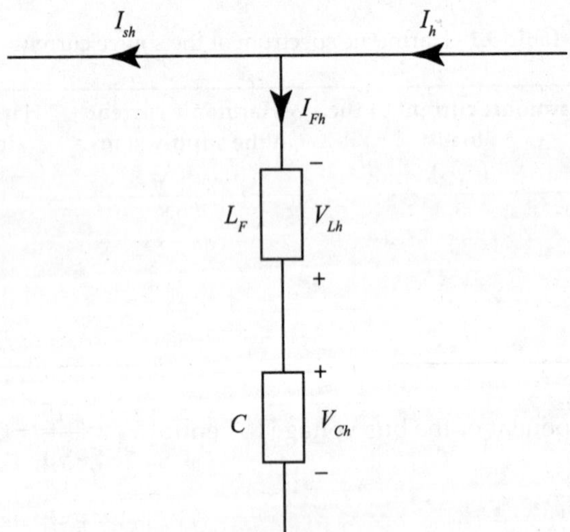

Figure 9.21 Equivalent circuit of the harmonic filter.

The reduction in the total harmonic distortion voltage is due to the fact that the resonant condition that was existing when the capacitors were connected has been shifted to a value much lower than the lowest harmonic order of the load current.

Another factor which needs to be considered when employing the harmonic filter, is the possibility of an over voltage across the capacitor unit. This over voltage is likely to cause excessive voltage stress on the capacitor and may reduce the useful life of the capacitor. Or else we may have to select low stress type of capacitors for such application. The presence of an over voltage can be analysed using the simplified equivalent circuit of the harmonic filter circuit. Figure 9.21 shows the simplified equivalent circuit of the filter on a single-phase basis.

$$\text{The harmonic voltage across the capacitor } V_{Ch} = \frac{I_{Fh}}{\omega_h C} \tag{9.26}$$

$$\text{The voltage across the inductance } V_{Lh} = I_{Fh} \times \omega_h L \tag{9.27}$$

EXAMPLE 9.11

For the harmonic filter of example 9.10, determine the effective rms voltage across the capacitor unit and the percentage increase in voltage across the capacitor.

Solution :

The harmonic component of the filter currents have been calculated and tabulated in Table 9.12. Using equation (9.25), the fundamental component of the filter current can be calculated. The voltage across the capacitor can be calculated using equation (9.26) and the results are tabulated in Table 9.15.

Table 9.15 Harmonic components of voltages across the capacitor

Harmonic order h	**Frequency** f_h	I_{Fh}	V_{Ch}
1	50	868.02	249.14
5	250	44.02	2.527
7	350	18.29	0.749
11	550	7.676	0.200
13	650	3.474	0.077
17	850	1.444	0.024

$$\text{The effective rms voltage across the capacitor} = \sqrt{V_{C1}^2 + V_{C5}^2 + V_{C7}^2 + V_{C11}^2 + V_{C13}^2 + V_{C17}^2}$$

$$= 249.15 \text{ V}$$

$$\text{the percentage rise in voltage across the capacitor} = \frac{249.15 - 239.61}{239.61} \times 100$$

$$= 3.98\%$$

Example 9.11 shows that an over voltage exists across the capacitor unit when it is used along with the tuning inductor. For this specific reason, it is a common practice that capacitors with a higher voltage ratings are recommended when they are used in harmonic filter circuits. In such cases the reactive power rating of the capacitors will have to be recalculated according to the following expression.

$$Q_{Cactual} = Q_{Crated}\left(\frac{f_{actual}}{f_{rated}}\right)\left(\frac{\text{Actual } V_{LL}}{\text{Rated } V_{LL}}\right)^2 \quad (9.28)$$

EXAMPLE 9.12

A three-phase capacitor bank rated for 100 kVAr, 415 V, 50 Hz is required to be used along with a tuning inductor. If the voltage across the capacitor is expected to rise to 425 volts and if the actual system frequency is 49.5 Hz, calculate the actual rating of the capacitor.

Solution :

Using equation (9.28) directly,

$$Q_{Cactual} = 100\left(\frac{49.5}{50}\right)\left(\frac{425}{415}\right)^2 = 103.83 \text{ kVAr}$$

9.8 IEEE STANDARD 519-1992

IEEE standard 519 recommended practices and requirements for harmonic control in electric power systems were first published in 1981. This document established levels of voltage distortion acceptable to distribution systems. It was widely accepted in establishing harmonic correction throughout the power industry. However, widespread use of variable speed drives, rectifiers and other forms of non-linear loads necessitated a rewriting of the IEEE 519.

Table 9.16 Current distortion limits for general distribution systems (120 V through 69,000 V)

Maximum harmonic current distortion in % of I_L						
Individual harmonic order (Odd harmonics)						
$\frac{I_{SC}}{I_L}$	**<11**	**11≤ *h* <17**	**17≤ *h*<23**	**23≤ *h*<35**	**35≤ *h***	***THD$_I$* (TDD)**
<20*	4.0	2.0	1.5	0.6	0.3	5.0
20 <50	7.0	3.5	2.5	1.0	0.5	8.0
50 <100	10.0	4.5	4.0	1.5	0.7	12.0
100 <1000	12.0	5.5	5.0	2.0	1.0	15.0
>1000	15.0	7.0	6.0	2.5	1.4	20.0

Even harmonics are limited to 25% of the odd harmonic limits. TDD refers to total demand distortion and is based on the average maximum demand current at the fundamental frequency measured at PCC.

* All power generation equipment is limited to these values of current distortion regardless of short circuit current ratio.

I_{SC}—Maximum short circuit current at PCC
I_L—Maximum demand load current at PCC
h—Harmonic order

The revised IEEE 519 was published in 1992 with special emphasis on the relationship of harmonic voltages to harmonic currents flowing within industrial plants. Limits for both harmonic voltages on the utility transmission systems and harmonic currents within the industrial distribution systems were established in this document. Since harmonic voltages are produced by the passage of harmonic currents passing through the distribution network, controlling the harmonic currents or system impedances within the industrial facility will have an impact on the reduction of harmonic voltages on the utility system. Table 9.16 is a representation of Table 10.3 of IEEE Standard 519-1992.

In order to understand the implication of IEEE 519 1992 document, it is very important to understand the meaning of the terms used in Table 9.16.

PCC : Point of common coupling is probably the most important and perhaps the most controversial term in the document. It is the electrical connecting point or interface between the power utility and the consumer's electrical system. Even though simple in concept, identification of this point is sometimes misunderstood, leading to confusion and wrong interpretation of the specifications. To avoid this confusion, it can be considered the energy metering point on the system.

I_{SC} **:** Is the available short circuit current at the PCC.

I_L **:** Is the load current corresponding to the maximum demand measured at PCC. For existing systems, this can be measured over a period of time and the average value is taken for computation. For new installations, this is to be calculated using the anticipated maximum demand.

EXAMPLE 9.13

An industrial power system is connected to the power utility through a step down transformer. The plant load consists of non-linear loads. The point of common coupling is the HT side of the transformer. The HT supply voltage is 11 kV and the short circuit level on the HT side of the transformer is 150 MVA. The maximum demand on the transformer is 630 kVA. The details of the harmonic spectrum of the load current are given in Table 9.17.

Table 9.17 Harmonic spectrum of the load current

Frequency *f*	**Harmonic order** *h*	**Harmonic current on the HT side** I_{Sh} **harmonic current**	**Percentage**
250	5	3.5	10.584
350	7	2.0	6.050
550	11	1.6	4.840
650	13	0.6	1.814
850	17	0.4	1.210

Determine whether the requirements of the IEEE 519 are satisfied or not.

Solution :

Fundamental component of the maximum demand load $I_L = \dfrac{630}{\sqrt{3} \times 11}$

$$= 33.07 \text{ A}$$

Short circuit current on the 11 kV side $I_{SC} = \dfrac{1,50,000}{\sqrt{3} \times 11}$

$$= 7873 \text{ A}$$

Short circuit ratio $\dfrac{I_{SC}}{I_L} = \dfrac{7873}{33.07}$

$$= 238.07$$

This corresponds to the fourth row of the Table 9.16.

Total harmonic distortion up to 11th harmonics, $THD_{<11} = \dfrac{\sqrt{3.5^2 + 2.0^2}}{33.07} \times 100$

$$= 12.19\%$$

From Table 9.16 (fourth row) the maximum permissible harmonic current distortion for harmonics less than 11 is only 12%. Therefore IEEE 519 requirement is violated in this range.

For the harmonic range of 11 to 17th harmonics,

$$THD_{11-17} = \frac{\sqrt{1.6^2 + 0.6^2}}{33.07}$$

$$= 5.167\%$$

From Table 9.16, we can see that the maximum permissible distortion is 5.5% . Hence the IEEE 519 requirement is met over this range of harmonics.

For the harmonic range of 17 to 23rd harmonics,

$$THD_{17-23} = \frac{\sqrt{0.4^2}}{33.07}$$

$$= 0.484\%$$

The maximum permissible distortion for this range is 5% and hence the requirements are met.

The total harmonic distortion THD_I taking into account all harmonics (TDD)

$$= \frac{\sqrt{3.5^2 + 2.0^2 + 1.6^2 + 0.6^2 + 0.4^2}}{33.07} \times 100$$

$$=13.295\%$$

The permissible THD or TDD is 15%. And so the situation is well within the requirements of the IEEE 519 standard. Since there is only a marginal excess in the harmonic distortion of frequencies below 11th harmonics, further correction is not immediately required.

EXAMPLE 9.14

If the rating of the transformer referred to in the example 9.13 is 100 kVA, 11 kV/415 V, delta/star connected, with a percentage reactance of 5% and if the point of common coupling (PCC) is assumed to be on the MV side of the transformer, check whether the IEEE 519 requirements are satisfied or not.

Solution :

Assume a base voltage of 11 kV and a base MVA of 100 MVA. The per unit reactance of the supply source

$$X_S = \frac{\text{Base MVA}}{\text{Short circuit MVA}}$$

$$= \frac{100}{150}$$

$$= 0.666 \text{ pu}$$

Per unit reactance of the transformer on the newly selected base

$$X_T = 0.05\left(\frac{11}{11}\right)^2 \left(\frac{100}{1}\right)$$

$$= 5 \text{ pu}$$

$$\text{The fault level on the MV side (415V)} = \frac{\text{Base MVA}}{X_T + X_S}$$

$$= \frac{100}{5.666} = 17.65 \text{ MVA}$$

Short circuit current corresponding to a short circuit MVA of 17.75 MVA

$$I_{SC} = \frac{17{,}675}{\sqrt{3} \times 0.415}$$

$$= 24{,}555 \text{ A}$$

Load current corresponding to a maximum demand of 630 kVA

$$I_L = \frac{630}{\sqrt{3} \times 0.415}$$

$$= 776.485 \text{ A}$$

Therefore short circuit ratio

$$\frac{I_{SC}}{I_L} = \frac{24{,}555}{776.485}$$

$$= 31.624$$

The current harmonic spectrum on the 415 V side can be computed using the percentage harmonic spectrum given in Table 9.17. The results are

I_5 = 10.584% of 776.45 = 82.18 A
I_7 = 46.98 A
I_{11} = 37.58 A
I_{13} = 14.09 A and
I_{17} = 9.40 A

The harmonic distortion for frequencies below 11th harmonics

$$THD_{<11} = \frac{\sqrt{82.18^2 + 46.98^2}}{776.485} \times 100$$

$$= 12.19\%$$

Similarly $$THD_{11-17} = \frac{\sqrt{37.58^2 + 14.09^2}}{776.485}$$

$$= 5.17\%$$

and $$THD_{17-23} = \frac{\sqrt{9.4^2}}{776.485}$$

$$= 1.21\%$$

The allowable limits from Table 9.16 are

$THD_{<11}$ = 7%
THD_{11-17} = 3.5%
THD_{17-23} = 2.5%

We see that the harmonic distortion levels for harmonic orders from 5 to 17 are in violation of the IEEE 519 requirements.

Total harmonic distortion THD_I = TDD $$= \frac{\sqrt{82.18^2 + 46.98 + 37.58^2 + 14.09^2 + 9.42^2}}{776.485} \times 100$$

$$= \frac{103.25}{776.485} \times 100 = 13.30\%$$

The permissible TDD or THD_I from Table 9.16 is only 8%. Comparing the results of examples 9.12 and 9.13, we have a conflicting situation. For the same installation, a change in the location of PCC results in major violation of the requirements of IEEE 519. When the PCC was on the 11 kV side of the system, it was almost in conformation to the requirements whereas when the PCC was identified on the 415 bus, the system violates the requirements of the harmonic orders up to 17th harmonics and will require installation of harmonic filters to bring down the harmonic distortions within the limits. For a new installation, one way to overcome this confusion is to increase the stiffness of the system by increasing the short circuit current ratio and moving to the next row of Table 9.16 with higher percentage distortion limits. This can be done by using a transformer with a lower percentage reactance value. Replacing the transformer with a new transformer in the case of an existing installations is not an economically viable solution. A more practical solution includes the use of detuned capacitors or harmonic filters and choice of the PCC on the HT side of the transformer. The second option is also the most accepted practice for industrial loads because of the other benefits like lower energy charges, better reliability, better voltage regulation etc.

9.9 IEEE 519-1992 VOLTAGE DISTORTION LIMITS

The voltage distortion limits in IEEE 519-1992 defines the quality of power on the power on the electric power system. Although, determining the combined voltage distortion on the power utility is a complex process, the overall combined distortion will remain under control if individual industrial power networks contribute distortions lower than the limiting values. Table 9.18 shows the voltage distortion levels as given in IEEE 519-1992.

Table 9.18 Recommended voltage distortion levels

Bus voltage at PCC	Individual voltage distortion%	Total Harmonic Distortion THD_V%
69 kV and below	3.0	5.0
69.001 through 161 kV	1.5	2.5
161.001 and above	1.0	1.5

These limits should be used as system design values for the worst case under normal operation (lasting longer than one hour). For shorter periods like starting-up or unusual conditions the limits may exceed by 50%.

9.10 SUMMARY OF DETUNED HARMONIC FILTER DESIGN

The basic problem involved in the design of detuned harmonic filter design can be summarised as follows.

Step 1: Determine the reactive kVA compensation required

Step 2: Select a capacitor bank with a voltage rating slightly above the normal rated voltage of the supply system or select a low stress type capacitor

Step 3: Determine the derated capacity of the capacitor bank using equation (9.28)

Step 4: Calculate the rating of the tuning inductor to tune the capacitor to the desired harmonic frequency

Step 5: Measure / calculate the harmonic spectrum of the load current

Step 6: Calculate the harmonic spectrum of the voltages and the effective rms bus voltage

Step 7: Calculate the harmonic spectrum of the source current and the effective rms source current

Step 8: Check for consistency with IEEE 519 standard.

It is to be noted that IEEE 519-1992 is meant to apply to harmonic distortion of the system at PCC level rather than to the harmonic distortion levels of the individual loads. Therefore, any attempt to apply these limits to individual loads is not in good spirit and such an interpretation will only involve additional expenses and should be avoided as far as possible.

9.11 QUESTIONS

1. Determine the total harmonic distortion and the effective rms load current of a load having the following harmonic spectrum

Frequency f_h	Harmonic order h	Load current I_h (A)
50	1	100.0
250	5	20.0
350	7	15.0
550	11	5.0
650	13	3.0
850	17	2.5
1150	23	0.5

2. Determine the total harmonic distortions and effective rms bus voltage of a voltage waveform having the following harmonic spectrum.

Frequency f_h	Harmonic order h	Load current I_h (A)
50	1	240.0
250	5	5.0
350	7	2.1
550	11	1.2
850	17	0.4

3. An industrial plant is supplied from the utility at 11 kV with a a short circuit level of 150 MVA on the 11 kV side of the transformer. The transformer rating is 1500 kVA, 11kV/ 415 V delta-star connected and with a percentage reactance of 4%. Determine the parallel resonant frequencies when the following capacitors are connected to the 415 bus of the industrial power net work.

   (a) 300 kVAr
   (b) 600 kVAr
   (c) 900 kVAr

4. The industrial plant referred to in problem 3 supplies a total load of 1200 kVA which include a non-linear load having the following harmonic spectrum.

Frequency f	Harmonic order h	Harmonic load current I_h
250	5	200
350	7	120
550	11	50

If it is proposed to install a 500 kVAr low stress type capacitor bank on the 415 volt side of the transformer to improve the power factor, determine:

(a) Effective rms value of the line current and total harmonic current distortion without the capacitor bank.

(b) Effective rms current, total harmonic current distortion, total harmonic voltage distortion when the capacitor bank is connected.

5. For the industrial power system of problem 4, if a harmonic filter is to be introduced, determine:

(a) Value of the tuning inductor to tune the capacitor bank to 4.7th harmonic order.

(b) Effective rms voltage, current and total harmonic distortions on voltage and current.

6. Determine whether IEEE 519 requirements are satisfied after the installation of the harmonic filter in the power system described in problem 5. Assume that the PCC is on the 11 kV side of the transformer.

7. Determine the typical harmonic current spectrum of a three-phase, 37 kW, 415 V, 50 Hz squirrel cage induction motor connected to a six-pulse variable speed drive. The full load current of the motor is 70 A. (Use Table 9.3 for percentage harmonic spectrum of six-pulse drives).

8. Determine the effective rms current and the total harmonic current distortion of the load current of problem 7.

CHAPTER 10

ENERGY ECONOMICS IN SYSTEM DESIGN

OBJECTIVES

At the end of this chapter you will

- Understand the importance of economic analysis in system design.
- Know about the time value of money.
- Understand the models used in time value analysis.
- Know the impact of depreciation on energy investments.
- Be able to solve problems relating to energy economics.

10.1 INTRODUCTION

Energy economics analysis is an important tool in justifying energy investment costs. A good knowledge of life cycle costing is required for carrying our energy economics analysis. The life cycle cost analysis evaluates the total owning and operating cost. It takes into account the "time value" of money and can incorporate fuel cost escalation into the economic model. This approach can also be used to evaluate competitive projects. In other words, the life cycle analysis considers the cost over the life of the system rather than just the initial or first cost.

10.2 TIME VALUE OF MONEY

In order to compare energy utilisation alternatives, it is essential to convert all cash flows for each alternative to an equivalent base. The 'life cycle cost' takes into account the time value of money. Because, a rupee in hand today is more valuable than one received at some time in the future. This is why time value must be placed on all cash flows into and out of a company.

In developing such a flow model, it is assumed that cash flows occur at discrete points in time as lump sum and that interest is computed and payable at discrete points in time.

10.3 SINGLE PAYMENT COMPOUND AMOUNT MODEL (SPCA)

This is perhaps the simplest form of cash flow model. Suppose that a sum of P invested for a period of 'n' years with an interest rate of 'i' percent yields a sum of S. Then,

Amount at the end of first year $= (1 + i)P$
Amount at the end of second year $= (1 + i)^2P$
Amount at the end of nth year $= (1 + i)^nP$

$$\text{i.e., } S = (1 + i)^n P \tag{10.1}$$

$$P = \frac{S}{(1 + i)^n} \tag{10.2}$$

EXAMPLE 10.1

What should be the sum invested now with an interest rate of 10% to yield a return of Rs. 1,00,000 at the end of 8 years.

Solution :

In this problem, S = Rs. 1,00,000, i = 10% and n = 8 years
Substituting in equation (10.2)

$$P = \frac{S}{(1 + i)^n}$$

$$= \frac{1{,}00{,}000}{(1 + 0.1)^8} = \text{Rs. } 46{,}650.50$$

10.4 UNIFORM SERIES COMPOUND AMOUNT MODEL (USCA)

A model which takes into account a uniform series of year end payments is called "uniform series compound amount" model. Figure 10.1 shows the details of USCA model.

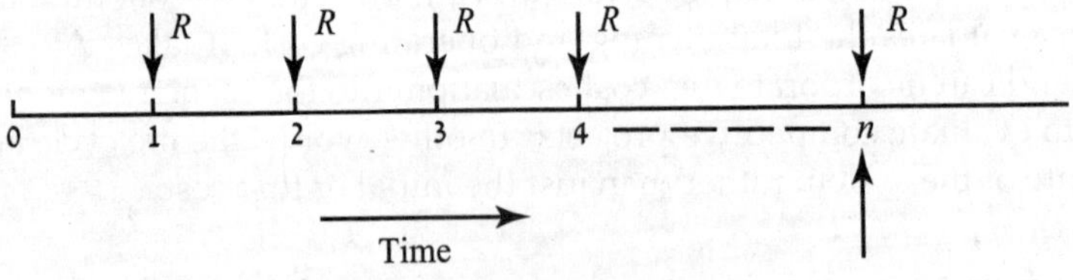

Figure 10.1 Uniform series compound amount model.

If R is the uniform year end payments and S is the future sum after 'n' years

$$S = R[1 + (1 + i) + (1 + i)^2 + \ldots (1 + i)^{n-2} + (1 + i)^{n-1}] \tag{10.3}$$

Multiplying both sides by $(1 + i)$

$$(1 + i)S = R[(1 + i) + (1 + i)^2 + \ldots (1 + i)n] \tag{10.4}$$

Subtracting equation (10.3) from equation (10.4)

$$(1+i)S - S = R[(1+i)^n - 1] \quad (10.5)$$

$$iS = R[(1+i)^n - 1]$$

$$\text{or } S = \frac{R[(1+i)^n - 1]}{i}$$

From the above equation we can evaluate the sinking fund payment model which gives the value of uniform series of deposits *R*, whose future value (*S*) is one rupee.

$$R = \frac{i}{(1+i)^n - 1}$$

EXAMPLE 10.2

Find the value of a uniform year end payment to obtain a sum of Rs. 5,00,000 at the end of 10 years. The interest rate is 9.5% per annum.

Solution :

Here, S = Rs. 5,00,000, n = 10 years and i = 9.5%

Substituting in equation 10.5, we get

$$R = \frac{iS}{(1+i)^n - 1}$$

$$= \frac{0.095 \times 5{,}00{,}000}{(1+0.095)^{10} - 1} = \text{Rs. } 32{,}133$$

10.5 UNIFORM SERIES PRESENT WORTH MODEL (USPW)

The model gives the present worth of a uniform series of year end payments. Figure 10.2 shows the details of USPW model.

From Equation (10.5) we have the future value *S* of a uniform series of deposits of *R*, as

$$S = \frac{R[(1+i)^n - 1]}{i}$$

From equation (10.2) the present worth of a future sum is given by $P = \dfrac{S}{(1+i)^n}$

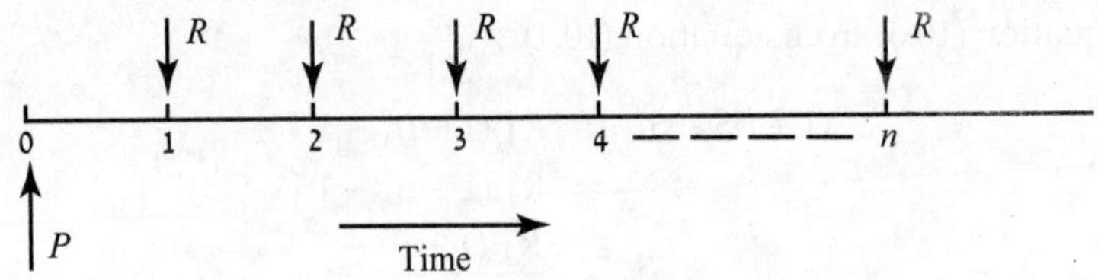

Figure 10.2 Uniform series present worth model.

Substituting for S, yields

$$P = R\frac{[(1+i)^n - 1]}{i(1+i)^n} \tag{10.6}$$

Present value of uniform investments of Rs. 1 USPW

$$= \frac{[(1+i)^n - 1]}{i(1+i)^n}$$

Or in other words the 'Capital recovery factor' (CR) is the uniform series of deposits whose present value is one rupee.

$$CR = \frac{i(1+i)^n}{(1+i)^n - 1} \tag{10.7}$$

EXAMPLE 10.3

Calculate the present worth of a uniform series of year end deposit of Rs. 10,000 for 5 years with an interest rate of 10% per annum.

Solution :

The details given are, R = Rs. 10,000, i = 10%, n = 5 years
Substituting in equation 10.6

$$\text{The present worth } P = \frac{10{,}000\left[(1+0.1)^5 - 1\right]}{0.1(1+0.1)^5} = 37{,}907.87$$

Table 10.12 gives the details of these models for 10% interest factor. In the table, each is defined when the number of periods (n) and interest rate (i) are specified. Another model used in life cycle costing is the "Gradient Present Worth model" which takes into account the escalation rates also.

The most commonly used methods in life cycle costing are the annual cost, present worth and rate of return analysis.

In the present worth method, a minimum return (i) is stipulated. All future expenditures are converted to a present value using this interest factor. The alternative with the lowest effective first cost is the most favoured choice.

Similar procedure is used in the annual cost method. The difference is that the first cost is converted into an annual expenditure. The alternative with the lowest effective annual cost is the favoured choice.

In many situations simple payback analysis is used instead of life cycle model. Payback is defined as "initial investment divided by annual savings". This method does not take into account the effect of interest or escalation. However, this method is popular because of its relatively simple calculations.

EXAMPLE 10.4

An electrical energy audit in a factory indicates that total energy consumption is 10^6 kWh per year. By upgrading the motor spares with high efficiency motors, a 10% saving in energy can be realised. The additional cost of energy efficient motors is estimated at Rs. 1,80,000. Assuming Rs. 3.50 per kWh energy charge and 20 year life cycle, is the expenditure justifiable on a minimum return of 20%. Solve the problem using present worth and annual cost method. Calculate also the simple pay back period.

Solution :

The various steps involved in the solution of the problem are recorded in Table 10.1 and Table 10.2.

Table 10.1 Present worth method

S.No.	Item	Existing cost (Rs.)	Proposed cost (Rs.)
1	First cost P	–	1,80,000
2	Annual cost R	3.5×10^6	3.15×10^6
3	Uniform series present worth USPW	4.87	4.87
4	$R \times$ USPW	17.05×10^6	15.341×10^6
5	Present worth (1+4)	17.05×10^6	15.521×10^6

Choose the alternative with the lowest total present worth. In this case the second option.

Table 10.2 Annual cost method

S.No.	Item	Existing cost (Rs.)	Proposed cost (Rs.)
1	First cost P	–	1,80,000
2	Annual cost R	3.5×10^6	3.15×10^6
3	Capital recovery factor CR -from table	0.20536	0.20536
4	$P \times CR$	–	0.037×10^6
5	Total cost (2+4)	3.5×10^6	3.187×10^6

Choose the alternative with the lowest total annual cost. In this case the proposed method has the lowest annual cost.

Simple payback analysis

Initial investment = Rs. 1,80,000/-

Annual saving = $(3.5 - 3.15)10^6$ = Rs. 0.35×10^6

$$\text{Therefore, payback period} = \frac{1,80,000}{3,50,000} \frac{\text{Initial investment}}{\text{annual savings}} = \frac{1,80,000}{3,50,000}$$

$$= 0.5142 \text{ years, say 6 months}$$

EXAMPLE 10.5

A floodlighting scheme covering an area of 9000 m^2 is designed for a luminance level of 300 lux. Calculate the economics of using HPS or MV lamps for the purpose. The details of the luminaires are given below. (Neglect light loss factors)

Luminaire type	List price Rs.	Lamp type	List price Rs.	Lumen output
HPMV/400W	1710	HPMV 400W	415	23,000
HPSV/250W	2150	SON 250W	731	27,000

Energy charges	Rs. 3.50/unit
Burning hours per day	12 hours
Installation charges/lamp	Rs. 500
Annual rate of return	10%
Average lamp life	15,000 burning hours

Solution :

The solution of the problem is recorded in Tables 10.3 and 10.4.

Table 10.3 Present worth method

S.No	Item	Option I using HPMV	Option II using HPSV
1	Illumination level in lux	300	300
2	Lumen output per lamp	23,000	27,000
3	Number of luminaires required	$\frac{9000 \times 300}{23,000} = 118$	$\frac{9000 \times 300}{27,000} = 100$
4	Cost of luminaires (Rs.)	118 × 1710 = 2,01,780	100 × 2150 = 2,15,000
5	Cost of lamps (Rs.)	118 × 415 =48,970	100 × 731 = 73,100
6	Installation charges (Rs.)	118 × 500 =59,000	100 × 500 = 50,000
7	Total investment—P = (4+5+6)	3,09,750	3,38,100
8	Power consumed per luminaire—W(includes ballast loss)	430	280
9	Burning hours per day—hours	12	12
10	Total power consumed = (3 × 8)-W	118 × 430 =50,740	100 × 280 = 28,000
11	Annual energy charges (Rs.) = $\frac{(9 \times 10)}{1000} \times 3.50 \times 365$	7,78,884.20	4,29,240.00
12	Lamp life—hrs	15,000	15,000
13	Replacement cycle	0.292	0.292

14	Quantity of lamp replacement per year-Nos	$118 \times 0.292 = 35$	$100 \times 0.292 = 30$
15	Cost of lamp replacement—(Rs.)	$35 \times 415 = 14{,}525$	$30 \times 731 = 21{,}930$
16	Annual cost = (11+15) = R (Rs.)	7,93,409.2	4,51,170
17	USPW = $\dfrac{(1+i)^n - 1}{i(1+i)^n}$ for $n = 20$ years	8.514	8.514
18	$R \times$ USPW	6755085.93	3841261.38
19	Present worth = (18+7)	7064835.93	4179361.38

Choose the alternative with the lowest present worth. In this case option II has the lowest present worth

Table 10.4 Annual cost method

S.No.	Item specification	Alternative I	Alternative II
1	Total investment (P)	3,09,750	3,38,100
2	Annual cost (R)	7,93,409.2	4,51,170
3	Capital recovery factor = $\dfrac{1}{\text{USPW}}$	0.1175	0.1175
4	$P \times CR$	36381.25	50756.63
5	Annual cost=(2+4)	829790.45	501926.63

Choose the alternative with the lowest annual cost. Here, Alternateive II has the lowest annual cost.

10.6 DEPRECIATION

Depreciation is the reduction in the value of the asset due to its usage, passage of time, wear and tear, technological obsolescence, depletion, inadequacy, decay or other factors. In accounting, depreciation is a term used to describe any method of attributing the historical or purchase cost of an asset across its useful life. Depreciation affects the accounting procedure for determining the profits and losses and the income tax of a company. For tax purposes, the expenditure for an asset like equipment cannot be fully expensed during its first year. The original investment must be charged off for the purpose of tax purposes over the useful life of the equipment. There are different types of depreciation calculations.

10.6.1 Straight Line Depreciation

This is the simplest method of calculating the annual depreciation rates

$$D = \frac{P - L}{n} \tag{10.8}$$

Where, D = Annual depreciation expense
P = Cost of fixed asset
L = Value of equipment at the end of its useful life (salvage value)
n = Life span of the equipment in years

EXAMPLE 10.6

A computer is purchased at a cost of Rs. 25,000 and its useful life is expected to be 5 years. The salvage value of the computer is Rs. 5000. Calculate the annual depreciation expense.

$$\text{Depreciation expense as given by equation 10.8} = \frac{P - L}{n} = \frac{25{,}000 - 5000}{5} = \text{Rs. } 4000.$$

Normally the depreciation is recorded in the form of book value. The book value at the beginning of the first year of depreciation is the original cost of the asset. At any other time the book value equals the book value minus the accumulated depreciation. Book value at the end of a year becomes the book value at the beginning of the next year. The asset is depreciated until its book value equals the scrap value (salvage value). Table 10.5 illustrates the book value of the computer over its useful life.

Table 10.5 Straight line depreciation

Book value - beginning of year Rs.	Depreciation expense Rs.	Accumulated depreciation Rs.	Book value - end of year Rs.
25,000	4000	4000	21,000
21,000	4000	8000	17,000
17,000	4000	12,000	13,000
13,000	4000	16,000	9000
9000	4000	20,000	5000 (scrap value)

10.6.2 Declining Balance Method

This method allows for larger depreciation charges in the early years. This method is also referred to as "fast write off method" or "accelerated depreciation method". Depreciation rate is calculated by taking a constant percentage of the declining non-depreciated balance. Common practice to calculate the declining balance, is to predetermine an accelerated rate. Sometimes a rate as high as twice the straight line depreciation rate may be used.

EXAMPLE 10.7

Solve example 10.6 using declining balance method.

Table 10.6 shows the method of double declining balance method of depreciation calculation of example 10.6. First the straight line depreciation is calculated and as the name suggests, double that value is used.

$$\text{Straight line depreciation rate in percentage} = \frac{100}{5} = 20\%$$

Double this percentage or 40% is used for the computation. Book value at the beginning of the first year of depreciation is the original cost of the asset. At any other time, book value equals original cost minus the accumulated depreciation. The asset is depreciated until the book value equals the salvage value.

Book value = Original cost – Accumulated depreciation

Table 10.6 Double declining balance method

Book value-beginning of year Rs.	Depreciation rate %	Depreciation expense Rs.	Accumulated depreciation Rs.	Book value - end of year Rs.
25,000	40	10,000	10,000	15,000
15,000	40	6000	16,000	9000
9000	40	3600	19,600	5400
5400	5400-5000	400	20,000	5000

10.6.3 Sum of Years Method

Sum of years is a depreciation method that results in a more accelerated write-off than straight line method, but less than the declining balance method. In this method the depreciation rate is determined by finding the sum of year's digits using the following formula.

$$N = \frac{n(n+1)}{2} \tag{10.9}$$

Where n = life of the equipment.

Each year's depreciation rate is determined as follows

First year $D = \dfrac{n(P-L)}{N}$

Second year $D = \dfrac{(n-1)(P-L)}{N}$ and, for the

nth year $D = (P-L)/N$

EXAMPLE 10.8

Solve the example 10.7 using the sum of year's digit method.

Table 10.7 shows the details of application of sum of year's method to example 10.8.

Table 10.7 Sum of year's method of calculating depreciation

Book value year beginning	$P-L$	Depreciation rate	Depreciation expense Rs.	Accumulated depreciation Rs.	Book value end of year Rs.
25,000.00	20,000	5/15	6666.67	6666.67	18,333.33
18,333.34	20,000	4/15	5333.33	12,000.00	13,000.00
13,000.00	20,000	3/15	4000.00	16,000.00	9000.00
9000.00	20,000	2/15	2666.67	18,666.67	6333.33
6333.33	20,000	1/15	1333.33	20,000.00	5000 (scrap value)

10.6.4 Units of Production Method

In this method, the useful life of the asset is expressed in terms of the total number of units expected to be produced. Annual depreciation is computed in three steps. The first step is to compute the depreciable cost. This is obtained by taking the difference between the original cost and the salvage value.

Depreciable cost = Original cost – Salvage value

The next step is to compute the depreciation per unit. This is computed by dividing the depreciable cost by the total number of units expected to be produced during the useful life of the assets.

$$\text{Depreciation per unit} = \frac{\text{Depreciable cost}}{\text{Total units of production}}$$

The next step is to compute the annual depreciation or the depreciation expense.

$$\text{Depreciation expense} = \text{Depreciation per unit} \times \text{Units produced during year}$$

$$\text{The book value} = \text{Original cost} - \text{Accumulated depreciation.}$$

EXAMPLE 10.9

The original asset of an industry is Rs. 70,00,000. The salvage value of the assets after 5 years is expected to be Rs. 10,00,000. If the industry is expected to produce 6000 units during the five years, calculate the annual depreciation over the five years. The yearly production schedule is:

First year	Second	Third	Fourth	Fifth
1000 units	1100	1200	1300	1400

Solution :

Table 10.8

Book value-beginning of year	Units of production	Depreciation cost per nit Rs.	Depreciation expense Rs.	Accumulated depreciation Rs.	Book value-end of year Rs.
70,00,000	1000	1000	10,00,000	10,00,000	60,00,000
60,00,000	1100	1000	11,00,000	21,00,000	49,00,000
49,00,000	1200	1000	12,00,000	33,00,000	37,00,000
37,00,000	1300	1000	13,00,000	46,00,000	24,00,000
24,00,000	1400	1000	14,00,000	60,00,000	10,00,000 Salvage value

$$\text{Depreciable cost} = \text{Rs. } 70{,}00{,}000 - 10{,}00{,}000 = \text{Rs. } 60{,}00{,}000$$

$$\text{Depreciation per unit} = \frac{60{,}00{,}000}{6000} = \text{Rs. } 1000$$

Table 10.8 shows the unit of production depreciation calculation.

10.7 TAX CONSIDERATIONS

When a company spends money for a service or anything that is short-lived, this expenditure is usually tax deductible and the company enjoys immediate tax benefits. To be eligible for depreciation benefits, an asset must have two features:

- It has a useful life beyond the taxable year and
- It wears out over a period of time, declining in its value due to natural causes or due to obsolescence.

Therefore, when a company buys an asset like equipment, vehicles, computers, buildings etc., that will last longer than one year, the company cannot enjoy an immediate tax benefit on the total investment. Instead, the company must depreciate the cost involved over the useful life of the asset, taking tax deductions for a part of the cost each year. Eventually the company gets full benefit for the entire cost of the asset, but this happens over a period of several years.

Expenses such as maintenance, energy costs, insurance, property tax etc., are also tax deductible expenses and these expenses reduce the income subject to the existing tax structure. For after tax life cycle analysis and payback analysis, the actual savings can be calculated as

$$AS = (1 - I)E + ID \tag{10.10}$$

Where AS = Annual after tax savings
E = Yearly annual energy savings
D = Annual depreciation rate
I = Income tax bracket

Equation (10.10) takes into account that the yearly annual savings is partially offset by additional taxes which must be paid due to reduced operating costs. On the other hand, the depreciation factors reduce the taxes directly.

10.8 AFTER TAX ANALYSIS

To compare the rate of return which account for taxes, depreciation, escalation etc., a cash flow analysis is required. Usually a computer aided analysis or trial and error method will be required.

Table 10.9 After tax analysis using single payment present worth method

Year	Investment	Tax credit (2)	After tax savings AS	SPPW factor	Present worth
0	$-P$				$-P$
1		$+TC$	AS_1	$SPPW_1$	$+P_1$
2			AS_2	$SPPW_2$	$+P_2$
3			AS_3	$SPPW_3$	$+P_3$
$n-1$			$AS_{(n-1)}$	$SPPW_{(n-1)}$	$+P_{(n-1)}$
					$\sum P$

The present worth factor tables will be used for the analysis. All money is converted to its present worth assuming an interest rate. The summation of all present worth terms should be zero when correct interest rate is selected. Table 10.9 indicates the steps involved in the after tax analysis.

When correct value of '*I*' is selected $\sum P = 0$

EXAMPLE 10.10

A heat recovery system installed to recover heat from the compressors provides a saving of 14 kW power. The additional power required for the operation of the heat recovery system is 1 kW. The other data are given below

Initial cost of the heat recovery plant	Rs. 12,50,000
Cost of electric energy	Rs. 3.50 per kWh
Escalation of energy cost	10% per year
Project life span	7 years
Investment tax credit	10%
Salvage value of the plant	Rs. 0.0
Interest rate	10%
Tax rate	30%

Determine the annual return on the investment using capital recovery method.

Solution :

The annual saving and the additional energy costs for the entire project period is calculated first.

Annual energy saving for the first year $= 14 \times 24 \times 365$
$= 1{,}22{,}640$ kWh

Cost of annual energy saved = Rs. $3.5 \times 1{,}22{,}640$
= Rs. 4,29,240

Cost of energy for the heat recovery plant = Rs. $1 \times 24 \times 365 \times 3.5$
= Rs. 30,660

Cost of energy saving and the additional cost of energy for the heat recovery plant can be calculated similarly using the escalated energy cost. The result is tabulated in Table 10.10.

Capital recovery factor corresponding to 10% interest and $n = 7$ years

$$\text{CRF} = \frac{i(1+i)^n}{(1+i)^n - 1}$$

$$= 0.20541$$

The total operating cost and the savings and profit after tax are tabulated in Table 10.11.

Table 10.10

	Year						
	1	**2**	**3**	**4**	**5**	**6**	**7**
Energy saving (Rs.)	4,29,240	4,72,164	5,19,380	5,71,318	6,28,450	6,91,295	7,60,425
Additional energy cost (Rs.)	30,660	33,726	37,099	40,808	44,889	49,378	54,316

From the table the following conclusion can be made:

$$\text{Return on investment before taxation} = \frac{\text{Gross savings}}{\text{Initial investment}} \times 100$$

$$= \frac{19{,}84{,}335}{12{,}50{,}000} \times 100$$

$$= 158\%$$

This is equivalent to an average return on investment of 22.68% per annum.

$$\text{Return on investment after taxation} = \frac{\text{Net savings}}{\text{Initial investment}} \times 100$$

$$= \frac{8{,}95{,}197}{12{,}50{,}000} \times 100$$

$$= 71.6\%$$

This corresponds to an average return of 10.23% per annum.
An analysis of the above problem brings into light the following factors:

- Cost escalation is a critical factor in economic analysis
- Annual costs are never the same
- Tax implication must be considered in all economic decisions
- Investment tax credit can be an important factor.

Table 10.11

Description	Year								Total
	0	1	2	3	4	5	6	7	
A. Operating costs.									
Installation cost (I).	12,50,000								
Capital recovery cost = $I \times$ CRF		2,56,723	2,56,723	2,56,723	2,56,723	2,56,723	2,56,723	2,56,723	
Additional energy cost		30,660	33,726	37,099	40,808	44,889	49,378	54,316	
Total operating costs		2,87,383	2,90,449	2,93,822	2,97,531	3,01,612	3,06,101	3,11,039	
B. Savings in energy costs		4,29,240	4,72,164	5,19,380	5,71,318	6,28,450	6,91,295	7,60,425	
C. Profit and taxes									
Gross profit (B-A) = C_1		1,41,857	1,81,715	2,25,558	2,73,787	3,26,838	3,85,194	4,49,386	19,84,335
Taxes (30% of B) = C_2		1,28,772	1,41,649	1,55,814	1,71,395	1,88,535	2,07,389	2,28,128	
Investment tax credit (10% on 12,50,000) = C_3		1,25,000							
D. Net Profit= $C_1 - C_2 + C_3$		1,45,629	40,066	69,744	1,02,392	1,38,303	1,77,805	2,21,258	8,95,197
E. Return on investment. % (D/I) 100		11.65	3.21	5.58	8.12	11.06	14.22	17.7	

10.9 CONCLUSION

Energy economic analysis is an important aspect of energy management programme. Every energy investment must be critically analysed for its effectiveness. Payback analysis is simple to calculate. Moreover, it permits the investors to recover the investment as rapidly as possible. However, this method should be used with other decision making tools. If used by itself, it may result in choosing less profitable investment which yields high initial returns for short periods, compared with more profitable investments which provide profits over longer period of time.

Table 10.12 10% Interest factor

Period n	Single payment compound amount (SPCA)	Single-payment present worth (SPPW)	Uniform series compound amount (USCA)	Sinking-fund payment (SFP)	Capital recovery (CR)	Uniform-series present-worth (USPW)
	Future value of one rupee	Present value of one rupee	Future value of uniform series of one rupee	Uniform series whose future value is one rupee	Uniform series with present value of one rupee	Present value of uniform series of one rupee
1	1.100	0.9091	1.000	1.00000	1.0000	0.909
2	1.210	0.8264	2.100	0.47619	0.57619	1.736
3	1.331	0.7513	3.310	0.30211	0.40211	2.487
4	1.464	0.6830	4.641	0.21547	0.31547	3.170
5	1.611	0.6209	5.105	0.16380	0.20380	2.791
6	1.772	0.5645	7.716	0.12961	0.22961	4.355
7	1.949	0.5132	9.487	0.10541	0.20541	4.868
8	2.144	0.4665	11.436	0.08744	0.18744	5.335
9	2.358	0.4241	13.579	0.07364	0.17364	5.759
10	2.594	0.3855	15.937	0.06275	0.16275	6.144
11	2.853	0.3505	18.531	0.05396	0.15396	6.495
12	3.138	0.3186	21.384	0.04676	0.14676	6.814
13	3.452	0.2897	24.523	0.04078	0.14078	7.103
14	3.797	0.2633	27.975	0.03575	0.13575	7.367
15	4.177	0.2394	31.772	0.03147	0.13147	7.606
16	4.595	0.2176	35.950	0.02782	0.12782	7.824
17	5.054	0.1978	40.545	0.02466	0.12466	8.022
18	5.560	0.1799	45.599	0.01955	0.11955	8.363
19	6.116	0.1635	51.159	0.01955	0.11955	8.365
20	6.727	0.1486	57.275	0.01746	0.11746	9.077
21	7.400	0.1351	64.002	0.012562	0.11562	8.649
22	8.140	0.1228	71.403	0.01401	0.11401	8.772
23	8.954	0.1117	79.543	0.01257	0.11257	8.883
24	9.850	01015	88.497	0.01130	0.11130	8.985
25	10.835	0.0923	98.347	0.10117	0.11017	9.077

10.10 QUESTIONS

1. A sum of Rs. 50,000 is invested in a bank for 10 years with an interest rate of 9.5% per annum. What is the future value of this investment after 10 years. Calculate the present worth of a sum of Rs. 75,000 that is available after 5 years, if the interest rate is 10%.
2. If a sum of Rs. 10,000 is deposited at the end of every year for 10 years with an interest rate of 12.5%, what will be the future worth of this investment at the end of 10 years.
3. If an amount of Rs. 500,000 is required at the end of 5 years, calculate the amount to be deposited annually with an interest rate of 10%.
4. If a sum of Rs. 20,000 is being deposited annually with an interest rate of 12.5% for 10 years, what will be the present worth of the sum that is available after 5 years?
5. If a sum of Rs. 1,00,000 is invested in an industry for 10 years, what is the value of capital recovery, if the interest rate is 10%
6. A three-phase, 3.7 kW, 415 volt, 50 Hz squirrel cage induction motor is required to drive a pump which is in operation for 3000 hours in an year. Two makes of motors with similar reliability and maintenance requirements are available in the market as detailed below.
   Motor A Cost = Rs. 6500, Full load efficiency = 85%
   Motor B Cost = Rs. 8250, Full load efficiency = 93%
   The cost of electricity is Rs. 5.50 per unit. The interest and depreciation may be taken as 10% per year. Determine which motor will have the minimum annual cost.
7. An industrial plant was established with an initial investment of Rs. 2,500,000. The salvage value of the plant at the end of its useful life of 20 years was estimated to be Rs. 2,00,000. Determine the book value of the plant at the end of five years using
   (a) Straight line depreciation
   (b) Double declining depreciation and
   (c) Sum of year's digit method.
8. If an industrial plant was established with an initial investment of Rs. 15,00,000 to produce 10,000 units of starters, and if the salvage value at the end of 8 years is Rs. 1,00,000, calculate the book value of the investment at end of each year. The production schedule is:

First year	Second	Third	Fourth	Fifth	Sixth	Seventh	Eighth
1000	1100	1200	1300	1400	1500	1250	1250

CHAPTER 11

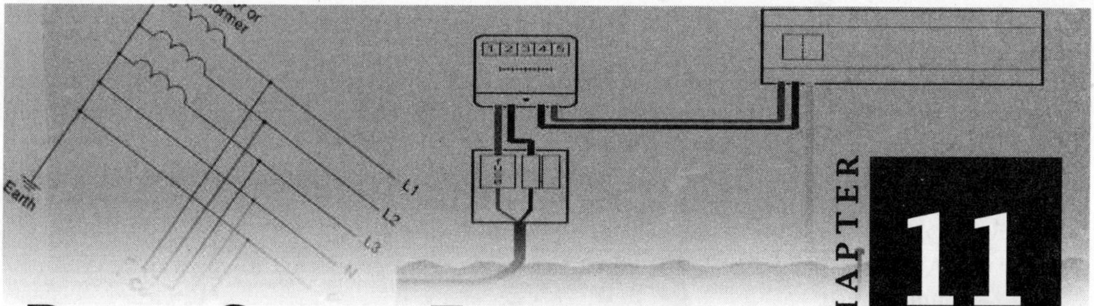

POWER SYSTEM EARTHING

OBJECTIVES

At the end of this chapter you will

- Understand the importance of earthing in electrical distribution systems.
- Know about the different types of power system earthing method.
- Be able to distinguish between IT, TN and TT systems of earthing.
- Know about the protective measures in different types of system.
- Understand the selection criteria for system earthing.

11.1 INTRODUCTION

Protective measures in an electrical system require the co-ordination of earth connection and the characteristics of the protective conductor and protective devices. There are different earthing schemes, each one characterised by the method of earthing the installation downstream side of a HV/MV transformer and the means used for earthing the exposed conductive parts of the MV installations supplied by the transformer. The choice of the earthing scheme governs the measures necessary for protection against indirect contact hazards. The basic requirements of protection against indirect contact is that all exposed conductive parts of the electrical equipment shall be connected to the protective conductor or to the earth under the specified conditions for each type of distribution system. Protective measures with protective conductors therefore necessitates coordination of the type of system earthing and the characteristics of protective conductors. IEC standard 60479 specifies the requirements for these protective measures and the disconnecting time.

11.2 TYPES OF SYSTEM EARTHING

Wiring regulations use a series of letter designations to precisely define the relationship between the source of supply (usually, the secondary of distribution transformers) including the neutral and of the exposed conductive parts and earth.

The first letter denotes the earthing arrangement of the source of energy. They are:

T = Direct connection of one or more points to earth (Terra).
I = All active parts isolated from earth or one point connected to earth through an impedance.

The second letter denotes the relationship of the exposed conductive parts of the installation to earth. These letters are:

T = Direct electrical connection of the exposed conductive parts to earth, independently of the earthing of any point of the source of energy.
N = Direct electrical connection of the exposed conductive parts to the earthed point of the source of energy, which for AC systems is the neutral point of the supply transformer.

The remaining letters describe the arrangement of the neutral conductor and the protective conductor.

S = Conductor which is separated for the neutral or the earthed conductor. (In DC system this is an earthed negative or positive conductor).
C = Neutral conductor and protective functions combined in a single conductor (PEN conductor).
PE = Protective earth conductor.
N = Neutral conductor.

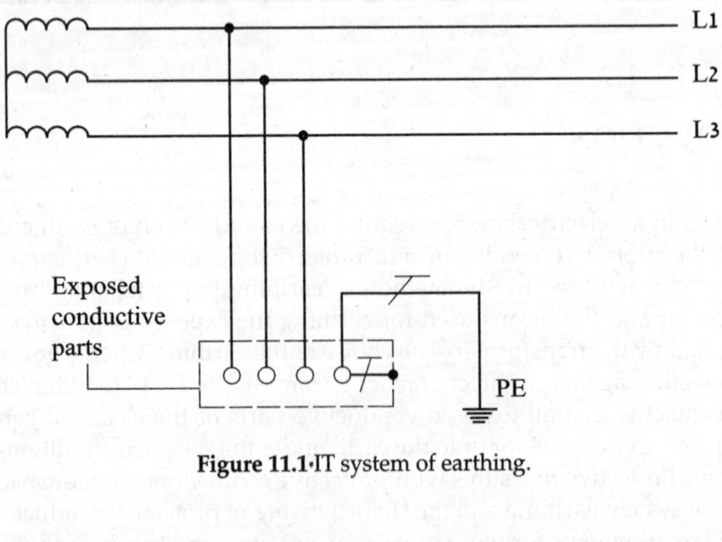

Figure 11.1 IT system of earthing.

According to the wiring regulations, the three main distribution systems are

- IT system
- TN system
- TT system

These systems of earthing are shown in Figures 11.1, 11.2 and 11.3.

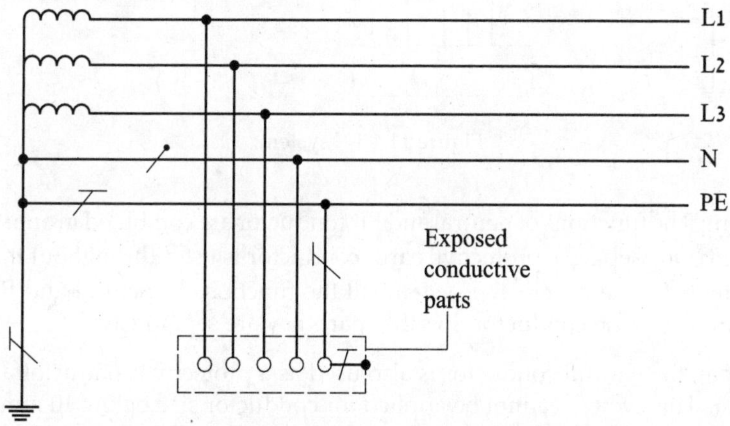

Figure 11.2 TN system of earthing.

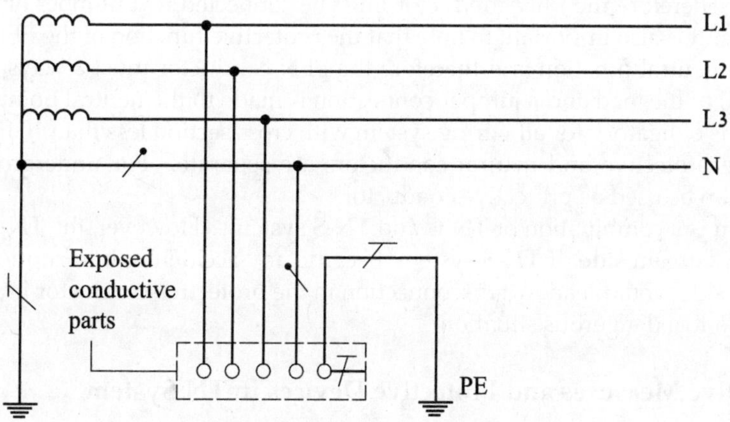

Figure 11.3 TT system of earthing.

11.3 TN SYSTEM

In TN system one point of the source is directly connected to the earth. The exposed conductive parts of the electrical installations are connected to this point by protective conductors. There are three types of TN system, based on the arrangement of the neutral and the protective earth (PE) conductor.

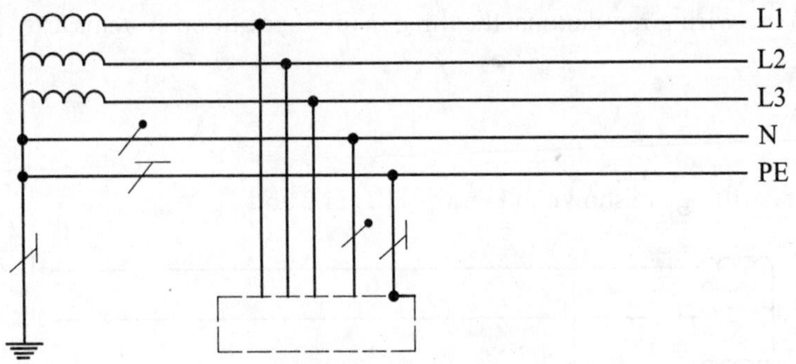

Figure 11.4 TN system.

- TN-C system: The functions of neutral and PE conductor are combined in one single conductor.
- TN-S system: One separate protective earth conductor is used throughout the system.
- TN-CS system: In one part of the system, all the functions of neutral and PE conductor are combined in one single conductor. In other part they are separate.

In TN-C system, the neutral conductor is also used as a protective conductor and is referred to as PEN conductor. This system cannot be applied for conductor size below 10 mm^2 or for portable equipment. This system requires an effective equipotential environment within the installation with disposed earth electrodes spaced as regularly as possible since the PEN conductor is both the neutral and at the same time carries phase unbalanced currents as well as third harmonics (and their multiples). Therefore the PEN conductor must be connected to a number of earth electrodes in the installation. It is also important to note that the protective function of the PEN conductor has priority over the neutral function and therefore the PEN conductor must always be connected to earthing terminal of the load and a jumper connection is made to the neutral point.

TN-S system is obligatory for all wiring system with cross section less than 10 mm^2 for portable equipment. The protective and neutral conductors are separate. For underground cables, the metallic sheath can be used as protective conductor.

TN-CS system is a combination of TN-C and TN-S systems. However, the TN-C should not be used on the downstream side of TN-S system, because any accidental interruption of the neutral on the upstream side would lead to a disconnection in the protective conductor in the downstream side and will lead to a dangerous situation.

11.3.1 Protective Measures and Protective Devices in TN System

All exposed conductive parts of the installation must be connected to the earthed point of the system by protective conductor which shall be earthed at or near each relevant transformer or generator. Generally the earthed point of the system will be the neutral point. If a neutral point is not available or accessible, a phase conductor shall be earthed. In no circumstances, phase conductor shall serve as PEN conductor. In TN systems the following protective measures are employed

- Over current protective devices
- Residual current devices.

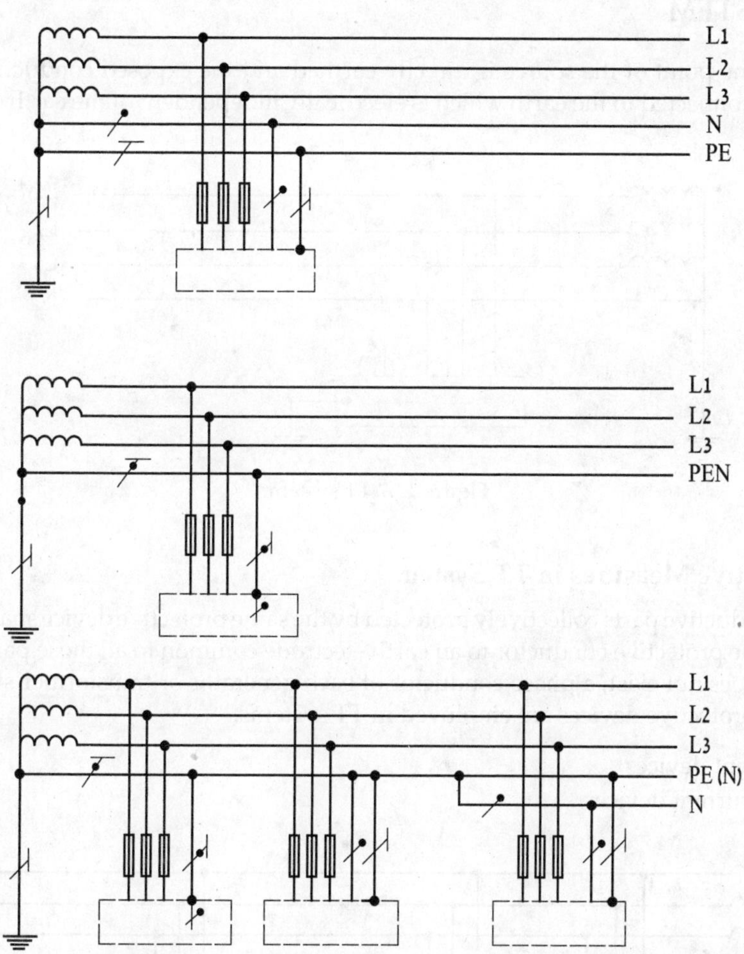

Figure 11.5 Over current protective devices in TN system.

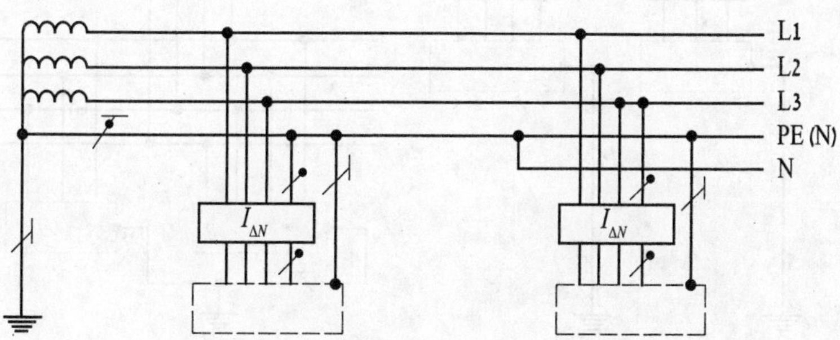

Figure 11.6 Residual current protective device.

11.4 TT SYSTEM

In TT system, one point of the source is directly earthed and the exposed conductive parts of the installation are connected to the earth which is electrically independent of the earth electrode of the power source.

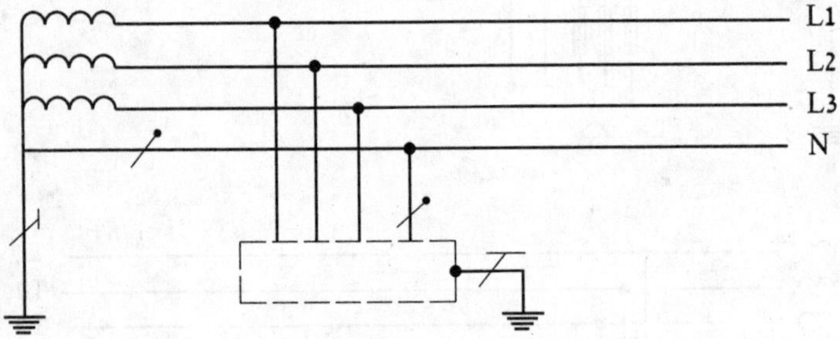

Figure 11.7 TT system.

11.4.1 Protective Measures in TT System

All exposed conductive parts collectively protected by the same protective device shall be connected, together with the protective conductor to an earth electrode common to all those parts. The neutral points, or if it does not exist, a phase conductor of each generator or transformer shall be earthed. The following protective devices are employed in TT systems.

- Over current devices
- Residual current devices

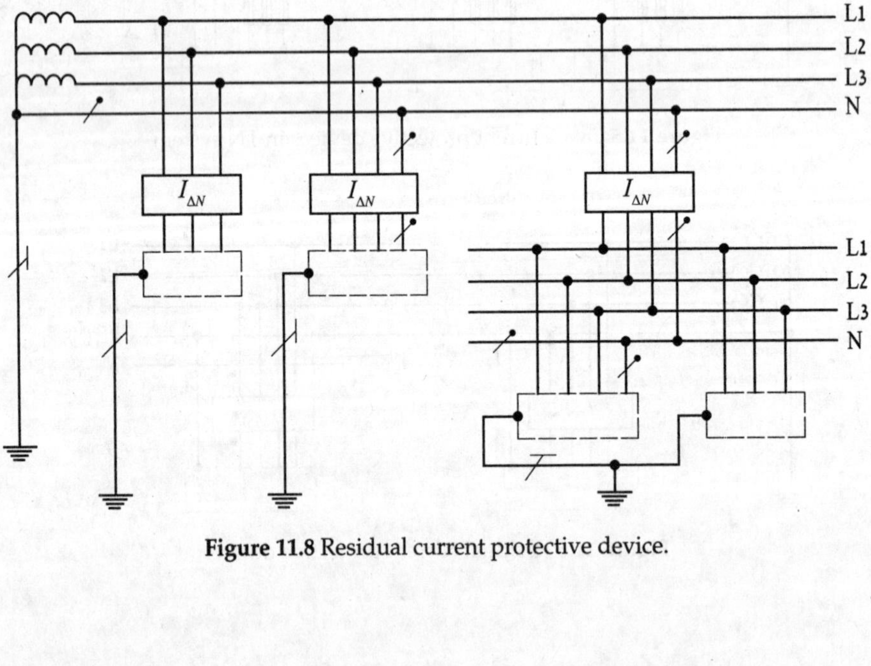

Figure 11.8 Residual current protective device.

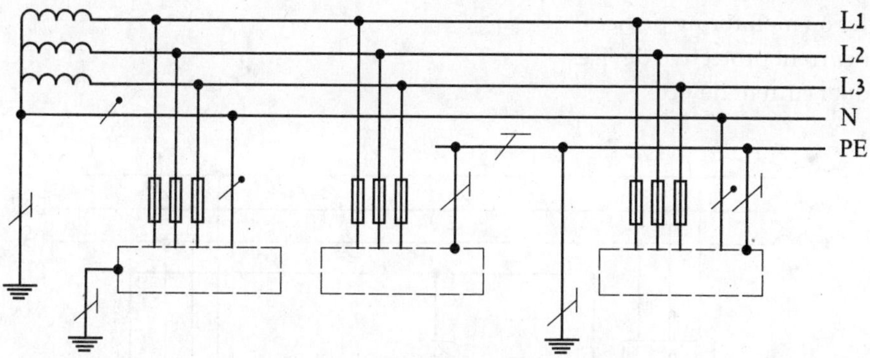

Figure 11.9 Over current protective device.

11.5 IT SYSTEM

In IT systems all parts are insulated from earth or one point is connected to earth through sufficiently high impedance. The exposed conductive parts are either

- Earthed individually
- Earthed in groups or
- Collectively connected to the earthing point of the system

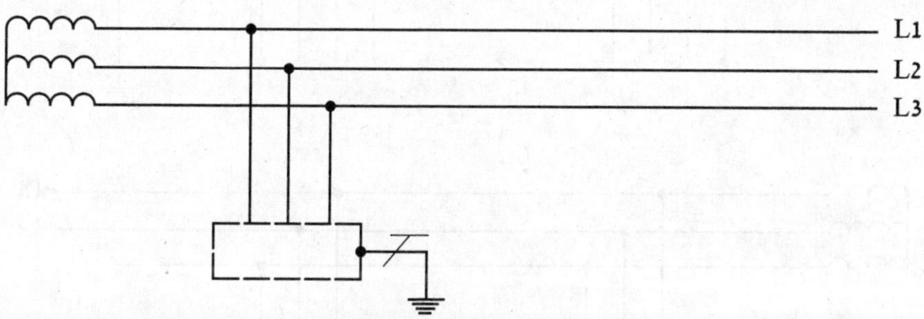

Figure 11.10 IT system.

11.5.1 Protective Measures and Protective Devices in IT System

In IT system all active parts shall be insulated from earth or connected to earth through sufficiently high impedance. This earth connection may be between the earth and the neutral point or an artificial neutral point. In the case of an artificial neutral point it can be connected directly to earth if the resulting zero sequence impedance is sufficiently high. When there is no neutral point, a phase conductor may be earthed through impedance. The following protective and monitoring devices are employed in IT systems.

- Insulation monitoring devices
- Over current protective devices
- Residual current devices

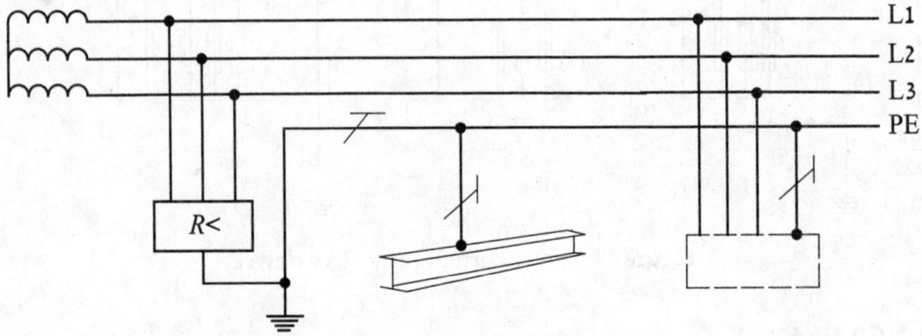

Figure 11.11 Insulation monitoring device.

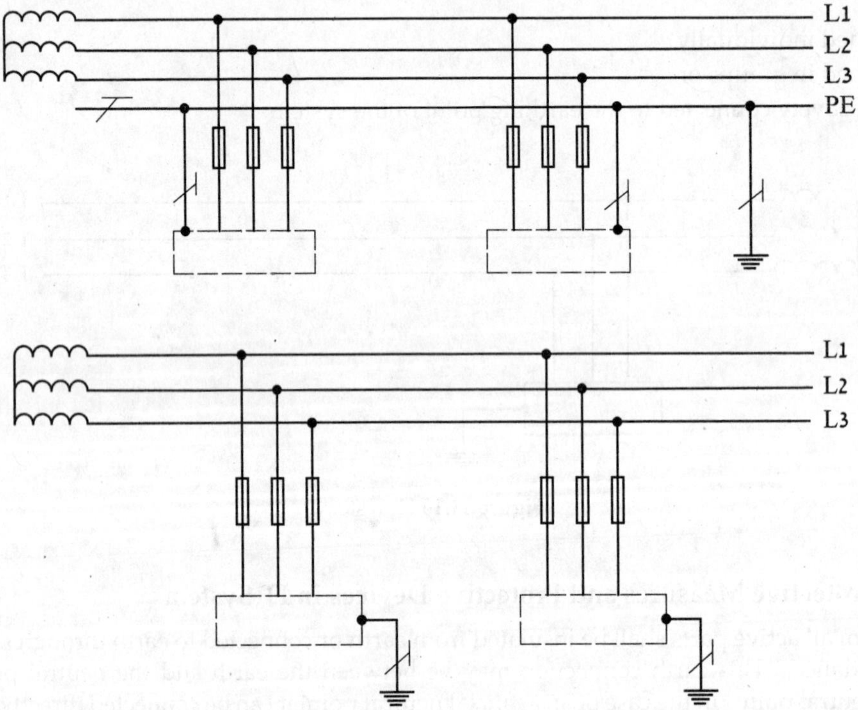

Figure 11.12 Over current protective device.

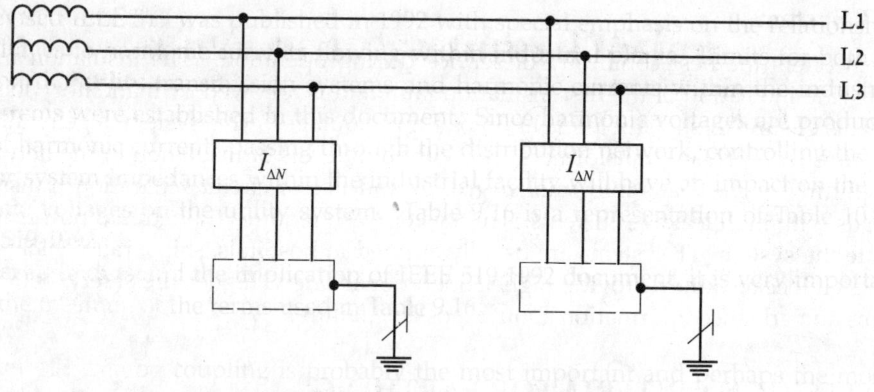

Figure 11.13 Residual current device.

11.6 MAIN CHARACTERISTICS OF THE EARTHING SYSTEMS

11.6.1 TT System

This is the simplest solution to design and install. They are widely used in installations supplied directly by the public MV distribution network. This system does not require continuous monitoring during their operation. An occasional check on the RCD is only required. The required level of protection is ensured by over current devices and RCDs. The residual current devices will also prevent the risk of fire if they are set to operate at ≤ 500 mA. The main disadvantage is that each insulation fault results in an interruption of supply. This can be overcome by installing RCDs in series (selective RCDs) or in parallel (selective circuits). Another advantage is the nuisance tripping of the RCDs because of the high leakage current of the loads or parts of the loads. This needs special measures like isolation transformers or the use of specific RCDs.

11.6.2 TN System

This system of earthing requires the installation of earth electrodes at regular intervals throughout the installation. The effectiveness of the protective devices should be analysed at the design stage and should be confirmed by mandatory measurements to confirm the effectiveness. Modifications to the system should be carried out only by a qualified person.

The TN-C system may appear to be less expensive system because of the absence of a separate protective conductor. However, this system is forbidden in situations like:

- Premises with risk of fire
- Non-linear loads with harmonic currents in the neutral

The TN-S can be used even with flexible conductors. They also provide a clean PE conductor because of the separation of neutral from PE conductor. They are ideal for computer systems and premises with special risks.

11.6.3 IT System

This system offers the best continuity of service during operation. The insulation monitoring devices provide indication of the first insulation fault and is followed by mandatory location and clearing. This ensures systematic prevention of supply outages. This system is generally suitable for installation with private transformers and with high levels of insulation in the network. This requires splitting of the loads into smaller groups and use of circuit separation transformers to supply loads with high leakage currents. This system also requires trained manpower to monitor and operate the system. IT systems are usually applied in hospitals, ships, rail vehicles furnaces, steel works etc because they provide higher operational safety, higher fire safety, better accident preventions and advance information about a possible failure.

11.7 SELECTION CRITERIA FOR EARTHING

The primary function of system earthing is to provide the required level of protection to humans against indirect contact. If this is the only consideration, the three systems of earthing are equivalent as long as the installation and operating rules are correctly followed.

Table 11.1 Influence of networks and loads on the selection of earthing system

Type of network	Advised	Possible	Not advised
Very large network with very good earth electrodes for exposed conductive parts(10 Ω max)	TN	TN-S	IT, TN-C
Very large network with low quality earth electrodes for exposed conductive parts(> 30 Ω)	TN	TT	IT
Network with high leakage currents (> 500 mA)	TN	IT,TT	
Network with over head lines	TT	TN	IT
Emergency standby generating set	IT	TT	TN
Loads sensitive to high fault currents (motors etc.)	IT	TT	TN
Loads with low insulation level (welding m/c, heating elements etc.)	TN	TT	IT
Single phase portable loads	TT, TN-S		IT, TN-C
Loads with sizable risks (conveyors, hoists etc.)	TN	TT	IT
Machine tools	TN-S	TN-C, IT	TT
Supply via star connected power transformer	TT	IT w/o neutral	IT with neutral
Premises with risk of fire	IT	TN-S, TT	TN-C
Installation with frequent modifications	TT		TN, IT
Electronic equipment	TN-S	TT	TN-C

The selection criterion for the best system earthing depends mainly on the wiring regulations, continuity of service, operating conditions and the type of network and loads. The order of priority of the different criteria of selection is:

1. Wiring regulations of the respective countries which impose certain types of system earthing.
2. Whether the consumer owns his transformer (e.g. HT consumer). In this case the owner gets a choice to decide the type of system earthing which is decided in consultation with the design engineer with due consideration to the level of continuity of service, operating condition and the characteristic of the loads.

A typical relation between the network and loads on the selection of earthing system is shown in Table 11.1.

11.8 EARTH LEAKAGE PROTECTION

The widespread use of electrical equipment and devices expose the people to two basic forms of hazards:

- The immediate effect of the electrical energy on the human body in the event of a current flow through the body and
- The indirect influence by the damaging events caused by the electrical energy such as fire caused by electric arcs etc.

When there is a current flow through the human body, the electric current itself is the source of danger which has a negative influence when it passes through the human body (muscle paralysis or ventricular fibrillation), or changes the body tissues (coagulation, burn). This means that there is to be a clear distinction between the hazards caused by the current flow through the human body and the other type of hazard caused by electrical energy. In the first case, the electrical energy acts directly on the human body, but in the second case, the electrical energy is only the indirect cause behind the fire damage, arc burning or spark which expose the body to damaging influence through heat and radiation. Along with this there is considerable risk of damage to property (due to fire caused by the electric arc).

Every electrical installation is subject to current leakages which can vary considerably depending on such factors as the installations condition, age, environment, etc. All electrical risks to people are the result of direct or indirect contact.

Direct contact

Direct contact caused by humans may be due to either carelessness, clumsiness or by accident. Direct contact happens when someone makes contact with a live electrical component of a device or installation. For example:

- A person inadvertently touching a live cable
- A child sticking a metal object into a power socket
- Using unprotected test cables or extensions.
- Someone touching a live bus bar in a distribution panel.

There are two ways of providing protection to people and livestock against direct contact. The first one is to provide protection via physical or electrical isolation of all live parts. This protection ensures that live parts cannot be touched under normal situations, even inadvertently. The methods used for this type of protection are use of barriers, enclosures, closed cabinets which physically or electrically isolate the live parts, shuttered sockets or proper insulation. Additional protection is provided by a residual current device rated at 30 mA. This additional protection provides a back-up protection if the basic protective measures fail.

Indirect contact

Indirect contact takes place when a person makes contact with a metallic earthed part which has accidentally been made live by an insulation failure. Touching of these parts, referred to as exposed conductive parts, can normally be avoided when one uses electrical equipment. This type of contact is very dangerous as, unlike direct contact, it is completely unexpected. For example, a person touching the metal frame of an electrical appliance with defective insulation may be electrocuted through no fault of their own if the appliance itself is not protected for such failures. Protection against indirect contact can be guaranteed by:

- Preventing access to potentially dangerous metal components through double insulations (Class II)
- Assuring good connection of all exposed conductive parts to an effective earth
- Using a residual current device depending on the type of system earthing.

As per the provisions of Indian Electricity Rules [Rule 61A, 71(1) and 73(1)] installation of a residual current circuit breaker (RCCB) is mandatory for all installations of 5kW and above, in all luminous signboards and X-ray installations. The Bureau of Indian Standards also recommends the use of RCCBs at construction sites, temporary installations, and agriculture and horticulture premises.

Risk of Electrocution

The dangerous effects of electricity depend on two factors:

- The time duration of the flow of electric current through the body and
- The magnitude of the current flowing through the body.

These two factors are independent and the severity of the risk varies in accordance with the level of each factor. The current value through the human body depends on the value of the voltage bridged by the person and the various resistances of the human body. The current value is defined using a standard safety voltage of 50 V. This voltage takes into account the maximum current which can be withstood by a human body with minimum internal resistance in given conditions. It also takes into account the maximum permissible time for the current to pass through the body without dangerous physio-pathological effects.

When subjected to such a voltage, the human body reacts like any other receiver with an internal resistance. An electric current passes through the body which causes three serious risks:

- Locking of the muscles or tetanisation: the muscles through which the current passes contract and remain contracted. If this includes the rib cage, breathing may be impaired.
- Action of the heart: The cardiac rhythm is completely disturbed and may cause ventricular fibrillation and
- Thermal effects may cause varying levels of damage to body tissues including severe burns in the case of large currents.

The effect of alternating current on human body as laid down in IEC standard 60479 is shown in Figure 11.14. There are four risk zones and the physiological effects for the various zones is indicated in Table 11.2.

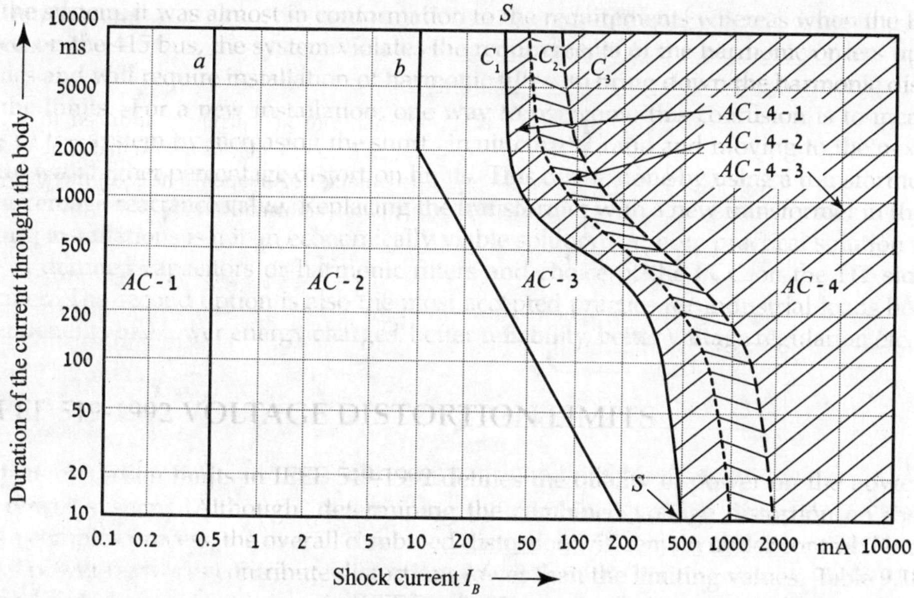

Figure 11.14 Effect of alternating current on human body (IEC Standard 60479).

Table 11.2 Physiological effects of shock current on human body

Current zones	Shock current	Duration of current flow	Physiological effects
AC-1	0-1 mA	Non-critical	Usually no reaction—no electrification or only slight electrification perceptible
AC-2	1-10mA	Non-critical	Strong, partly painful effects on muscles, fingers, feet and joints of arms and legs
AC-3	10-500 mA	Minutes up to split seconds	Strong and painful effects on arms, legs and shoulder muscles, rise in blood pressure, difficulty in breathing. Reversible interferences in the heart, unconsciousness in the upper range.
AC-4-1	Between C_1 and C_2		Ventricular fibrillation unlikely to occur (probability blow 5%)
AC-4-2	Between C_2 and C_3		Possible ventricular fibrillation (probability up to 50%)
AC-4-3	Via C_3		High risk of ventricular fibrillation (above 50%)

Residual current devices (RCD) A residual current device continuously measures the difference between the value of input and output currents. If the value is not equal to zero, it indicates the presence of a leakage current. When the leakage current reaches the level at which the differential is set (sensitivity), the device trips and breaks the circuit. The principle of operation of a residual current device is shown in Figure 11.15.

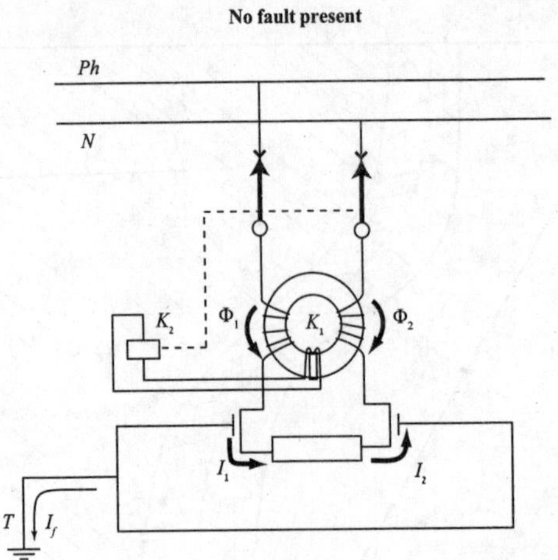

Therefore no current is induced in coil K_1, and coil K_2 is not excited. The contacts do not open. The equipment operates normally

$I_f = 0$, thus

$I_1 = I_2$

$\Phi_1 = \Phi_2$

$\Phi_1 - \Phi_2 = 0$

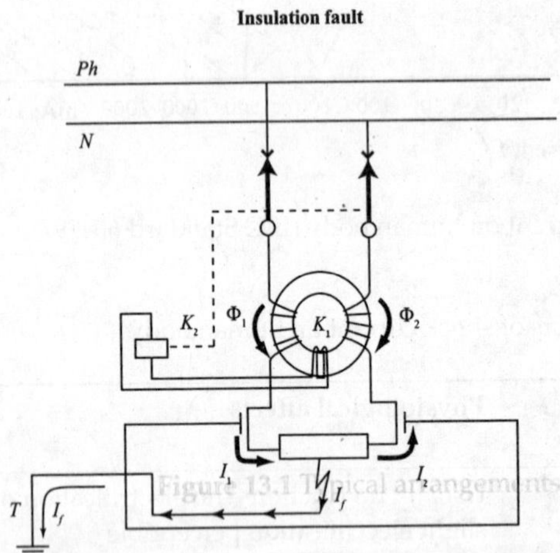

A current is thus induced in coil K_1...

$I_f \neq 0$,

$I_1 > I_2$, thus

$\Phi_1 > \Phi_2$, thus

$\Phi_1 - \Phi_2 \neq 0$

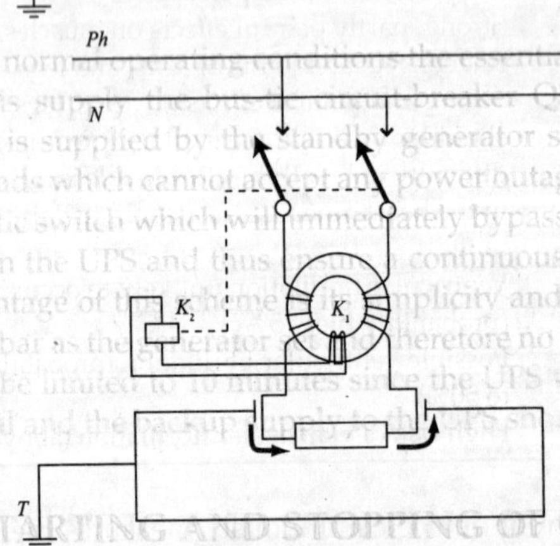

...coil K_2 is excited, the contacts open and the equipment is automatically switched OFF.

Figure 11.15 Operating principle of a residual current device.

The sensitivity of a residual current device $I_{\Delta n}$ is the current level at which the device will surely trip. Standard stipulation for the tripping is that the tripping must occur between 50% of $I_{\Delta n}$ and $I_{\Delta n}$. All residual current devices are provided with a test function knob. This facility is essential to make sure that these devices are in healthy condition. The test knob is to be activated regularly to assure that the devices will function satisfactorily.

Selection of residual current devices

The selection of a residual current device is based on the following requirements:

- To protect against direct or indirect contacts
- To protect against over current and short circuits and
- To carry the normal current in the circuit without interruption.

If protection against direct contact is required, a residual current device with a sensitivity of 30 mA is to be selected. On the other hand, if protection is required for indirect contact, residual current devices with sensitivity of 30, 100 or 300mA may be selected depending upon the installation characteristics. The current carrying capacity of the RCD is selected according to the nature of the load (25A, 40A, 63A, etc.)

RCCBs are the residual current devices which provide two basic functions: 1) Fault current detection , measurement and cut off, and 2) Isolation of an installation.

RCBOs are devices which provide three basic functions: 1) fault current detection, measurement and cut-off, 2) Protection against overload and short circuit, and 3) Isolation of an installation.

Types of residual current device

The different types of residual current device are:

- Type A
- Type AC
- Type S and
- Type Hpi.

Type A : These devices are sensitive to residual alternating currents and residual currents with a DC component and are used for special applications where it is possible that the fault current is not purely sinusoidal.

Type AC : These devices are sensitive to residual alternating currents and are used for general purpose applications.

Type S : These devices are provided with a delayed trip for coordination with other residual current devices and are used with downstream devices.

Type Hpi : Residual current devices provided with enhanced immunity to unwanted tripping in environments with disturbances. They are capable of detecting fault currents with DC components, e.g. computer circuits, printers, diesel generators etc.

Typical technical specifications of residual current devices are shown in Table 11.3.

Table 11.3 Typical specifications of residual current devices (Source: Legrand India)

Specification		RCCB			RCBO		
		Type AC	Type A-S	Type Hpi	Type AC	Type AC 2 modules	Type Hpi
No. of modules	2 pole	2	2	42	4	2	2
	4 pole	4	4		8	–	4
Nominal rating (A)	2 pole	25,40,63	63	25,40,63,80	10,16,25,40,63	6,10,16,25,40,63	25,32,40
	4 pole	25,40,63	63	25.40.63.80	16,25,40,63		25,32,40
Rated sensitivity (mA)	2 pole	30,100,300	300	30	30,100,300	30,300	30
	4 pole	30,100,300	300	30	30,100,300		
Rated making and breaking capacity	Upto 63 A	500 A	500 A	500 A	10,000 A	6000 A	6000 A
	≥63 A	630 A	630 A	630 A	10,000 A		
Rated operating voltage (V)	2 pole	240	240	240	240	240	240
	4 pole	415	415	415	415	415	415
Minimum operating voltage (V)	2 pole	170	170	170	170	170	170
	4 pole	196	196	196	196	196	196
Rated insulation voltage (V)	2 pole	250	250	250	250	250	250
	4 pole	500	500	500	500	500	500
Testing		By pressing test button	By pressing test button	By pressing test button	By pressing test button	By pressing test button	By pressing test button

11.9 QUESTIONS

1. What is the importance of earthing of electrical installations?
2. How is system earthing different from equipment earthing?
3. What are the different types of system earthing?
4. Explain briefly the protective measures that can be employed in different types of system earthing.
5. What are the selection criteria for system earthing?

GENERAL REQUIREMENTS OF INSTALLATION OF MV STANDBY GENERATORS [2]

1. Minimum 1 m clearance shall be provided on three sides of the generator set. When two generators sets are installed side-by-side, minimum 2 m clearance shall be provided between them.
2. Fuel tank of DG sets shall be installed outside the generator room.
3. Exhaust pipe of DG sets shall maintain a minimum height of 1.8 m clearance from the floor level.
4. Voltmeters and frequency meter shall be connected before the circuit breaker in the generator control panel.
5. Watt-hour meter and ammeters in each phase shall also be provided in GCP. For generators of 500 kVA and above, kVA/KW meter and PF meter shall also be provided.
6. Change over switches of approved makes is permitted for capacities up to and including 800A.
7. Above 800A capacity, circuit breakers with castle key interlock shall be provided.
8. U/V coil may be provided at grid side ACB/MCCB on respective switchboards at installations where generators are installed for preventing chances of back feeding from generators to grid side.
9. Separate isolators shall be provided before the change over switch (COS), if it has no OFF position.
10. When generator is directly connected to the main bus it may not be possible sometimes to maintain grading of outlets. Stipulation with respect to maximum single motor rating is also difficult to insist. Generator capacity can be deemed as sufficient even if it is called up on to start induction motors of 60% of its rating, provided suitable starters are employed for limiting the starting current, and grading is affected. However in such cases, it should be ensured that generator is not getting overloaded which can be achieved by providing thermal O/L protection in the generator circuit.
11. Generator room shall be made of non-inflammable materials.
12. Three-pole interlocked circuit breakers/isolators are permitted for supply changeover, provided neutral of all sources is earthed at neutral bus in the panel board.
13. [illegible] Pollution Control Board (CPCB) approved type acoustic enclosure shall be provided [illegible] generators. For generators without in built enclosure arrangements, room acoustic treatment [illegible] be provided. Autotransfer switches of adequate ampere rating may be permitted [illegible] changeover.
14. Standard size of cables and earthling conductors for generators are given in the Table 13.1.
15. Protections required for generators are given in the Table 13.1.
16. Protection [illegible] control wiring diagram shall be forwarded along with scheme for approval by

CHAPTER 12

SOLAR PV SYSTEMS FOR HOUSEHOLD APPLICATIONS

OBJECTIVES

At the end of this chapter you will

- know the students to the basics of Solar PV systems and their relevance to the current energy scenario.
- be familiar with the different types of PV modules available in the market
- know the different components of a solar PV system
- be able to make simple designs for domestic applications
- be able to provide hands on examples to design simple solar PV systems for homes.

12.1 INTRODUCTION

Photovoltaic systems (PV) have been in use world over for more than 50 years generating electricity from the solar energy as renewable electrical energy converted directly from sunlight. Even though it was a very costly proposition originally, the advancements in technology and manufacturing have brought down the costs to an affordable level and solar PV systems are being seen as a way to reduce the huge gap between the supply and demand in the energy sector. This locally generated electricity can be used directly or can be stored in batteries for use at night or can even be fed into the existing power grid when it is not being consumed by the appliances within the house, thereby earning credits against our regular energy consumption. It is most suitable for homes with a flat roof or with slanting roof facing south (in India).

Though blessed with abundant sunshine, averaging approximately 3000 hours a year, our country is yet to utilise this advantage in terms of electric power generation. The country is still heavily dependent on fossil fuels, which are fast depleting and unviable in the long run. The highest contribution of electric energy currently generated comes from coal, which is highly polluting and a non-renewable source. It is estimated that India's coal reserves are fast depleting and will last only up until 2040 or so. Apart from coal, the other sources are hydro, gas and nuclear. A significant portion of oil used to generate energy in India is imported. For economic stability it is imperative that the country reduces its dependence on imported crude oil. Hydroelectric generation also comes with certain disadvantages such as human displacement, soil erosion, diminished forest cover and wildlife habitats.

Sunlight is the most dependable and viable source of energy available to us. However, India's solar energy installations today constitute only a small percentage of installed electricity generation capacity in the country. There is an urgent need to develop cost-effective technologies to harness sunlight in order to generate a significant percentage of clean energy and thus, gradually reduce dependence on environmentally degrading, non-renewable and expensive resources.

With its ever growing electricity demand, Government of India has initiated steps to develop its large potential for solar energy based power generation. In November 2009, the Government of India launched Jawaharlal Nehru National Solar Mission under the National Action Plan on Climate Change. The Central Government of India plans to generate 1 GW of power by 2013 and up to 20 GW grid-based solar power by the year 2020.

Ministry of New and Renewable Energy (MNRE) has recently released a news update on grid-connected solar photovoltaic (PV) technologies, reporting that 1.23 GW of solar power generation capacity has been connected to the grid in the previous three years, as of January 31st, 2013. By far the state with the highest capacity added is Gujarat at 824 MW, followed by Rajasthan at 251 MW, with Maharashtra a distant third at 25 MW. However, solar makes up only 10% of the 12.4 GW of renewable capacity added during this time, with wind taking up 2/3 of the total capacity.

The Government of Kerala has released its draft solar policy for 2013, calling for 500 MW of new solar projects by 2017 and 1.5 GW by 2030. "Kerala Solar Energy Policy 2013" covers both solar electric and solar thermal technologies, calling for mandatory use of solar water heating in many buildings. The policy has been implemented from April 2013. This policy seeks to evaluate in a realistic manner the possibilities of harnessing solar energy to optimal levels and to put in place the necessary framework for achieving the targets.

With the Government's resolve to kickstart the development and implementation of home solar PV installations on a large scale and with considerable subsidies, solar power options for residential users will become very attractive in the years to come. The Government of Kerala has initiated a program to provide ten thousand 1 kW home PV systems (off grid) with considerable subsidies. The response from the general public is very encouraging and the program is in an advanced stage of implementation. The ANERT has shortlisted 22 PV system providers for the speedy implementation of this project (for more details of this scheme please log on to www.anert.gov.in)

This chapter discusses some of the basic technical, economic and regulatory information one should know before buying a photovoltaic (PV) solar electric generation system for homes. This is not a comprehensive technical or economic guide on photovoltaic systems. For that information, please consult an experienced photovoltaic system designer, authorized retailer or supplier.

12.2 TYPES OF PV ELECTRICAL SYSTEMS

There are two general types of designs for PV power systems for homes. The two principal classifications are:

- **Grid-connected or Utility-interactive** systems and
- **Stand-alone or Off grid photovoltaic** systems.

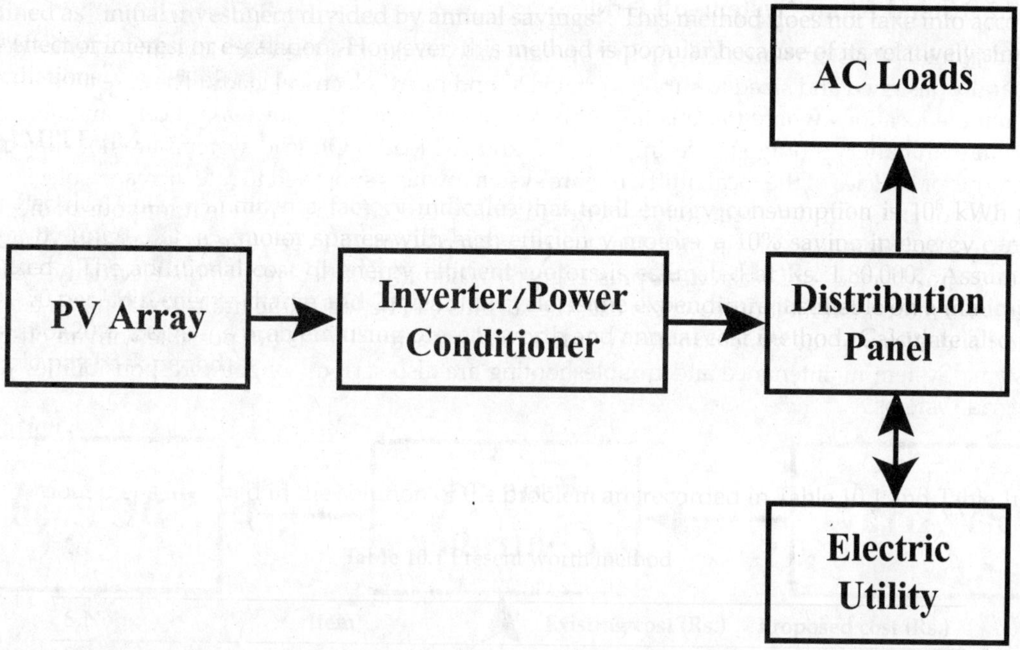

Figure 12.1 Grid-connected photovoltaic system

12.2.1 Grid Connected System

Grid-connected or utility-interactive PV systems are designed to operate in parallel with and interconnected with the electric utility grid (Fig. 12.1). The primary component is the inverter, or power conditioning unit (PCU). The inverter converts the DC power produced by the PV array into AC power consistent with the voltage and power quality required by the utility grid. The inverter automatically stops supplying power to the grid when the utility grid is not energized (islanding protection). A bidirectional interface is made between the PV system AC output circuits and the electric utility network, typically at service entrance. This allows the power produced by the PV system to either supply on-site electrical loads, or to feed the grid when the PV system output is greater than the on-site load demand. During periods when the electrical demand is greater than the PV system output (at night-time), the balance of power required is drawn from the electric utility. Additional safety features are required in all grid-connected PV systems, to ensure that the PV system will not continue to operate and back feed onto the utility grid when the grid is down for service or repair.

12.2.2 Stand Alone System–off Grid

Stand-alone PV systems are designed to operate independent of the electric utility grid, and are generally designed and sized to supply certain DC and/or AC electrical loads. They are most suited for remote locations where there is no utility supply. Figure 12.2 shows a typical standalone PV system with battery storage powering both DC and AC loads. Off-grid systems are not subject to the terms or policies of the local utility, nor are system owners subjected to rate increases, blackouts, or brownouts. Off-grid systems use batteries to store electricity and provide it for your home, but batteries don't last forever. In fact, they will need replacement every five to six years. Another serious drawback is the wasted surplus energy. When a grid-tied renewable electricity system generates more than the homeowners use, the surplus can be fed to the utility, creating an energy credit and allowing the system to always run at full capacity. This is not possible with an off-grid system. System maintenance and troubleshooting are also serious, ongoing responsibilities with off-grid systems.

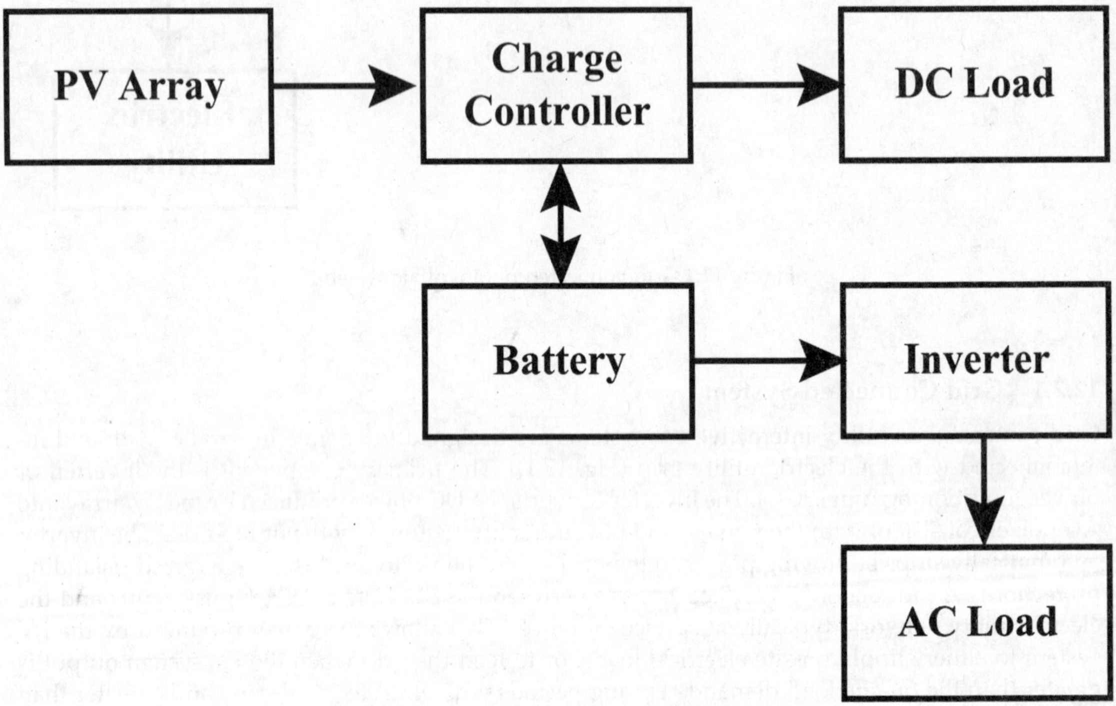

Figure 12.2 Stand-alone PV system-off grid photovoltaic system

12.3 SYSTEM COMPONENTS

Pre-engineered solar power systems can be purchased that come with all the components, right down to the nuts and bolts. Any good company/PV provider can size and specify systems for you, given a description of your site and needs. Nevertheless, familiarity with system components, the different types that are available, and criteria for making a selection is important.

Basic components of grid-connected PV systems with and without batteries are:

- Solar photovoltaic modules
- Array mounting racks
- Grounding system (protective earthing)
- Junction box
- Surge protection (often part of the junction box)
- Inverter
- Meters–system meter and kilowatt-hour meter etc.
- Disconnects:
  - Array DC disconnect
  - Inverter DC disconnect
  - Inverter AC disconnect
  - Exterior AC disconnect

If the system includes batteries, it will also require:

- Battery bank with cabling and housing structure
- Charge controller
- Battery disconnect

12.3.1 Solar and Modules

Unlike a solar hot water system that uses the sun's energy to heat water, solar electric or photovoltaic technology uses the sun's energy to make electricity. The basic building block of PV system is the solar "cell." Many such cells are wired together to produce a PV "module," and many modules are linked together to form a PV "array." PV modules are available in the range of power output from about 10 watts to 300 watts, and produce a direct current like that from an automobile battery. Rating of a module is the maximum power (peak power–W_P) the module can produce under standard test conditions (STC) with 1000 watts of sunlight per square meter at a module temperature of 25 °C in still air. Actual conditions will rarely match standard conditions and so actual power output will almost always be less.

Caution-1: A 1000 Wp (watt peak) solar plant will not generate 1000 watts of power under normal conditions of operation

PV modules are made from several processes or technologies. They all do the same job of producing electricity from sunlight. There are mainly three types of PV module manufacturing technology and a fourth one called hybrids (modules that use a combination of two technologies). The three main types come from the three different forms of crystalline materials.

(1) Mono-crystalline - High performance solar cells (efficiency 15–19%)
(2) Poly-crystalline - Standard solar cells (efficiency 11–15%)
(3) Amorphous - Thin film solar cells (efficiency 5–8%)
(4) Hybrid solar modules

As a general rule, thin film modules have lower costs than crystalline silicon modules for modules of similar powers. Two figures that are useful in comparing modules are the modules' price per watt and the rated power output per unit area (or efficiency). When looking through a manufacturer's catalog of solar modules, you will often find the rated power, the overall dimensions of the module, and its price. Find the cost per watt by dividing the module's price by its rated output in watts. Find the watts per area, by dividing the rated output of the panel by its surface area.

Caution-2: Since module costs and efficiencies continue to change as technology and manufacturing methods improve, it is difficult to provide general recommendations that will be true into the future regarding, for example, which type of module is cheapest or the best overall choice. It is best to make comparisons based on current information provided by manufacturers, combined with the specific requirements of our application.

12.3.2 Array Mounting Racks

A well-designed PV system needs clear and unobstructed access to the sun's rays for most or all of the day, throughout the year. The orientation of PV arrays (the compass direction that the array faces) affects its performance. The best location for a PV system is a south-facing roof, but roofs that face east or west may also be acceptable. Flat roofs also work well for solar electric systems, because PV modules can be mounted flat on the roof facing the sky or bolted on frames tilted toward the south at an optimal angle. On roof-mounted systems, the PV array is typically mounted on fixed racks, parallel to the roof for aesthetic reasons and lifted off several centimeters above the roof surface to allow airflow that will keep them as cool as practical. Mounting racks are adjustable, allowing setting of the angle of the PV modules. Solar modules can also be placed on the ground, either on a fixed pole or a "tracking" mount that follows the sun to orient the PV modules. Tracking devices allow the array to automatically follow the sun. Tracked PV arrays can increase the system's daily energy output by 25 percent to 40 percent. Despite the increased power output, tracking systems usually are not justified by the increased cost and complexity of the system.

Caution-3: Tracking systems are not recommended for home solar applications.

12.3.3 Grounding System

Grounding system provides a well-defined, low-resistance path from the PV system to the ground to protect the system from current surges, lightning strikes or equipment malfunctions. Grounding also stabilizes voltages and provides a common reference point. The grounding requirements shall normally be as per the Electrical Inspectorate standards. Equipment grounding provides protection from shock caused by a ground fault. A ground fault occurs when a current-carrying conductor comes into contact with the frame or chassis of an appliance or electrical box. All system components and any exposed metal, including equipment boxes, receptacles, appliance frames and PV mounting equipment should be properly grounded.

12.3.4 Junction Box

Output wires from individual PV modules or strings are run to the junction box, typically located on the roof. These wires may be single conductor pigtails with connectors that are pre-wired onto the PV modules. The output of the junction box is a larger two-wire conductor laid in conduit. A junction box typically includes a safety fuse or circuit breaker for each string and may include a surge protector.

12.3.5 Surge Protection

Surge protectors help to protect the system from power surges that may occur if the PV system or nearby power lines are struck by lightning. A lightning surge is a sudden increase in voltage significantly above the design voltage.

12.3.6 Inverter

Because all photovoltaic panels produce electricity in DC, an inverter is required for all solar power systems to make the electricity usable in domestic environment. Its purpose is to convert the direct current (DC) electricity that is generated from a photovoltaic panel into an alternating current (AC) that can be used by in home appliances. Inverters take care of four basic tasks of power conditioning:

- Converting the DC power generated by the PV modules or battery bank to AC power required for practical use.
- Ensuring that the frequency of the AC power is 50 Hz, the standard frequency of electricity supplied by the utility.
- Reducing voltage fluctuations.
- Ensuring that the shape of the AC wave is appropriate for the application, i.e., a pure sine-wave for grid-connected systems

12.3.6.1 Criteria for selecting a grid-connected inverter

The following factors should be considered for a grid-connected inverter:

- A UL1741 listing and IEEE1547 compliance of the inverter for use in a grid-interactive application
- The DC voltage of the electricity generated by the solar array or battery bank
- The DC power window of the PV array
- Characteristics indicating the quality of the inverter, such as high efficiency and good frequency and voltage regulation
- Anti islanding protection
- Additional inverter features such as meters, indicator lights, and integral safety disconnects
- Manufacturer warranty, which is typically 5–10 years
- Maximum Power Point Tracking (MPPT) capability, which maximizes power output.

Caution-4: The solar inverter is most often the critical component of a photovoltaic installation. Many people fall into the trap of underestimating the systems needs for a solar inverter and purchase one that handles a smaller capacity than necessary in an attempt to save money. This is a mistake that can be costly in the long run. A better way to cut costs is to purchase an inverter with a slightly larger capacity than we plan to initially install and add more PV panels to boost production later on down the line. This leaves the system with a flexibility to upgrade the capacity of the PV system with minimum cost at a later stage.

12.3.7 Metering

Metering provides an easy access to the various parameters of the system and allows us to check whether the system is operating properly or not. Most of the modern PV systems are provided with multifunction digital instruments with soft touch controls.

12.3.8 Disconnects (Isolators and Circuit Breakers)

Automatic and manual safety disconnects protect the wiring and components of a PV system from power surges and other equipment malfunctions. They also ensure that the system can be safely shut down and system components can be removed for maintenance and repair. For grid connected systems, safety disconnects ensure that the generating equipment is isolated from the grid, which is important for the safety of utility personnel. In general, disconnect is needed for each source of power or energy storage device in the system.

12.3.9 Battery Bank

Batteries store direct current electrical energy from the PV modules generated during the day for later use. This energy storage comes at a cost, however, since batteries reduce the efficiency and output of the PV system, typically by about 10 percent for lead-acid batteries. Batteries also increase the complexity and cost of the system. The different types of batteries used in solar applications are:

- Lead-acid batteries
  - Flooded (Liquid vented)
  - Sealed (Valve-Regulated Lead Acid-VRLA)
    Absorbent glass mat
    Gel cell
- Alkaline batteries
  - Alkaline batteries
  - Nickel-iron

12.3.9.1 Lead-acid batteries

Sealed lead acid batteries are most commonly used in grid-connected systems. Sealed batteries are spill-proof and do not require periodic maintenance. Flooded lead acid batteries are usually the least expensive but require adding distilled water at least monthly to replenish water lost during the normal charging process. There are two types of sealed lead acid batteries: sealed absorbent glass mat (AGM) and gel cell. AGM lead-acid batteries have become the industry standard, as they are maintenance free and particularly suited for grid-tied systems where batteries are typically kept at a full state of charge. Gel-cell batteries, designed for freeze-resistance, are generally a poor choice because any overcharging will permanently damage the battery.

12.3.9.2 Alkaline batteries

Because of their relatively high cost, alkaline batteries are only recommended where extremely cold temperatures are anticipated or for certain commercial or industrial applications requiring their advantages over lead-acid batteries.

12.3.9.3 Sizing battery banks

For grid-connected systems, batteries are usually sized for relatively short time periods with 8 hours being typical. Size may vary, however, depending on the particular needs of a facility and the length of power outages expected. For comparison, battery banks for off-grid systems are usually sized for providing the back-up for one to three cloudy days.

Caution-5: Lead acid AGM batteries are the industry standard for grid connected systems and are maintenance free.

12.3.10 Charge Controller

A charge controller is only necessary in systems with battery back-up. The primary function of a charge controller is to prevent overcharging of the batteries. Most charge controllers also include a low voltage disconnect that prevents over-discharging batteries. In addition, charge controllers prevent charge from draining back to solar modules at night. Some modern charge controllers incorporate maximum power point tracking, which optimizes the PV array's output, increasing the energy it produces.

12.3.10.1 Selection

Charge controllers are selected based on:

- PV array voltage – The controller's DC voltage input must match the nominal voltage of the solar array.
- PV array current – The controller must be sized to handle the maximum current produced by the PV array.

12.3.10.2 Interaction with inverter

Since the majority of charge controllers are installed in off-grid systems, their default settings may not be appropriate for a grid-connected system.

Caution-6: The charge controller must be selected such that it does not interfere with the proper operation of the inverter. In particular, the controller must be set up such that charging the batteries from the PV array takes precedence over charging from the grid.

Anyone considering installing a solar PV system for their home should consider hiring a professional to assist in making sure the system is built properly. There are many variables to take into account, including the size of the home, the electricity use of the home, possible future increase in the energy use of the home, as well as the types and sizes of the photovoltaic cells and solar inverters necessary to keep the system running smoothly. A professional installer can make sure that your system is designed to handle the electric load of your home efficiently and economically. However please make sure that the person or company with whom you are entrusting the work of designing and installing your solar PV system provides satisfactory answers to the following questions:

- Has the company/PV provider installed any similar PV systems previously?
- How many years of experience does the company/PV provider have in installing similar PV systems?
- Is the company/PV provider properly licensed to carry out such work?
- Do the brands of equipment and accessories offered by them satisfy the MNRE regulations?
- Does the company/PV provider have any pending judgments or liens against it?
- Does the company/PV provider have proper service facility/experienced technicians in the locality?
- Can the company/PV provider provide a list of customers for whom they have installed similar PV systems?
- Will the company/PV provider provide warranty for the entire system (not for individual components) for the stipulated period at no extra cost?

If the answers to the above questions are convincing, go ahead and have sunny days ahead.

12.4 TYPICAL DESIGN FOR HOME APPLICATIONS

The design process involves the following steps:

1. Determine the load to be served in watt-hours/day.
2. Determine the average solar energy available on at least a month by month basis.
3. Calculate the size of solar panel that will be needed to meet the required load under the worst month conditions.
4. Calculate the size and type of battery that will be needed to provide the required reliability of power.
5. Determine the type of charge controller.
6. Determine the inverter capacity.

System sizing is the process used for determining the minimum PV panel, inverter, battery and charge controller sizes needed to deliver the required electrical energy under the solar conditions that exist at the site. It balances the output from the system with the solar input while taking into consideration losses in the system. In order to do the system sizing, we need to know the following in formations:

- The solar energy available in kWh/m^2/day at the site for the lowest solar energy month of the year.
- The average Wh/day required by the user to operate the desired appliances and any special needs for power that go much beyond the average.
- The losses that occur in the PV system that reduces the energy available to the user.

12.4.1 Determining the Load to be Served in Watt-hours per Day

The steps involved in the estimation of total watt-hours to be supplied are:

- Determine the power rating in watts of each of the appliances to be used in the household
- Estimate the number of hours per day that each appliance will be used.
- For each appliance multiply the power rating in watts by the hours of use to get watt hour/day
- Calculate the total watt hour per day for all appliances taken together.

EXAMPLE 12.1
A residential building has the following electrical loads:

1. Three numbers 18 W CFL used 4 hours per day
2. Two numbers 60 W ceiling fans used 8 hours per day
3. One refrigerator of 125 W used 24 hours with 50% duty cycle
4. One 5 W night lamp used 7 hours per day
5. One 125 W washing machine used for 1 hour per day

Calculate the total watt-hour per day for all appliances taken together.

Answer

Watt-hours used by CF lamps	$= 3 \times 18 \times 4$
	$= 216$
Watt-hours used by ceiling fans	$= 2 \times 60 \times 8$
	$= 960$
Watt-hours used by refrigerator	$= 125 \times 24 \times 0.5$
	$= 1500$
Watt-hours used by night lamp	$= 5 \times 7$
	$= 35$
Watt-hours used by the washing machine	$= 125 \times 1$
	$= 125$
Total watt-hour required per day	$= 216 + 960 + 1500 + 35 + 125$
	$= 2836$ watt-hour.

12.4.2 Determining the Average Solar Energy Available on at least a Month by Month Basis

Actual measurements at the site are the best but at least the date for one full year is needed and if possible the data for several years is preferred. Measurements taken with instruments tilted at the same angle as that of the solar panels are best but horizontal "meteorological" measurements are also acceptable. NASA satellite measurements are better than "sunshine hours" recorded for the site. "Sunshine hour" measurements indicate the solar variation over the year but are not a good measure of actual solar energy in kWh/m^2/day but are better than having nothing. It will be advisable to choose the average value of solar irradiation (solar insolation) for the lowest month as the design basis.

Table 12.1 Table Monthly average Solar Irradiation (kWh/m^2/day) in Trivandrum, Kerala, India (Source: www.synergyenviron.com)

January	**February**	**March**	**April**	**May**	**June**
5.80	6.46	6.83	6.24	5.57	4.83
July	**August**	**September**	**October**	**November**	**December**
4.91	5.26	5.74	5.24	4.94	5.30

Annual average value = 5:59 kWh/m^2/day

12.4.3 Calculating the Size of Solar Panel that is Needed to Meet the Required Load

In order to find out the size of the solar panel, we need first to find out the energy that should be produced by the solar panel to meet the requirement of the household. For a system with batteries (that is the case with most of the household PV systems) the main components of losses are the losses happening in the interconnections and that in the battery. Wiring and connection losses will amount to about 10% of our requirement and that in the battery will be about 20%. Hence, the total losses will be about 30% of our requirement. In other words, the solar panels need to produce

30% more energy than that is required by the user. Therefore, the watt-hour/day to be generated by the solar panel is about 130% of the watt-hour needed by the equipment in the household.

EXAMPLE 12.2

Determine the watt-hour per day to be generated by the solar panels for the example 12.1

Answer

Total watt-hour required per day = 2836

Therefore, watt-hour to be produced by the panels = 1.3 × 2836

= 3686.8

Panel generation factor (PGF) is an important component used in deciding the solar panel size. The PGF is computed using the solar irradiation data for a particular location. Usually, the starting point is the lowest monthly average solar irradiation. Typical value of daily average solar irradiation is between 5 and 6 kWh/m^2/day in India. The monthly average solar irradiation in Trivandrum is shown in Table 12.1 for the example shown above the lowest average solar irradiation is 4.83 kWh/m^2/day for the month of June in Trivandrum. This is the same as the total energy as would come from the sun shining at 1000 W/m^2 each day for the number of hours equal to the kWh/m^2/day figure. Normally, the rating of the solar panel in watt-peak is based on a solar irradiation of 1000W/m^2.

In our case for example the lowest average monthly solar irradiation is 4.83 kWh/m^2/day.

This can be interpreted as (4.83 hours × 1000W/m^2/day = 4:83 kWh/m^2/day). That is equivalent to 4.83 hours of 1000 W/m^2 sunlight every day. Therefore, each W_p of the panel would deliver an energy equivalent to 4.83 Wh/day if all other conditions were perfect. The conditions are normally not perfect and so we have to correct for the variations from standard conditions. Theoretically, we can say that each W_p of the solar panel will generate an energy equal to 4.83 Wh/day at the minimum. However, there are several factors that influence the performance of the solar panels and we need to apply corrections for the same.

Corrections include:

- 15% for ambient temperature above 25 °C (85% derating)
- 5% for losses due to sunlight not striking the panel straight on (caused by glass having increasing reflectance at lower angles of incidence) (95% derating)
- 10% for losses due to not receiving energy at the maximum power point (not present if there is a MPPT charge controller) (90% derating)
- 5% allowance for dirt collected on the solar panel (95% derating)
- 10% allowance for the panel being below specification and for ageing (90% derating)

Therefore, the rating of the solar panel gets derated by a factor which is equal to

$$\text{Derating factor} = 0.85 \times 0.95 \times 0.90 \times 0.95 \times 0.90$$
$$= 0.62$$

This is equivalent to saying that 62% of the rating will only be actually available for use.

In case MPPT charge controller is used, the derating factor will be

$$\text{Derating factor} = 0.85 \times 0.95 \times 0.95 \times 0.90$$
$$= 0.69$$

Therefore, the panel generation factor in Wh/Wp/day (PGF) is obtained by multiplying the lowest watt-hour/day by the derating factor.

$$\text{The PGF for the example} = 4.83 \times 0.62 = 2.99 \text{ (without MPPT) or}$$
$$= 4.83 \times 0.69 = 3.33 \text{ (with MPPT)}$$

Hence, we can conclude that every Wp capacity of the solar panel would produce a minimum of 2.99 Wh/day (without MPPT) or 3.33 Wh/day (with MPPT) during the lowest solar month.

EXAMPLE 12.3

Calculate the size of the PV panel and that of the solar plant for the example 12.1, if the location of the building is in Trivandrum.

Answer

Minimum solar irradiation for Trivandrum = 4.83 kWh/m^2/day

Panel generation factor (with MPPT controller) = $4.83 \times 0.69 = 3.33$ Wh/Wp/day

Watt hour/day for all appliances taken together = 2836 Wh/day

Wh/day needed from the panel = $1.3 \times 2836 = 3686.8$ Wh/day

Panel size required $= \dfrac{3686.8}{3.33} = 1107.15$ Wp

Therefore, number of PV modules $= \dfrac{\text{Panel size required}}{\text{Rated Wp of the PV modules}}$

From table 12.2, we may choose 110 Wp module,

$$\text{Number of PV modules} = \frac{1107.15}{110}$$
$$= 10.065 \text{ (say 10 modules)}$$

The capacity of the solar plant is therefore $10 \times 110 = 1100$ Wp. The panels are connected as two in series and five in parallel.

Thus, if we choose 110 Wp modules, we will require 10 solar modules rated at 110 Wp. The surface area of each module from table 12.2 is 0.85 m^2. Therefore, the roof area required for installing the solar PV system will be about $10 \times 0.85 = 8.5$ m^2.

Table 12.2 Roof area requirement of Solar PV systems

Power in watts	Open circuit voltage (V_{oc}) in volts	Short circuit current (I_{sc}) in Amps	Voltage at maximum Power (V_{mp}) in volts	Current at maximum power (I_{mp}) in Amps	Area.sq.m
40	21	2.54	17	2.35	0.35 (3.76)
50	21	3.17	17	2.94	0.43 (4.63)
60	21	3.81	17	3.53	0.50 (5.38)
70	21	4.44	17	4.12	0.56 (6.03)
74	21	4.7	17	4.35	0.61 (6.66)
75	21	4.75	17	4.41	0.61 (6.66)
80	21	5.08	17	4.71	0.62 (6.67)
90	21	5.71	17	5.29	0.70 (7.53)
100	21	6.35	17	5.88	0.78 (8.40)
110	21	6.98	17	6.47	0.85 (9.25)
120	21	7.62	17	7.06	1.0009 (10.86)
125	21	7.94	17	7.35	
130	21	8.25	17	7.65	
140	21	8.89	17	8.24	
150	43	4.65	35	4.29	1.163 (12.52)
160	43	4.96	35	4.57	

Table 12.3 Technical specifications of PV modules (Courtesey: Indsolar Pvt Ltd)

PV Module Efficiency (%)	PV Capacity (kilo watt)							
	0.1	0.25	0.5	1	2	4	10	100
	Approximate Roof area needed in sq m							
4	3	7.5	15	30	60	120	300	3000
8	1.5	3.8	7.5	15	30	60	150	1500
12	1	2.5	5	10	20	40	100	1000
16	0.8	2	4	8	16	32	80	800

12.4.4 Battery Sizing

The battery recommended for a solar PV system is a deep cycle battery. Automobile batteries are not generally suitable for solar applications. They are specifically designed to discharge to low energy level and rapid recharge or cyclic charge and discharge day after day for years. The battery shall be large enough to operate the appliances at night and on cloudy days. Battery size also shall take into account rainy season with no solar input for a few days (about 2–4 days).

12.4.4.1 Typical specifications for a deep cycle battery for solar system

4000 cycles to 10% depth of discharge
3300 cycles to 30% depth of discharge
2500 cycles to 50% depth of discharge
1500 cycles to 80% depth of discharge

Therefore, we can see that the life expectancy of the battery will be more if the battery is subjected lesser depth of discharge. From the above information we can determine the life expectancy: A solar system is discharged on an average to a depth of 50% per day, then, it could be expected to have a life of around 2500 days or 6.8 years. On the other hand if the depth of discharge permitted is only 30%, the life expectancy will be 3300 cycles or about 9 years.

12.4.4.2 Calculating the battery size

The following assumptions are made for selecting the capacity of the battery bank:

1. Battery efficiency is assumed to be 85%
2. Depth of discharge is assumed to be 50%
3. Cloudy days assumed to be 2 (battery supports 2 days of energy requirements)
4. Battery voltage levels are chosen based on the wattage capacity of the plant

   12 V for up to 500 W capacity
   24 V for up to 1000 W capacity
   48 V for up to 2000 W capacity

Tables 12.4, 12.5 show the technical specifications of low maintenance tubular floated and tubular GEL VRLA batteries.

EXAMPLE 12.4

Determine the size of the battery for use with the solar PV system described in example 12.1.

Answer:

Wh/day required by the load	= 2836 (from Example 12.1)
Battery must supply 2836/0.85	= 3336 Wh/day (based on assump-1)
Batter bank must supply 3336/0.5	= 6672 Wh (based on assump-2)

Battery bank must be sized6672 $\times$ 2 = 13344 Wh (based on assump-3)

= (say 13400 Wh)

With 24 V battery, Ah capacity = 13400/24 = 558 Ah (say 600Ah)

Therefore, we can choose 8 Nos. of 150 Ah, 12 V batteries connected as two in series and four in parallel (4 $\times$ 150 = 600 Ah at 24 volts) and this will be able to supply the load for 2 days without interruption. As 12 V batteries are more popular, two 12 V batteries are connected in series to get 24 V DC. Since the solar plant capacity is 1100 Wp, the operating voltage is selected as 24 volts DC. From table 12.4b, we can choose 6LMS150 L, 12 V 150 Ah low maintenance tubular floated – LMS batteries.

12.4.5 Selection of Charge Controller

A charge controller is an essential element of PV solar system with storage batteries. The basic functions of the charge controller are:

1. Block the reverse flow of current
2. Prevent battery over charging
3. Prevent battery over discharge
4. Protect the battery from overload and
5. Display battery status and flow of power.

12.4.5.1 Charge controller specifications

The charge controller is specified in terms of current and voltage.

Charge controller is selected to match the voltage of the PV array and batteries.

Make sure that the charge controller has enough capacity to handle the current from the PV array.

The standard practice is to size the charge controller to operate at 30% more than the short circuit current Is_c of the array.

EXAMPLE 12.5

Determine the size of the charge controller for the problem specified in example 12.1.

Answer

From example 12.3, the number of PV modules is equal to 10, each rated for a peak power of 110 watts. From example 12.4, the battery voltage is selected as 24 volts. Therefore, the PV modules shall be connected as two numbers in series and five numbers in parallel. So that the open circuit voltage of the PV modules will be around (21 $\times$ 2) = 42 volts. From Table 12.2 the short circuit current of each module is 6.98 ampere.

Therefore the.total short circuit current from all the five PV modules connected in parallel will be equal to 6.98 $\times$ 5 = 34.9 A.

Hence, the current rating of the charge controller = 130% of the short circuit current

$$= 1.3 \times 34.9$$

$$= 45.37 \text{ A (say 50A)}$$

Therefore, we can choose a charge controller rated for 24 volt 50 A capacity.

Additional specifications that can be included in the specifications are voltage, current, power and energy display. State of the charge on the battery can also be displayed on the charge controller.

Table 12.4a Charging characteristics of solar batteries (Courtesey: Exide batteries) 12 V low maintenance Tubular floated LMS series

Model of operation	Voltage setting per mono-block unit for ambient-temperature		Current setting
	12V mono-block	6V mono-block	
Bulk voltage	15.6V ± 0.1V	7.80V ± 0.1V	Maximum-20% of battery Ah capacity Minimum-10% of battery Ah capacity
Float voltage	13.8 ± 0.1V	6.90V ± 0.1V	
Low voltage disconnect	11.4V ± 0.1V	5.70V ± 0.1V	
Load reconnect voltage	12.96V ± 0.1V	6.48V ± 0.1V	
Recharge factor	110% of discharge Ah		

Table 12.4b Technical Data

Type of Battery	Nominal Voltage (V)	Capacity @c10 to 1.80 v.p.c at 27 °C (Ah)
6LMS20	12	20
6LMS20L	12	20
6LMS40	12	40
6LMS40L	12	40
6LMS75	12	75
6LMS75L	12	75
6LMS100L	12	100
6LMS120L	12	120
6LMS150L	12	150
6LMS200L	12	200
3LMS300	6	300

Table 12.5a Tubular Gel VRLA Batteries (Courtesey: Exide batteries) Recharging Characteristics during Operations

	Recommended parameters for ambient temperature of 25–30°C	**Temperature compensation factor at 25 °C**
Charging current	Maximum-20% of battery Ah capacity Minimum-10% of battery Ah capacity	
Bulk voltage	2.40 ± 0.02V/cell	Float: –3mV/°C/2 V cell
Float voltage	2.28 ± 0.02V/cell	
Equalizing voltage	2.40 ± 0.02V/cell	
Load reconnect voltage	2.20 ± 0.02V/cell	
Low voltage disconnect	1.90 ± 0.02V/cell	Cyclic: –5mV/°C/2 V cell
Recharge factor	105% of discharge Ah	

Table 12.5b Technical Data

Type of Battery	**Nominal Voltage (V)**	**Capacity @10 hr/1.75 V (Ah)**
SG 200	2V	200
SG 300	2V	300
SG 400	2V	400
SG 500	2V	500
SG 600	2V	600
SG 800	2V	800
SG 1000	2V	1000

12.4.6 Selection of Inverters

Inverters take care of four basic tasks of power conditioning in the solar PV systems:

- Converting the DC power coming from the PV modules or battery bank to AC power
- Ensuring that the frequency of the AC cycles is 50 cycles per second,
- Reducing voltage fluctuations,
- Ensuring that the shape of the AC wave is appropriate for the application, i.e., a pure sine-wave for grid-connected systems.

12.4.6.1 Criteria for selecting the inverters

The following factors should be considered for a grid-connected inverter:

- A UL1741 listing and IEEE1547 compliance of the inverter for use in a grid-interactive application,
- The voltage profile of the incoming direct current from the solar array or battery bank
- The DC power window of the PV array
- Characteristics indicating the quality of the inverter, such as high efficiency and good frequency and voltage regulation

- Additional inverter features such as meters, indicator lights, and integral safety disconnects
- Manufacturer's warranty, which is typically 5–10 years
- Maximum Power Point Tracking (MPPT) capability, which maximizes power output.

There are essentially two types of grid-interactive inverters: those designed for use with solar PV systems with batteries and those designed for a system without batteries. Some of the basic qualities of inverters for PV applications are explained below.

Power Quality – Inverters for grid-connected systems produce better than utility-quality power. For grid-connection, the inverter must have the words "Utility-Interactive" printed directly on the listing label.

Voltage Input – The inverter's DC voltage input window must match the nominal voltage of the solar array, usually 235V to 600V for systems without batteries and 12, 24 or 48 volts for battery-based systems.

AC Power Output – Grid-connected systems are sized according to the power output of the PV array, rather than the load requirements of the building. This is because any power requirements above what a grid-connected PV system can provide is automatically drawn from the grid.

Surge Capacity – The starting surge of equipment such as motors is not a consideration in sizing grid-connected inverters. When starting, a motor may draw as much as six times its rated wattage. For grid-connected systems, this start-up surge is automatically drawn from the grid.

Frequency and Voltage Regulation – Inverters of high quality will produce near constant output voltage and frequency.

Efficiency – Modern inverters commonly used in residential and small commercial systems have peak efficiencies of 92 percent to 94 percent, as rated by their manufacturers. Actual field conditions usually result in overall efficiencies of about 88% to 92%. Inverters for battery-based systems have slightly lower efficiencies.

Integral Safety Disconnects – The AC disconnect in most inverter models may not meet requirements of the electric utility. Therefore, a separate exterior AC disconnect may be required even if one is included in the inverter. All inverters that are UL listed for grid-connection include both DC disconnects (PV input) and AC disconnects (inverter output). In better inverters, the inverter section can be removed separately from the DC and AC disconnects, facilitating easy repair.

Maximum Power Point Tracking (MPPT) – Modern non-battery based inverters include maximum power point tracking. MPPT automatically adjusts system voltage such that the PV array operates at its maximum power point. For battery-based systems, this feature has recently been incorporated into better charge controllers.

Inverter-Chargers – For battery-based systems, inverters are available with a factory integrated charge controller, referred to as inverter-chargers. Be sure to select an inverter-charger that is rated for grid-connection.

In the event of a grid power outage, use of an inverter-charger that is not set up for grid-connection would result in overcharging and damaging the batteries, known as "cooking the batteries."

Automatic Load Shedding – For battery-based systems, the inverter can automatically shed any unnecessary loads in the event of a utility power outage. Dedicated loads, i.e., the loads that will be kept powered up during the grid outage, are connected to a separate electrical sub-panel. A battery-based system must be designed to power these critical loads.

Warranty – Inverters typically must carry warranties of 5 years, although the industry is moving toward a 10-year warranty. The transformer and solid state components of an inverter are both susceptible to overheating and damage from power spikes, reducing its useful life.

EXAMPLE 12.6

Determine the size and specification of the grid connected inverter for example 12.1

Answer

Grid-connected systems are sized according to the power output of the PV array, rather than the load requirements of the building. This is because any power requirements above what a grid-connected PV system can provide is automatically drawn from the grid. In this case the power output of the PV array is 1100 watt. Hence, the capacity of the inverter can be fixed as 1200 VA.

The final specification of the inverter can be 24 V DC inputs, 240 V AC output, 1200 VA grid smart inverter, conforming to UL 1741 and IEEE1547 standards.

12.5 ESTIMATING THE WH/DAY THAT CAN BE USED FOR A PARTICULAR SIZE OF PANEL

There are situations when we may have to compute the available energy from a particular size of solar panels. The following steps are useful for computing the maximum available energy inWh/day that can be served by a particular size of panel:

Multiply the minimum average value of $kWh/m^2/day$ by 0.69 to get the local panel generation factor.

Multiply the local generation factor by the rating of the solar panel in Wp. This will give the estimated Wh/day from the panel.

Divide the estimated Wh/day from the panel by 1.3 to get the estimated appliance Wh/day that can be served by that panel.

Eample 12.7

A 240 Wp solar panel panel is installed at a site having a low month solar value of 5.2 $kWh/m^2/day$. What is the maximum Wh/day of appliance load that this panel can serve?

Answer

Panel generation factor = $5.2 \times 0.69 = 3.59$

Wh/day of the panel = $240 \times 3.59 = 861.6$

Wh/day of appliance use is = $861.6 = 1.3 = 662$ Wh/day

Typical specifications for solar panel, solar panel mounting structure and solar inverters are given in Tables 12.6, 12.7 and 12.8.

Table 12.6 Specifications of Solar PV Modules

Type	**Polycrystalline Silicon**
Origin	Manufactured in India
Efficiency	> = 13%
Fill factor	> = 70%
Degradation Warranty	Panel output (Wp) capacity to be > = 90% of design nominal power after 10 years and > = 80% of design nominal power after 25 years
Module Frame	Non-corrosive and electrolytically compatible with the mounting structure material
Termination	Box Thermoplastic, IP 65, UV resistant
Blocking diodes	Schottky type
Module Minimum rated Power	The nominal power of a single PV module shall not be less than 74Wp
RF Identification tag for each solar module	Shall be provided inside the module and must be able to withstand environmental conditions and last the lifetime of the solar module
RF Identification tag data	(a) Name of the manufacturer of PV module (b) Name of manufacturer of solar cells (c) Month and year of manufacture (separately for solar cells and module) (d) Country of origin (separately for solar cells and module) (e) I-V curve for the modules (f) Wm, Im, Vm and FF for the module (g) Unique serial No and model No of the module (h) Date and year of obtaining IEC PV module qualification certificate (i) Name of the test lab issuing IEC certificate (j) Other relevant information on traceability of solar cells and module as per ISO 9000 standard
Power output rating	To be given for standard test conditions (STC). I-V curve of the Sample module shall be submitted.
Compliance with standards and codes	IEC 61215/IS 14286 IEC 61730 Part 1 and 2
Salt Mist Corrosion Testing	As per IEC 61701

Table 12.7 Solar PV Modules Mounting Structure

Wind velocity withstanding capacity	150 km/hour
Structure material	Hot dip galvanized steel with a minimum galvanization thickness of 120 microns or aluminium alloy
Bolts, nuts, fasteners, panel mounting clamps	Stainless steel SS 304
Mounting arrangement for RCC-flat roofs	With removable concrete ballast made of pre-fabricated PCC (1:2:4), M15
Mounting arrangement for metal sheet roofs	Mounting directly on the sheet metal ensuring stability and wind withstanding capacity, or penetrating the sheet metal and fixing to the substructure, ensuring that the roof remains waterproof and ensuring stability and wind withstanding capacity
Mounting arrangement for elevated structures	The elevated structure has to be securely anchored to the supporting surface. Concrete foundations of appropriate weight and depth for elevated structures mounted directly on the ground; bolted with anchor bolts of appropriate strength for elevated structures mounted on RCC surfaces
Mounting arrangement for ground installations	With removable concrete ballast made of pre-fabricated PCC (1:2:4), M15; assuring enough ground clearance to prevent damage of the module through water, animals and other environmental factors
Installation	The structures shall be designed for simple mechanical on-site installation. There shall be no requirement of welding or complex machinery at the installation site
Minimum distance between roof edge and mounting structure	0.6 m
Access for panel cleaning and maintenance	All solar panels must be accessible from the top for cleaning and from the bottom for access to the module junction box
Panel tilt angle	North-south orientation with a fixed tilt angle of 11–13 degrees (depending on location), south facing

Table 12.8 Specification for solar grid interactive Inverter

Total output power (AC)	To match solar PV plant capacity while achieving optimum system efficiency
Input DC voltage range	As required for the solar grid inverter DC input
Maximum power point (MPPT) tracking	Shall be incorporated
Number of independent MPPT inputs	1 or more
Operation AC voltage	Single phase 230 V or three phase 415 V (+12.5%; –20%)
Operating frequency range	47.5–52.5 Hz
Nominal frequency	50 Hz
Power factor of the inverter	> 0.98 at nominal power
Total harmonic distortion	Less than 3%
Built-in protection	AC high/low voltage; AC high/low frequency
Anti-islanding protection	As per VDE 0126-1-1, IEC 60255.5 / IEC 60255.27
Operating ambient temperature range	–10 °C – + 60 °C
Humidity	0-95% Rh
Inverter efficiency Inverter weighted efficiency	> = 95% > = 94%
Protection degree	IP 65 for outdoor mounting, IP 54 for indoor mounting
Communication interface	RS 485/RS 232/RJ 45
Safety compliance	IEC 62109-1, IEC 62109-2
Environmental testing	IEC 60068-2 (1, 2, 14, 30)
Efficiency measurement procedure	IS/IEC61683
Cooling	Convection
Display type	LCD for data display. LCD / LED for status display
Display parameters to include	Output power (W), cumulative energy (Wh), DC voltage (V) DC current (A), AC voltage (V), AC frequency (Hz), AC current (A), Cumulative hours of operation (h).

12.6 QUESTIONS

1. What is the importance of solar energy in the current economic scenario?
2. How are solar PV systems classified?
3. What are the different components of a solar PV system?
4. Explain briefly the important features of a grid interactive inverter?
5. Explain the different types of batteries used in solar PV systems with their relative merits?
6. What are the important factors that must be looked into before placing an order for a solar PV system?
7. A residential building has the following electrical loads to be supported by the solar PV system. Design a suitable solar PV system assuming that the number of cloudy days will be 3:
    a. 4 Nos. of 15 W CF lamps for 4 hours
    b. 3 Nos. of 65 W ceiling fans for 6 hours
    c. One refrigerator of 125 W used 24 hours with 40% duty cycle
    d. One 5 W night lamp used 7 hours per day
    e. One 150 W washing machine used for 1 hour per day
    f. One number 100 W LCD TV for 4 hours
    g. One number Personal computer 125 W for 1 hour

CHAPTER 13

EMERGENCY AND STANDBY GENERATION

OBJECTIVES

At the end of this chapter you will be

- Provided with the basic information about the significance of standby power supply system in an electrical system.
- Able to explain the different types of ratings for standby generating sets.
- Able to understand the need for UPS systems along with standby generating systems in certain applications.
- Familiar with the basic requirements of standby generator installations.

13.1 INTRODUCTION

Emergency and standby power systems are intended to provide an alternate source of power if the normal source of power, most often the utility fails. As such, reliability of these types of systems is critical for most of the operations and good design practices are essential for the smooth functioning of the system. The emergency and standby generation system is classified as follows:

1. **Emergency Power System:** As defined in IEEE Standard. 446–1995 [1], is "an independent reserve source of electric energy that, upon failure or outage of the normal source, automatically provides reliable electric power within a specified time to critical devices and equipment whose failure to operate satisfactorily would jeopardize the health and safety of personnel or result in damage to property
2. **Standby Power System:** As defined in [1], is "an independent reserve source of electric energy that, upon failure or outage of the normal source, provides electric power of acceptable quality so that the user's facilities may continue in satisfactory operation.

Emergency and standby power systems are generally designed into the overall electrical system for one of the following two reasons:

Legal Requirements: As required by the National Electrical Code (NEC), and other local, state codes and requirements. These are mainly concerned with the safety of human life, protection of the environment, etc.

Economic Considerations: Continuous process applications very often require a continuous source of electrical power to avoid significant production and economic loss. In some cases even a momentary loss of power can be disastrous. Co-generation systems which are used to sell power back to the utility as part of an energy management strategy are not discussed in this chapter.

Engine driven alternating current generator sets are often used in commercial/industrial sites as a prime source of electrical energy. They are also extensively used in both industry and commercial buildings as a source of back-up power. This chapter discusses some of the aspects which have to be handled when implementing engine driven alternating current generator sets. The main types of prime movers used in engine driven generator sets for industrial sites and commercial buildings are diesel engines, gas turbines, and steam turbines. Turbines are used mainly for production sets whereas diesel engines can be used for both production and standby sets. Gas-turbine generator sets are generally lighter in weight than diesel engine-generator sets, run more quietly, and generally require less cooling and combustion air, leading to lower installation costs. However, gas-turbine generator sets are more expensive than diesel engine-generator sets, and require more starting time (normally around 30 seconds compared to the 10–15 seconds for diesel). The long starting time requirement and lack of available small sizes (¡500 kW) makes the gas-turbine generators infeasible in many applications.

Generator installations must also take into consideration the combustion and cooling air required by the generator and prime mover, as well as the provisions for the removal of exhaust gasses. Noise abatement is also an important consideration. These considerations increase the installation costs, especially for reciprocating engine units such as diesel or gasoline engines. Further, the fuel supply must be considered. National Building Code and insurance considerations may force the fuel storage tank to be away from the generator(s), usually forcing the addition of a fuel transfer tank near the generator(s).

The choice of the prime mover is determined by such considerations as the availability and type of fuel and is not covered in this chapter. Since diesel engines are very often used, some specific information about diesel generator sets will be given. The power output requirement for the generator set is probably the most important criterion to be defined. The active power output depends on the type of fuel used, and on site conditions including ambient temperature, cooling medium temperature, altitude, and relative humidity. It also depends on load characteristics such as possible overloading and load variations over time.

13.2 RATINGS OF THE STANDBY GENERATING SETS

Care must also be taken when sizing engine-generator sets for a given application since several ratings exist for the output capability of a given machine. The different power ratings are:

13.2.1 Continuous Power Rating

The engine can supply 100% rated power for an unlimited time. This rating is normally used for production sets.

13.2.2 Prime Power Rating

The engine can supply a base load for an unlimited time, and 100% rated power for a limited time. The base load and acceptable time for 100% rated power are different for each manufacturer. Typical values are a base load of 70% of the rated power, 100% rated power during 500 hours per year.

13.2.3 Standby Power Rating

This is the maximum power that the engine can deliver and is limited in time, typically less than 500 hours per year. This rating should only be applied to generator sets which are used **exclusively for emergency power**. Since the engine is incapable of supplying more power, a security factor of at least 10% should be used when defining the standby power rating.

The standard overload capacity is defined as 10% more power during 1 hour for every 12 hours of operation. There is no overload capacity with a standby power rating.

A typical example of a diesel engine set is as follows:

Continuous power rating of 15.50 kW,
Prime power rating of 17.60 kW, and
Standby power rating of 18.80 kW.

A second alternative for emergency or standby system power is a second utility power source. However, the availability of a second utility source which is sufficiently independent from the normal utility service may be economically infeasible.

One common application for standby generators is to supply UPS (uninterrupted power supply) equipment during power outages. Since the generator has relatively high impedance as compared to a utility supply, voltage waveform distortion can occur due to harmonic currents generated by the UPS. Generator manufacturers normally derate their machines by up to 60% to ensure correct voltage waveforms when loads are on UPS equipment without harmonic current filtering.

When generator sets are used as a prime source of electrical energy the following points should be considered:

Provide for parallel operation with other sets and/or with utility
Allow for long maintenance periods (overhaul)
Ensure black-start capabilities, black start: The capability of starting generator sets without the
presence of a utility supply.
Use low speed equipment for long life (maximum 750 rpm for diesel engines).

When used as a standby source:

Ensure quick and reliable start-up and loading,
Implement reliable load shedding to avoid overloading or stalling,
Allow for periodic testing under load,
Provide for parallel operation with utility if set is used during peak loads.

The typical arrangement for supply of essential loads including critical loads for commercial buildings, small industrial sites, etc., is shown in Figure 13.1.

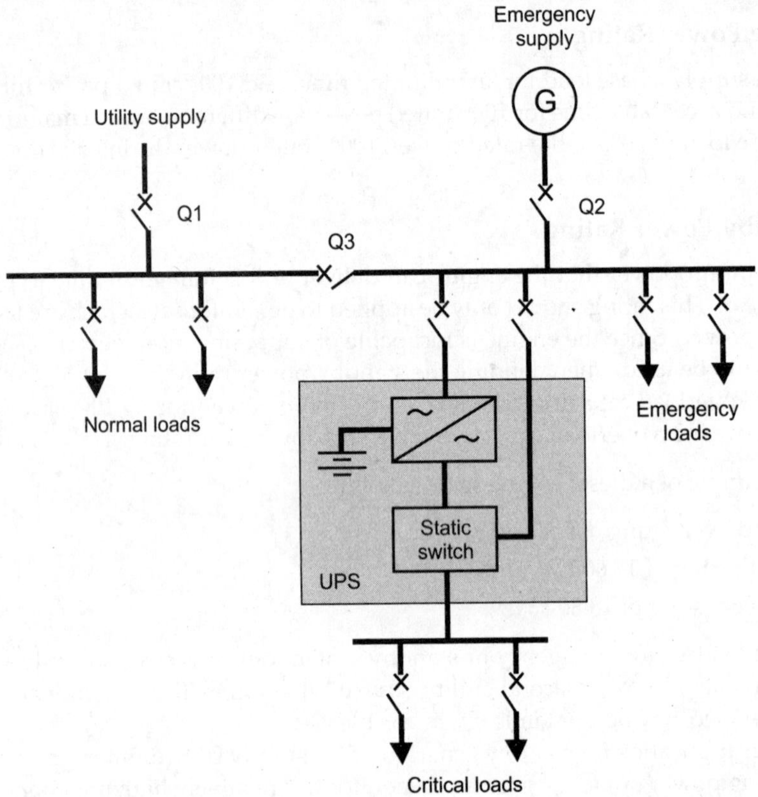

Figure 13.1 Typical arrangements of standby power systems

Under normal operating conditions the essential load is supplied from the utility supply. Upon loss of this supply the bus-tie circuit-breaker Q3 is tripped, the generator set is started, and then load is supplied by the standby generator set by closing the generator circuit-breaker, Q2. Critical loads which cannot accept any power outage are supplied by the UPS. The UPS is equipped with a static switch which will immediately bypass the rectifier/inverter module in case of an internal fault in the UPS and thus ensure a continuous supply of electrical power to the critical loads. The advantage of this scheme is its simplicity and clarity. All essential loads are connected to the same bus bar as the generator set and therefore no load shedding is required. UPS backup time can normally be limited to 10 minutes since the UPS will be supplied by the emergency supply. Both the normal and the backup supply to the UPS should be taken from the essential bus bar.

13.3 STARTING AND STOPPING OF GENERATOR SETS

Since diesel generator sets are often used for emergency power, it is necessary that steps be taken to ensure that the set will start correctly and quickly when required.

A starting time of 15 seconds from the start order to the closing of the generator circuit-breaker will be guaranteed by manufacturers. Specifying shorter starting times should be avoided since the decrease in starting time will be small and could increase the cost of the set. Critical equipment must be supplied by an UPS in any case.

13.3.1 Starting

Two techniques are commonly used for starting. These are compressed air and battery, compressed air is generally used for larger sets. The starting equipment should be designed for a minimum of 3 consecutive starts. It should be carefully monitored in order to enable preventive maintenance to be carried out prior to a failure during an attempted start. Failure to start is most often due to a problem with the starting battery. Where reliable starting is essential, consideration should be given to using compressed air.

13.3.2 Loading the Generator

When a generator set is operating alone, the load shall be applied in one or more steps. The variation in frequency and voltage will depend upon the magnitude of the step loads. As an example, step loads of 90% can be applied to a diesel generator set without the frequency varying more than 10% and the voltage more than 15%. Should specific limits on frequency and voltage variations be required, they should be specified together with the type of load which is to be connected. This information should include motor starting characteristics such as the starting current, and the type of starting.

13.3.3 Stopping the Generator

When stopping a generator set, the power output should be reduced to zero by transferring the load to other sources, and the circuit-breaker then tripped. The generator set should be run for several minutes to allow it to cool down prior to shutdown. In some cases the cooling system should continue to operate after shutdown in order to remove latent heat from the machine. Manufacturer's recommendations for shutdown should be followed in all cases.

13.4 STAND-ALONE OPERATION

Generator sets are often designed to operate independently (isochronous mode). In such cases the system frequency will be controlled by the engine governor. Overloads exceeding the maximum power output of the set will cause the system frequency to decrease and this can be used for initiating load shedding.

13.5 PARALLEL OPERATION WITH UTILITY SUPPLY

Since the utility supply is much stronger it will determine the system frequency and the system voltage. The governor will therefore be used to control the active power output of the engine, and the voltage regulator will control the reactive power output of the generator. The generator set must know in which configuration it is operating in order to be able to switch the governor and voltage regulator operation from frequency and voltage control (isochronous operation) to active and reactive power control (parallel operation).

The following figures indicate the different types of emergency/standby power supply arrangements. Figure 13.2 is the basic arrangement of a simple emergency/standby supply arrangement. Emergency supply arrangements for hospitals require special attention because of the life saving aspect of the equipment. National codes give specific guidelines on emergency supply arrangements for hospitals. Figure 13.3 is a typical arrangement of emergency power supply system for a hospital with a connected load below 150 kVA. A more complex form of emergency/standby supply system applicable for a high voltage installation is shown in Figure 13.4. In this case the system is supplied from two independent grid power sources and one standby generator. This is a highly flexible system for easy maintenance of system components.

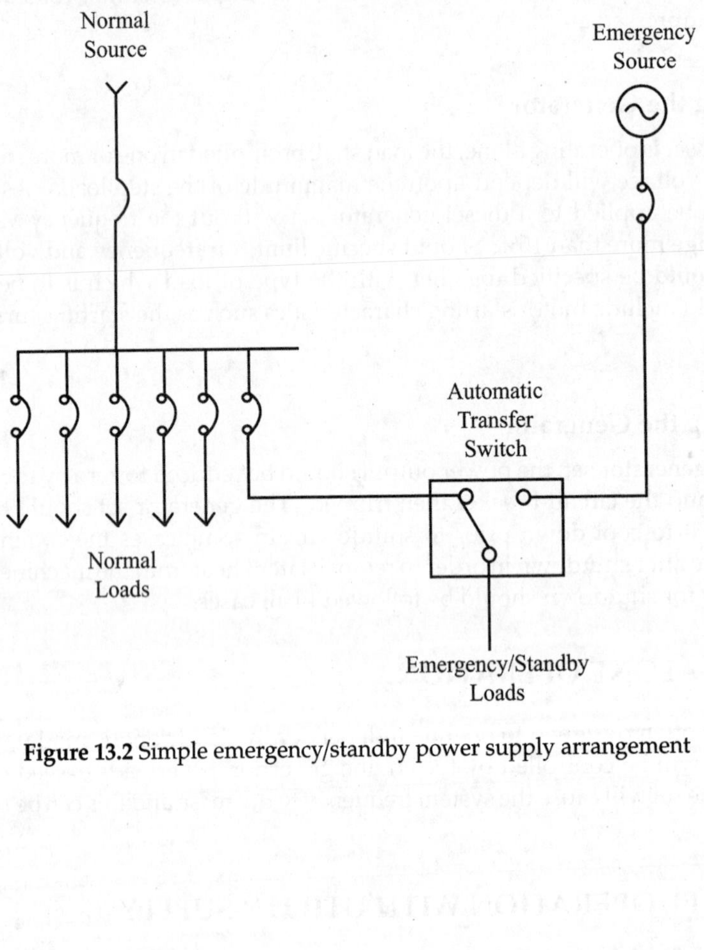

Figure 13.2 Simple emergency/standby power supply arrangement

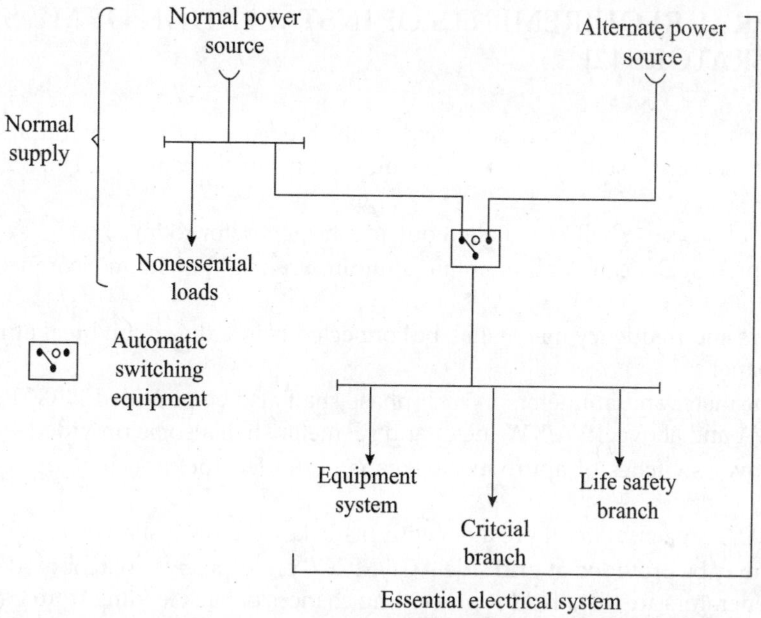

Figure 13.3 Emergency/standby power supply for hospitals with connected load below 150 kVA

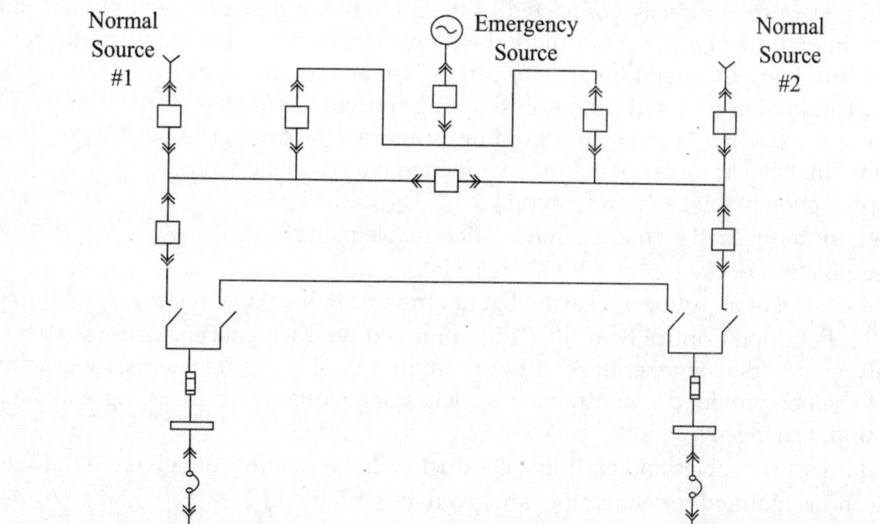

Figure 13.4 Emergency/standby power supply arrangement for a high voltage installation

13.6 GENERAL REQUIREMENTS OF INSTALLATION OF MV STANDBY GENERATORS [2]

1. Minimum 1 m clearance shall be provided on three sides of the generator set. When two generators sets are installed side-by-side, minimum 2 m clearance shall be provided between them.
2. Fuel tank of DG sets shall be installed outside the generator room.
3. Exhaust pipe of DG sets shall maintain a minimum height of 1.8 m clearance from the floor level.
4. Voltmeters and frequency meter shall be connected before the circuit breaker in the generator control panel.
5. Watt-hour meter and ammeters in each phase shall also be provided in GCP. For generators of 500 kVA and above, kVA/KW meter and PF meter shall also be provided.
6. Change over switches of approved makes is permitted for capacities up to and including 800A.
7. Above 800A capacity, circuit breakers with castle key interlock shall be provided.
8. U/V coil may be provided at grid side ACB/MCCB on respective switchboards at installations where generators are installed for preventing chances of back feeding from generators to grid side.
9. Separate isolators shall be provided before the change over switch (COS), if it has no OFF position.
10. When generator is directly connected to the main bus it may not be possible sometimes to maintain grading of outlets. Stipulation with respect to maximum single motor rating is also difficult to insist. Generator capacity can be deemed as sufficient even if it is called up on to start induction motors of 60% of its rating, provided suitable starters are employed for limiting the starting current, and grading is affected. However in such cases, it should be ensured that generator is not getting overloaded which can be achieved by providing thermal O/L protection in the generator circuit.
11. Generator room shall be made of non-inflammable materials.
12. Three-pole interlocked circuit breakers/isolators are permitted for supply changeover, provided neutral of all sources is earthed at neutral bus in the panel board.
13. Central Pollution Control Board (CPCB) approved type acoustic enclosure shall be provided for all generators. For generators without in built enclosure arrangements, room acoustic treatment shall be provided. Autotransfer switches of adequate ampere rating may be permitted for supply changeover.
14. Standard size of cables and earthling conductors for generators are given in the Table 13.1.
15. Protections required for generators are given in the Table 13.1.
16. Protection and control wiring diagram shall be forwarded along with scheme for approval by the Electrical Inspectorate.
17. Electricity duty at the rate specified by the Government is payable for generators up to and including 10 kVA rating.

Table 13.1 Size of cables, earthing conductors and protection for standby generators (Source:KSEIE handbook)

Generator capacity kVA	Full load current rating Amps	Cable size AYFY mm^2	Earth conductor size mm^2/SWG	Protection	Panel instruments
5	7	4	8.3/10	MCB	AM, VM & FM
7.5	10.5	4	8.3/10	MCB	
10	14	4	8.3/10	MCB	
12.5	17.5	6	8.3/10	MCB	AM, VM, FM & EM
15	21	10	8.3/10	MCB	
20	28	10	8.3/10	MCB	
25	35	16	18.6/6	MCB	
30	42	16	18.6/6	MCB	
35	49	25	18.6/6	MCB	
40	56	35	27.27/4	MCB	
45	63	35	27.27/4	MCB	
50	70	35	27.27/4	MCCB	
63	88	50	27.27/4	MCCB	
82.5	115	95	25 × 3	MCCB	
100	140	120	25 × 3	MCCB+ Stand by low Set earth fault Relay using CT in neutral Earthing conductor	AM, VM, FM & EM
125	175	150	25 × 3		
160	224	185	25 × 3		
180	252	300	25 × 3		
200	280	2 × 120	25 × 3		
225	315	2 × 225	25 × 3		
250	350	2 × 185	25 × 3		
320	448	2 × 300	25 × 3		
400	560	2 × 400	25 × 3		
500	700	3 × 400	25 × 3	ACB with Overload and E/F release	AM, VM, FM EM, PFM, kVAM
625	875	4 × 400	25 × 3		
750	1050	Bus-Bar Trunking	25 × 3		

For 1000 kVA and above, ACB with thermal O/L, voltage controlled O/C relay, over voltage, under voltage, negative sequence, low set stand by earth fault relays and REF/Different relay with fuel shutoff facilities. Overspeed protection shall be provided for the engine.

13.7 GUIDELINES FOR GENERATORS RUNNING IN PARALLEL

There are several specific guidelines when standby generators are running in parallel. The following are some of the specific guidelines for generators running in parallel.

1. P.F. meters shall be provided in generator control panels. For higher capacity generators kVA and kVAR meters shall be provided.
2. Reverse power relays shall be provided for generators. For higher capacity generators reverse reactive power relays shall also be provided.

3. (a) Neutral switching facility shall be provided.
   (b) Neutral of largest capacity generators shall only be connected to the system.
   (c) Neutral of largest capacity generator shall only be earthed. Neutrals of other generators, running in parallel, shall be in floating condition.
   (d) Also ensure that generator breakers can be made ON only if functional neutrals are earthed and closed.
4. Synchronizing breakers and neutral switches/contactors shall be electrically operating type having remote switching facility.
5. There shall be facility for remote control of generator voltage and engine speed.
6. Double frequency meter and double voltmeter shall be provided in the synchronizing panel.
7. For generators of one MVA and above, synchro check relay shall also be provided in synchronizing panel.
8. Switch ratings, cable sizes, earthing conductor sizes, etc., shall be designed considering fault current contribution by all generators running in parallel.
9. Bus bars shall be designed for maximum current during parallel running.

13.8 GUIDELINES FOR GENERATORS INSTALLED ABOVEGROUND LEVEL

The following list shows the special requirements for those generators which are installed above ground level.

1. Structural fitness certificate from a competent structural engineer shall be submitted and should contain the load bearing capacity of the structure, rating of generator, weight of generator, etc. The certificate shall state the structure can bear the loading addition to any other load already in existence in the concerned structure.
2. Separate room shall be provided for generator set.
3. Generator set shall be mounded on antivibration pads.
4. Proper oil drainage facility shall be provided.
5. Maximum fuel tank capacity shall not exceed 100 litres or fuel required for 8 hours of continuous operation whichever is minimum.
6. Emergency engine stop facility shall be provided outside the generator room.
7. For apartment buildings, no objection certificate from apartment owners shall be produced, to install the generator above ground level.
8. No objection certificate from the owner of the building shall be produced if a tenant installs generator

13.9 GUIDELINES FOR PORTABLE GENERATORS

1. Generators up to 4kVA in rating are treated as portable generators.
2. A residual current device (ELCB) having an operating time of 20 ms at a residual current of 30 mA shall be provided.
3. (a) In single-phase generators one terminal shall be connected to the earth and designated as the neutral.

(b) Three phase generators should have their windings connected in star, with the star connection made available and connected to earth.

4. All the exposed metal work of the generator should be adequately connected to the earth terminal.
5. The supply from a single-phase generator can be changed over to a 3-phase supply system through a 4-pole changeover switch subject to the following conditions.
   (a) The neutral conductors of the load side and generator side should be of adequate capacity to carry the total current in the neutral.
   (b) The 3 poles in the 4-pole changeover switch shall be linked by using rigid conductors of adequate short circuit and continuous current rating capacity as detailed in Table 13.2.

Table 13.2 Ratings of changeover switches for generators

Capacity of single-phase generator	Changeover-switch rating	Minimum copper area of conductors used for linking the poles
Upto 3 kVA	32 A	20 sq.mm
3 to 6 kVA	63 A	40 sq.mm
6 to 10 kVA	100 A	50 sq.mm

   (c) The copper strip/rigid conductors used for linking the poles should be insulated by using tapes or sleeves.
   (d) Shielding should be provided when the generator is kept exposed to weather.
   (e) Portable generator shall be kept at a place, sufficiently ventilated so as to avoid pollution.

REFERENCES

1. IEEE Recommended Practice for Emergency and Standby Power Systems for Industrial and Commercial Applications, IEEE Std. 446-1995, December 1995.
2. State Electrical Inspectorate Engineers Handbook-2014.

13.10 QUESTIONS

1. Give a brief description of salient features of emergency and standby power systems.
2. What are the different power ratings of standby generators? Explain briefly.
3. With a neat single line diagram, explain the components of a typical standby power system.
4. Explain the procedure for starting and stopping of standby generators.
5. What are the special precautions to be taken for standby generators installed above ground level?
6. Discuss the general requirements of installation of standby generators of medium voltage (MV) class.

APPENDIX

TABLE I: Recommended levels of illumination for different parts of domestic buildings *(National Electrical Code 1985)*

S. No.	Location		Illumination level-lux
1	Entrances, hall ways		100
2	Living room		300
3	Dining room		150
4	Bedroom	General	300
		Dressing, bed heads	200
5	Kitchen		200
6	Kitchen sink		300
7	Bathroom		100
8	Sewing		700
9	Workshop		200
10	Stairs		100
11	Garage		70
12	Study		300

TABLE II: Recommended levels of illumination for different parts of recreational /assembly buildings *(National Electrical Code 1985)*

S. No.	Location	Illumination level-lux
1	**Assembly and Concert Halls**	
	Foyers, auditoria	100-150
	Platforms	450
	Corridors	70
	Stairs	100
2	**Cinema Halls**	
	Foyers	150
	Auditoria	50
	Corridors	70
	Stairs	100
3	**Theatres**	
	Foyers	150
	Auditoria	70
	Corridors	70
	Stairs	100

TYPICAL LAMP DATA

TABLE III: Incandescent lamps

Rating (Watts)	Life (hours)	Initial lumens
60	1000	870
100	750	1750
150-PAR	2000	1740
200	2000	2300
500	2000	6500

TABLE IV: Compact fluorescent lamps

Rating (Watts)	Life (hours)	Initial lumens
5	10,000	220
7	10,000	380
11	10,000	560
13	10,000	680
15	10,000	810
18	10,000	1050
23	10,000	1500
26	10,000	1800
32	10,000	2400

TABLE V: Mercury vapour lamps

Rating (Watts)	Life (hours)	Initial lumens
100	18,000	3700
175	24,000	8600
250	24,000	12,100
400	24,000	22,500
1000	24,000	57,000

TABLE VI: Metal halide lamps

Rating (Watts)	Life (hours)	Initial lumens
50	15,000	3400
70	15,000	5600
100	15,000	9000
150	10,000'	13,500
175	10,000	15,000
250	10,000	20,500
400	20,000	36,000
1000	12,000	1,10,000

TABLE VII: High pressure sodium vapour lamps

Rating (Watts)	Life (hours)	Initial lumens
35	16,000	2250
50	24,000	4000
70	24,000	5800
100	24,000	9500
150	24,000	16,000
250	24,000	27,500
400	24,000	47,500
1000	24,000	1,40,000

TABLE VIII: Pulse start metal halide lamps

Rating (Watts)	Life (hours)	Initial lumens
50	15,000	3400
70	15,000	5600
100	15,000	9000
150	15,000	15,000
175	15,000	17,500
200	15,000	21,000
250	15,000	26,300
320	20,000	34,000
400	20,000	44,000
450	20,000	50,000

COMPARISON OF LIGHT SOURCES

Characteristics	Incandescent	Quartz halogen	Mercury vapour	Metal halide	Pulse start MH	HP SV	LP SV	Compact Fluorescent lamp
Initial cost	Low	Low	High	High	High	High		Lower
Initial lumen per watt	15-24	15-24	30-6	69-115	60-111	79-130	62-150	60-80
Efficiency	Poor	Poor	Fair	Good	Best	Better	Best	Fair
Average lamp life (hours)	1000	2000	24,000	15,000	30,000	24,000	20,000	10,000
Annual operating cost	High	Medium	Low	Low	Low	Low	Very low	Low
Long burning hours	Fair	Fair	Good	Good	Good	Good	Good	Good
Short burning hours	Good	Good	Good	Good	Good	Good	Good	Good
Colour acceptability	Good	V good	Fair	V good	V good	Fair	Fair	Good
Degree of light control	V good	V good	Fair	Good	Good	Good	Poor	Fair
Maintained lumen output	Good	V good	Poor	Good	Good	Good	V good	Good
Hot restart time	Instant	Instant	10 m	10-15 m	2-4 m	1-3 m	1 m	Instant

TABLE IX: Alphanumeric notation, graphical symbols and colours (as per *National Electric Code 1985*)

Designation of conductor	Alphanumeric notation	Graphical symbol	Colour
Supply system AC			
Phase 1	L1		Red
Phase 2	L2		Yellow
Phase 3	L3		Blue
Neutral	N		Black
Apparatus AC			
Phase 1	U		Red
Phase 2	V		Yellow
Phase 3	W		Blue
Neutral	N		Black
Supply system DC			
Positive	L+	+	Red
Negative	L–	–	Blue
Mid conductor	M		Black
Supply system 1ϕ			
Phase	L		Red
Neutral	N		Black
Protective earth	PE		Green and Yellow
Earth	E		No colour
Clean earth	TE		
Chasis or frame	MM		
Equipotential terminal	CC		

TABLE X: Size of cables, earthing conductors and protection for generators

Generator capacity kVA	Full load current rating Amps	Cable size AYFY mm^2	Earth conductor size mm^2/SWG	Protection	Panel instruments
5	7	4	8.3/10	MCB/MCCB	
7.5	10.5	4	8.3/10	MCB/MCCB	AM, VM and FM
10	14	4	8.3/10	MCB/MCCB	
12.5	17.5	6	8.3/10	MCB/MCCB	
15	21	10	8.3/10	MCB/MCCB	
20	28	10	8.3/10	MCB/MCCB	
25	35	16	18.6/6	MCB/MCCB	AM, VM,
30	42	16	18.6/6	MCB/MCCB	FM and EM
35	49	25	18.6/6	MCB/MCCB	
40	56	35	27.27/4	MCB/MCCB	
45	63	35	27.27/4	MCB/MCCB	
50	70	35	27.27/4	MCCB	
63	88	50	27.27/4	MCCB	
82.5	115	95	25 × 3	MCCB	
100	140	120	25 × 3	MCCB+ stand	
125	175	150	25 × 3	by low set	
160	224	185	25 × 3	earth fault	AM, VM, FM
180	252	300	25 × 3	Relay using	and EM
200	280	2 × 120	25 × 3	CT in neutral	
225	315	2 × 225	25 × 3	earthing	
250	350	2 × 185	25 × 3	conductor	
320	448	2 × 300	25 × 3		
400	560	2 × 400	25 × 3		
500	700	3 × 400	25 × 3	ACB with	AM, VM, FM
625	875	4 × 400	25 × 3	Overload and	EM, PFM,
750	1050	Bus - Bar Trunking	25 × 3	E/F release	kVAM

For capacities 1000 kVA and above, ACB with thermal O / L, voltage controlled O / C relay, over voltage, under voltage, negative sequence, low set stand by earth fault relays and REF/ Different relay with fuel shutoff facilities. Over speed protection shall be provided for the engine.

TABLE XI: Standard size of cables and earthing conductors for motors

Motor rating		Approx. full load current	Backup fuse rating		Cable size AYFY	Earth conductor		
						Copper		Al
kW	HP	Ampere	DOL	Assisted	mm^2	mm^2	SWG	mm^2
0.75	1	1.8	6	-	4	3.251	10	25
1.50	2	3.5	10	-	4	3.251	10	25
2.25	3	5.0	16	-	4	3.251	10	25
3.75	5	7.5	25	-	4	3.251	10	25
5.50	7.5	11.0	32	25	6	3.251	10	25
7.50	10	14.0	32	25	6	3.251	10	25
9.30	12.5	18.0	50	35	10	3.251	10	25
11.20	15	21.0	50	35	10	3.251	10	25
15.00	20	28.0	63	50	16	3.251	10	25
18.70	25	36.0	80	63	25	3.251	10	25
22.50	30	40.0	100	63	25	4.877	6	35
26.00	35	47.0	120	100	35	4.877	6	35
30.00	40	55.0	120	100	35	4.877	6	35
37.50	50	66.0	160	100	50	4.877	6	35
45.00	60	80.0	200	120	70	5.893	4	50
55.00	75	95.0	200	120	95	5.893	4	50
59.00	80	105.0	250	200	150	6.910	2	70
67.50	90	118.0	250	200	185	6.910	2	70
75.00	100	135.0	250	200	185	6.910	2	70
90.00	125	165.0	320	250	240	6.910	2	95
110.00	150	200.0	350	320	400	7.620	1 or 0	120
131.00	175	230.0	400	320	2 × 150	7.620	1 or 0	120
150.00	200	270.0	500	350	2 × 185	7.620	1 or 0	120
185.00	250	3250.0	600	500	2 × 240	25 × 3		185

For motors of rating 185 kW and above, breaker shall be provided and following protection shall also be incorporated.

(a) Thermal O/L
(b) Short circuit
(c) Earth fault
(d) Negative sequence
(e) Under voltage
(f) Locked rotor

TABLE XII: Maximum rating of a motor that can be connected to a transformer

Method of starting	Starting current of motors in multiples of full load current	Motor rating permissible in HP
DOL	7.0	$(2S_T - L)/7.0$
ATS with 80% tap	5.5	$(2S_T - L)/5.5$
ATS with 40% tap, Star Delta	2.5	$(2S_T - L)/2.5$
Soft starter	2.0	$(2S_T - L)/2.0$
Slip ring	2.0	$(2S_T - L)/2.0$

Where, S_T is the rating of transformer in kVA and L the maximum demand excluding the motor in question. When the number of starts/hour is more than 6 or loading cycle is more than 6, maximum rating shall be limited to 2% of the fault level.

PROPERTIES OF ACSR CONDUCTORS

Code Name	No. of Aluminium Strands	Dia in mm	No. of Steel Strands	Dia in mm	Dia of Cable in mm	Gross area of Aluminium. mm^2	Gross area of Cable mm^2	Ultimate strength in kg	Resistance at 20 °C in ohm/km	Current for 30° rise above ambient	Modulus of elasticity kg/sq.cm $\times 10^{-6}$
Squirrel	6	2.11	1	2.11	6.33	20.71	24.44	770	1.37	70	0.81
Easel	6	2.59	1	2.59	7.77	31.63	36.90	1135	0.91	100	0.81
Rabbit	6	3.35	1	3.35	10.15	52.95	61.78	1860	0.54	148	0.81
Mink	6	3.66	1	3.66	10.98	63.06	73.57	2205	0.45	167	0.81
Dog	6	4.72	7	1.57	14.15	105.20	118.80	3310	0.22	254	0.74
Tiger	30	2.36	7	2.36	16.52	131.50	162.10	5790	0.22	296	0.79
Wolf	30	2.59	7	2.59	18.13	158.10	195.00	6875	0.18	343	0.79
Lynx	30	2.79	7	2.79	19.53	183.90	226.80	7945	0.16	385	0.79
Panther	30	3.00	7	3.00	21.00	211.70	261.20	9095	0.14	427	0.79
Kundah	42	3.49	7	1.94	26.76	402.90	423.57	9550	0.07	726	0.65
Finch	54	3.64	19	2.18	32.83	564.00	635.50	18235	0.05	1000	-

TABLE XIII: Flexible house wiring single core cables (with copper conductor) 250/440 V

Conductor nominal area in mm^2	No. and size of wire in mm	Resistance in ohms per km	Current carrying capacity in (amps)	Thickness (mm)		Overall dia (mm)		Equivalent SWG
1.0	14.0/0.3	18.89	11	0.9	0.65	3.00	4.55	1/18
1.6	23.0/0.3	11.50	15	0.9	0.65	3.50	4.80	3/22
2.0	28.0/0.3	9.42	17	0.9	0.65	3.90	5.40	3/20
3.0	44.0/0.3	6.00	22	1.0	0.75	4.15	5.55	7/22
4.5	65.0/0.3	4.06	26	1.0	0.75	4.90	6.10	7/20

Colours

Single core unsheathed—Red, Yellow, Blue, Black, Green, and White.
Single core sheathed. Core colour black. Sheath White.
Packing: 91.44 metres in convenient to handle corrugated boxes.

PROPERTIES OF CIRCULAR AND SHAPED COPPER AND ALUMINIUM CONDUCTORS

Nominal area of conductor (mm^2)	No. and size(mm)	Conductor diameter mm.	Maximum resistance at 20 °C			
			Copper		Aluminium	
			1 core ohm/km	Multicore ohm/km	1 core ohm/km	Multicore ohm/km
1.5	1/1.38	1.38	11.9000	12.9000	19.1700	20.0000
1.5	7/0.50	1.50	13.3000	13.3000	–	–
2.5	1/1.78	1.78	7.1400	7.2800	11.8000	12.0000
2.5	7/0.67	2.01	7.2700	7.4100	12.1000	12.4000
4.0	1/2.25	2.25	4.4700	4.5600	7.3900	7.5400
4.0	7/0.85	2.55	4.5200	4.6100	7.5500	7.7000
6.0	1/2.76	2.76	2.9700	3.0300	4.9100	5.0100
6.0	7/1.04	3.12	3.0200	3.0800	4.9900	5.0900
10.0	1/3.57	3.57	–	–	2.9400	3.0000
10.0	7/1.35	4.05	1.7800	1.8300	2.9600	3.0200
16.0	7/1.70	5.10	1.1300	1.1500	1.8700	1.9100
25.0	7/2.14	6.42	0.7120	0.7270	1.1800	1.2000
35.0	7/2.52	7.56	0.5140	0.5240	0.8510	0.8680
50.0	19/1.78	8.90	0.3790	0.3870	0.6280	0.6410
70.0	19/2.14	10.70	0.2620	0.2680	0.4350	0.4430
95.0	19/2.52	12.60	0.1890	0.1930	0.3130	0.3200
120.0	37/2.03	14.21	0.1530	0.1530	0.2480	0.2530
150.0	37/2.25	15.75	0.1220	0.1240	0.2020	0.2060
185.0	37/2.52	17.64	0.0972	0.0991	0.1610	0.1640
240.0	61/2.25	20.25	0.0740	0.0754	0.1220	0.1250
300.0	61/2.52	22.68	0.0590	0.0601	0.0976	0.1000
400.0	61/2.85	25.65	0.0461	0.0470	0.0763	0.0778
500.0	61/23.20	28.80	0.0366	0.0378	0.0605	0.0617
630.0	91/2.98	32.78	0.0283	0.0289	0.0469	0.0478
800.0	91/3.37	37.07	0.0221	0.0221	0.0226	0.0367
1000.0	91/3.78	41.58	0.0176	0.0179	0.0291	0.0297

BIBLIOGRAPHY

1. National Electrical Code 2011; Bureau of Indian Standards Publication, 2011.
2. National Building Code 2005; Bureau of Indian Standards Publication, 2005.
3. The Electricity Act, 2003.
4. CEA Regulations, 2010.
5. IES Lighting Handbook (10th edition); Illumination Engineering Society.
6. Rapid Lighting Design and Cost Estimation: P.C. Sorcar, McGraw Hill Publishing Co., New York, 1979.
7. Electrical Systems Design and Specification Handbook for Industrial Facilities: S.J. Marrano et al.; Fairmont Press, 1998.
8. Electrical Systems Design: T.R. Bosela; Prentice Hall, New Jersey, 2003.
9. Electrical Systems Design Data Handbook, M.K. Giridharan, IK International Publishing House; 2011.
10. Official websites of:
    (a) M/s Havells India Ltd.
    (b) M/s Crompton Greaves Ltd.
    (c) M/s Schneider India Ltd.
    (d) International Electrotechnical Commission.
    (e) M/s Leonics Company Ltd., Thailand.
    (f) M/s Indsolar Ltd.
    (g) M/s Exide Batteries.

Index

A

After tax analysis 353
Alkaline batteries 384
Alternating current systems, voltage limits for 10
Aluminium conductors 226
Area lighting 272
Array mounting racks 382
Automatic transfer switches 188
Automatically switched capacitor banks 281
Average lumen method 46

B

Back-to-back switching 293
Ballast factor 50
Ballasts 32
Battery bank 384
Battery sizing 391
Bedding and serving materials 231
Building services, classification of 17
Burning position 28

C

Cable laying, methods of 251
Cables
    materials used in 226
    selection of 225, 256
Cabling 95
Capacitor control devices 288
Capacitors, location of 282
Central Electricity Authority (CEA)[4] 3
Changeover switches 187
Charge controller 384, 393
Circuit breakers 201
    selection of 207
    selective discrimination of 211
Climate control, design aspects of 22
Colour appearance 52
Colour characteristics 27
Colour rendering 52
Conductor size 89
Conduit wiring 89
Connected load and maximum demand 78, 81
Control switchgear 182
Copper conductors 226
Crest factor 300
Current carrying capacity 229
Current limiting circuit-breakers 206
Current rating calculations 230

D

Declining balance method 350
Depreciation 349
Design procedures 52
Detuned capacitor bank, design of (harmonic filter) 325
Detuned harmonic filter, design summary of 339
Diagrams, graphic symbols for 12
Direct glare 50
Disconnects (isolators and circuit breakers) 383
Distortion power 300
Distribution architecture, selection of 113
Distribution board 80, 95
    types of 86
Disturbance sensitivity 112
Disturbances to sensitive loads 311
Domestic installations, earthing of 90
Double-ended power supply 124
Drives, selection of 138
Dry type transformers 130

E

Earth leakage protection 369
Earthing conductor size 179
Earthing, criteria for 368
Earthing systems, characteristics of 367
Electrical isolation 182
Electrical resistivity 230
Electrical system design safety, aspects of 11
Electricity act, 2003[3] 1
Electricity supply industry, 2
Emergency power supply 95
Emergency power system 401
Energy efficient fluorescent lamps 31
Energy saving incandescent lamps 29
Exterior lighting design 261

F

Factor of maximum utilisation 77
Factor of simultaneity 77
Fault current calculations 172
Fault location 175, 176
Fixed capacitors 281
Fixtures, arrangement of 55
Fluorescent lamps 29
Full voltage starters 141
Fuses 190
  current limiting property of 194
  selective coordination with 197

G

Generator sets, starting and stopping of 404
Generators installed above ground level, guidelines for 410
Generators running in parallel, guidelines for 409
Grid connected system 379
Grid-connected inverter criteria for selecting a 383
Grounding system 382

H

Harmonic distortion 300
Harmonics 297
  sources of 302
Harmonics disturbances 305
Harmonics methods to reduce the impact of 322
High intensity discharge lamps 31
High mast lighting 274
High pressure sodium vapour lamp (HPS) 35
High-rise apartment buildings, features for 93
Home applications, design for 386

I

IEEE 519-1992 voltage distortion limits 339
IEEE standard 519-1992 334
Impregnated paper 228
Incandescent lamps 27
Indirect glare 51
Individual compensation 283
Industrial buildings, classification of 108
Industrial installations 107
  power distribution architecture of 109
Installation flexibility 111
Insulating materials 227
Interconnected switchboards 124
International Electrotechnical Commission (IEC) 8
Inverter 383
  selection of 394
Isolated bank switching 291
Isolators (switch disconnectors) 184

J

Junction box 382

L

Lamp blackening 28
Lamp burn out 49
Lamp industry, development of 26
Lamp lumen depreciation 49
Lateral light distribution 264
Lead-acid batteries 384
LED, advantages 43
LEDs for lighting 42
LED colours 40
Light emitting diodes 39
Light fittings and fans 90
Light loss factor (LLF) 48
Light sources 25
  selection of 268
Lighting calculations 44
Lighting services, design aspects of 18
Load break switches 184
Load requirements, estimation of 75
Long hand method 53